Advanced and Applied Geophysics

Advanced and Applied Geophysics

Edited by **Karl Seibert**

R CALLISTO
REFERENCE

New York

Published by Callisto Reference,
106 Park Avenue, Suite 200,
New York, NY 10016, USA
www.callistoreference.com

Advanced and Applied Geophysics
Edited by Karl Seibert

© 2015 Callisto Reference

International Standard Book Number: 978-1-63239-012-7 (Hardback)

Contents

Preface

Earth, as it is now, is the result of continuous change that has taken place since its formation millions of years ago. Geology is the science which deals with the study of earth as a composition of rocks forming land. Geophysics on the other hand is the study of nature of the earth through application of physics; it is the combination of branches of science of geology and physics. Earth's magnetic field deflects high energy particles from the sun because of which life on earth is possible. However, it is the because of the moon's gravitational force that oceanic tides are formed. Geophysics therefore includes not only the phenomena of the earth such as its magnetic field, heat flow, gravity and seismic waves, but also the outer-space factors, such as moon's gravity, that have an impact on earth.

Geophysics is a broad term which incorporates various fields for the physical study of the earth. It includes the study of solid earth physics, terrestrial magnetism, gravity and tides, seismology, hydrology, volcanology, terrestrial electricity and atmospheric phenomena.

This book traces the development of the science of geophysics and looks at the field of geophysical surveys to help predict the behaviour of earth and its materials in relation to various geological and other phenomena.

I thank all the researchers and scientists who have contributed to this book. I would also like to thank my family and my publisher for their continued support and trust in me.

Editor

Subsurface and Petrophysical Studies of Shaly-Sand Reservoir Targets in Apete Field, Niger Delta

P. A. Alao,[1] A. I. Ata,[2] and C. E. Nwoke[3]

[1] *Department of Geology, Institute of Earth and Environmental Sciences, University of Freiburg, Albertstraße 23b, 79104 Freiburg, Germany*
[2] *Department of Geology, University of Malaya, Kuala Lumpur, Malaysia*
[3] *School of Earth, Atmospheric and Environmental Sciences, The University of Manchester, Oxford Road, Manchester, Greater Manchester M13 9PL, UK*

Correspondence should be addressed to P. A. Alao; peter.alao@merkur.uni-freiburg.de

Academic Editors: Y.-J. Chuo and A. Tzanis

Conventional departures from Archie conditions for petrophysical attribute delineation include shaliness, fresh formation waters, thin-bed reservoirs, and combinations of these cases. If these departures are unrecognized, water saturation can be overestimated, and this can result in loss of opportunity. Wireline logs of four (4) wells from Apete field were studied to delineate petrophysical attributes of shaly-sand reservoirs in the field. Shale volume and porosities were calculated, water saturations were determined by the dual water model, and net pay was estimated using field-specific pay criteria. Ten sand units within the Agbada formation penetrated by the wells were delineated and correlated and their continuity was observed across the studied wells. The reservoirs had high volume of shale (Vcl), high hydrocarbon saturation, low water saturation, and good effective porosity ranging 12.50–46.90%, 54.00–98.39%, 1.61–46.0%, and 10.40–26.80%, respectively. The pay zones are relatively inhomogeneous reservoirs as revealed from the buckle's plot except in Apete 05. The direction of deposition of the sands was thus inferred to be east west. Empirical relationships apply with variable levels of accuracy with observation of the porosity-depth, water saturation-depth, and water saturation-porosity trends. Core data is recommended for better characterization of these reservoirs.

1. Introduction

Shales can cause complications for the petrophysicist because they are generally conductive and may therefore mask the high resistance characteristic of hydrocarbons [1]. Several factors are to be considered when delineating petrophysical attributes for shaly-sand reservoirs because clay minerals add conductivity to the formation especially at low water saturations.

Clay minerals attract water that is adsorbed onto the surface, as well as cations (e.g., sodium) that are themselves surrounded by hydration water. This gives rise to an excess conductivity compared with rock, in which clay crystals are not present, and this space might otherwise be filled with hydrocarbon. Using Archie's equation in shaly sands results in very high water saturation values and may lead to potentially hydrocarbon bearing zones being missed. Moreover, in clean sands, the irreducible water volume is a function of the surface area of the sand grains and therefore the grain size, but for shaly sands the addition of silt and clay usually decreases effective porosity due to poorer sorting and increases the irreducible water volume with the finer grain size [2]. Archie's equation was developed for clean rocks, and it does not account for the extra conductivity caused by the clay present in shaly sands. Therefore, Archie's equation would not provide accurate water saturation in shaly sands. In fact, water saturations obtained from Archie's equation have a tendency to overestimate the water in shaly sands. Several models have been proposed by many researchers for shaly-sand analysis such as Juhasz model, dual water model, Indonesian model, Waxman and Smits model, and so forth.

2. Synopsis of the Geology

The stratigraphic sequence of the Niger Delta comprises three broad lithostratigraphic units, namely, (1) a continental shallow massive sand sequence, the *Benin Formation*, (2) a coastal marine sequence of alternating sands and shales, the *Agbada Formation*, and (3) a basal marine shale unit, the *Akata Formation* (Figure 2). Outcrops of these units are exposed at various localities (Figure 1). The Akata Formation consists of clays and shales with minor sand intercalations. The sediments were deposited in prodelta environments. Petroleum in the Niger Delta is produced from these unconsolidated sands in the Agbada Formation. Characteristics of the reservoirs in the Agbada Formation are controlled by depositional environment and by depth of burial. The sand percentage here is generally less than 30%. The Agbada Formation consists of alternating sand and shales representing sediments of the transitional environment comprising the lower delta plain (mangrove swamps, floodplain, and marsh) and the coastal barrier and fluviomarine realms. The sand percentage within the Agbada Formation varies from 30 to 70%, which results from the large number of depositional offlap cycles [3]. A complete cycle generally consists of thin fossiliferous transgressive marine sand, followed by an offlap sequence which commences with marine shale and continues with laminated fluviomarine sediments followed by barriers and/or fluviatile sediments terminated by another transgression [4, 5].

The Benin Formation is characterized by high sand percentage (70–100%) and forms the top layer of the Niger Delta depositional sequence. The massive sands were deposited in continental environment comprising the fluvial realms (braided and meandering systems) of the upper delta plain. The Niger Delta time-stratigraphy is based on biochronological interpretations of fossil spores, foraminifera, and calcareous nonnoplankton. The current delta-wide stratigraphic framework is largely based on palynological zonations labeled with Shell's alphanumeric codes (e.g., P630, P780, and P860). This allows correlation across all facies types from continental (Benin) to open marine (Akata). There have been concerted efforts, within the work scope of the stratigraphic committee of the Niger Delta (STRATCOM), to produce a generally acceptable delta-wide biostratigraphic framework [9] but not much again has been accomplished after several data gathering exercise by the committee. The sediments of the Niger Delta span a period of 54.6 million years during which, worldwide, some thirty-nine eustatic sea level rises have been recognized [10]. Correlation with the chart of Galloway [11] confirms the presence of nineteen of such named marine flooding surfaces in the Niger Delta. Eight of these are locally developed. Adesida et al. [10] defined eleven lithological mega sequences marked at the base by regional mappable transgressive shales (shale markers) that are traceable across depobelt boundary faults and proposed these as the genetic sequences that can be used as the basis for lithostratigraphy of the Niger Delta.

3. Methodology

Composite wireline log data from four well logs were interpreted. The basic analysis procedure used involves the following steps; each of which is described in the following sections.

3.1. Import and Well Log Data. The well data was imported into the software used and well log correlation (Figure 3) was done after which the petrophysical attributes were delineated. Well correlation helped in determining the direction of thickness of sand being mapped and the lateral continuity of these reservoirs.

3.2. Zoning and Point Selection. Zoning is of vital importance in the interpretation of well logs. The logs were split into potential reservoir zones and nonreservoir zones. Hydrocarbon bearing intervals were identified and differentiated based largely on the readings from the shallow and deep reading resistivity tools. However, hydrocarbon typing (oil and gas differentiation) was based on density-neutron logs overlay.

3.3. Compute Shale Volume from the Gamma Ray. This was derived from the gamma ray log first by determining the gamma ray index I_{GR} [12]:

$$I_{GR} = \frac{\left(GR_{log} - GR_{min}\right)}{\left(GR_{max} - GR_{min}\right)}, \tag{1}$$

where I_{GR} = gamma ray index; GR_{log} = gamma ray reading of the formation; GR_{min} = minimum gamma ray reading (sand baseline); GR_{max} = maximum gamma ray reading (shale baseline).

For the purpose of this research work, Larionov's [13] volume of shale formula for tertiary rocks was used:

$$Vsh = 0.083 \left(2^{3.7*I_{GR}} - 1\right) \tag{2}$$

Vsh: volume of shale and I_{GR}: gamma ray index.

3.4. Compute Total Porosity and Shale-Corrected (Effective) Porosity. Total and effective porosity was estimated from the density, neutron, and sonic logs using Archie's equation.

3.5. Compute Water Saturation. Water saturation was estimated using Archie's water saturation formula and Schlumberger's dual water model.

3.6. Estimate Net Pay. Calculate net pay using field-specific net pay cutoffs. Cutoff criteria used are water saturation < 50%, porosity > 10%, and volume of clay < 50%.

3.7. Use of Crossplots. Pickett, buckles, and neutron-density crossplots were generated to understand reservoir properties. Pickett plot which is a resistivity-porosity plot generated was used to determine saturation values alongside Archie parameters *a* and *m*. Porosity is calculated from the neutron porosity and density porosity logs and is plotted against the resistivity data obtained from the deep resistivity log. Porosity is plotted on the *y*-axis with a logarithmic scale ranging from 0.1% to 100%, while the resistivity is plotted on the *x*-axis with a logarithmic scale ranging from 1 to 100 ohm meter. In order to properly characterize the reservoir sands delineated and correlated across the studied wells, Buckles plot, a plot of Sw

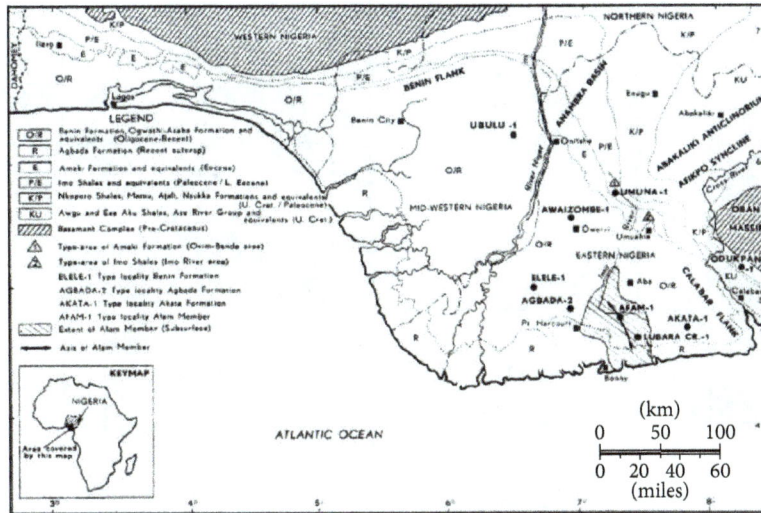

FIGURE 1: Map of Southern Nigeria showing outcrops of cretaceous and tertiary formations and type localities of subsurface stratigraphic units. After Short and Stauble [6].

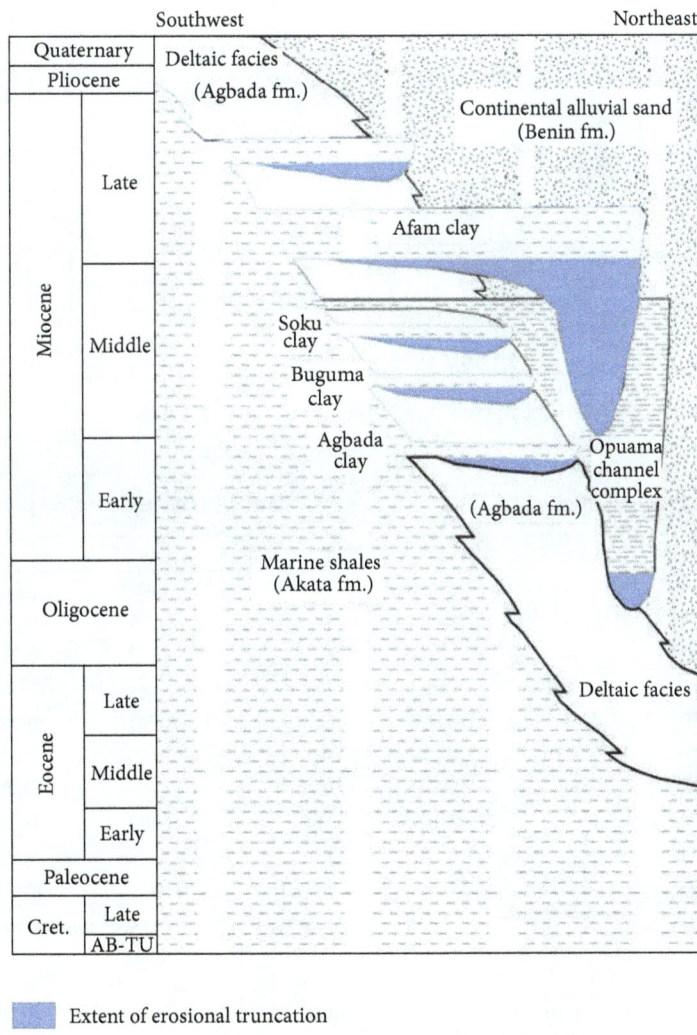

FIGURE 2: Stratigraphic column of the Niger Delta. After Shannon et al. [7] and Doust et al. [8].

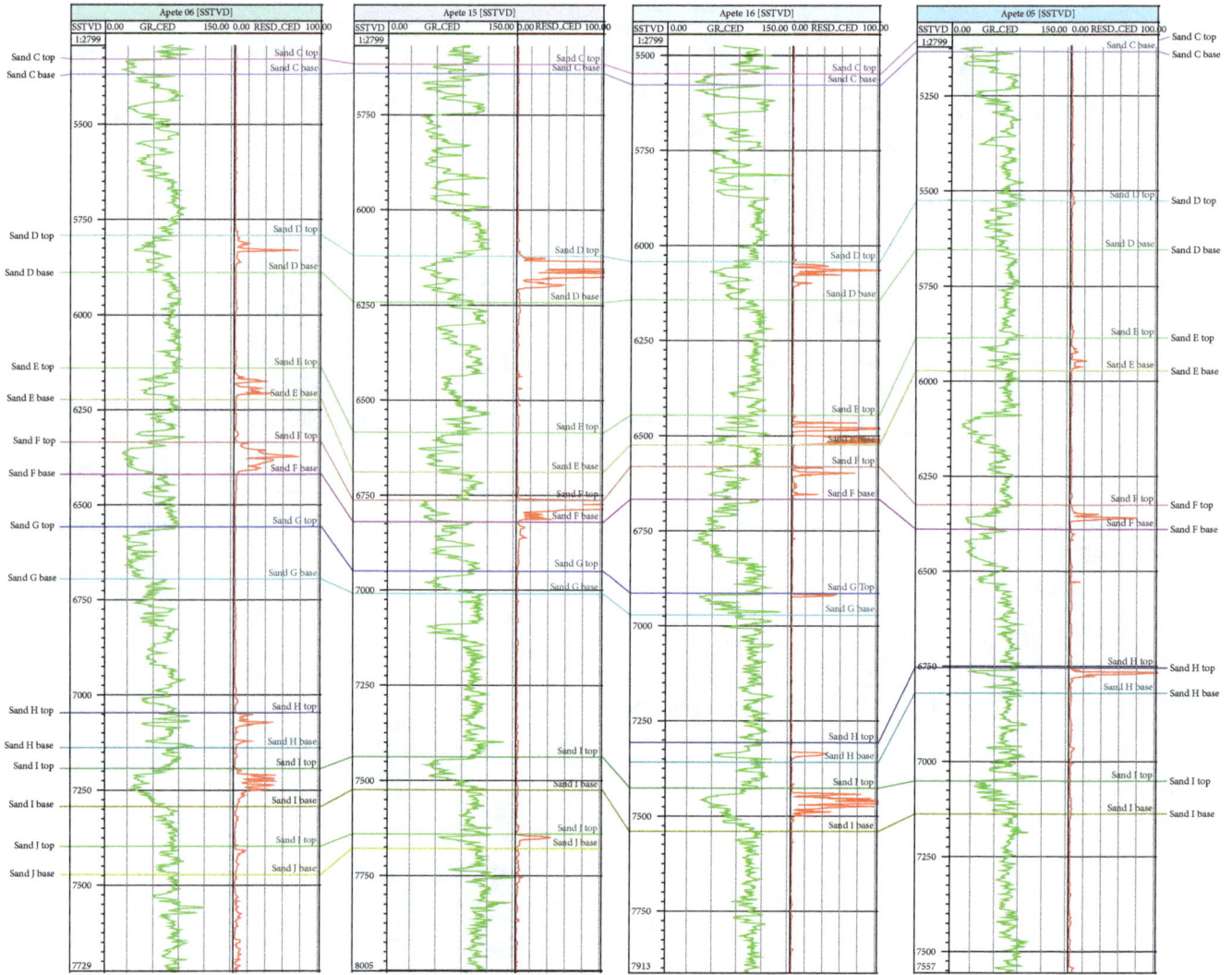

FIGURE 3: Well correlation of all reservoir sands.

TABLE 1: Average total porosity for all reservoir sands.

Reservoir	Apete 05	Apete 06	Apete 15	Apete 16
A	—	0.254	—	0.250
B	0.223	0.225	—	0.224
C	0.280	0.234	—	—
D	—	0.196	0.174	0.122
E	0.214	0.230	—	0.104
F	0.272	0.248	0.239	0.154
G	—	0.232	—	0.178
H	0.210	0.177	—	0.129
I	—	0.204	—	0.141
J	—	0.126	0.172	—

Blank spaces mean the zone is not a reservoir.

TABLE 2: Average effective porosity for all reservoir sands.

Reservoir	Apete 05	Apete 06	Apete 15	Apete 16
A	—	0.247	—	0.218
B	0.219	0.218	—	0.203
C	0.259	0.214	—	—
D	—	0.167	0.158	0.111
E	0.201	0.207	—	0.101
F	0.250	0.221	0.193	0.145
G	—	0.211	—	0.169
H	0.205	0.160	—	0.113
I	—	0.200	—	0.137
J	—	0.123	0.167	—

Blank spaces mean the zone is not a reservoir.

versus Φ, was generated to depict whether or not the sands are at irreducible water saturation. Porosity is plotted on the y-axis with a scale ranging from 0 to 40% porosity (shown in decimals), while water saturation is plotted on the x-axis with a scale ranging from 0 to 100% (shown in decimals) water saturation. The scale for bulk volume water lines (grey lines) ranges from 0.01 to 0.25 and is shown as a secondary y-axis. The implicit assumption in the Buckles plot approach

TABLE 3: Average net-to-gross ratio for all reservoir sands.

Reservoir	Apete 05	Apete 06	Apete 15	Apete 16
A	—	0.793	—	0.826
B	0.857	0.771	—	0.851
C	0.515	0.933	—	—
D	—	0.712	0.622	0.449
E	0.654	0.615	—	0.125
F	0.797	0.931	0.886	0.785
G	—	0.981	—	0.692
H	0.389	0.278	—	0.289
I	—	0.853	—	0.654
J	—	0.328	0.279	—

Blank spaces mean the zone is not a reservoir.

TABLE 4: Average volume of shale for all reservoir sands.

Reservoir	Apete 05	Apete 06	Apete 15	Apete 16
A	—	0.217	—	0.308
B	0.292	0.339	—	0.243
C	0.162	0.122	—	—
D	—	0.363	0.290	0.350
E	0.415	0.346	—	0.274
F	0.245	0.192	0.288	0.312
G	—	0.162	—	0.276
H	0.327	0.360	—	0.288
I	—	0.258	—	0.279
J	—	0.469	0.357	—

Blank spaces mean the zone is not a reservoir.

TABLE 5: Average water saturation from dual water model for all reservoir sands.

Reservoir	Apete 05	Apete 06	Apete 15	Apete 16
A	—	0.333	—	0.238
B	0.071	0.298	—	0.106
C	0.200	0.436	—	—
D	—	0.118	0.169	0.231
E	0.223	0.002	—	0.046
F	0.353	0.071	0.164	0.246
G	—	0.460	—	0.176
H	0.107	0.023	—	0.079
I	—	0.017	—	0.087
J	—	0.011	0.016	—

Blank spaces mean the zone is not a reservoir.

TABLE 6: Average water saturation from Archie's model for all reservoir sands.

Reservoir	Apete 05	Apete 06	Apete 15	Apete 16
A	—	0.377	—	0.324
B	0.164	0.406	—	0.189
C	0.221	0.449	—	—
D	—	0.241	0.232	0.384
E	0.355	0.115	—	0.140
F	0.417	0.113	0.257	0.339
G	—	0.492	—	0.244
H	0.192	0.151	—	0.157
I	—	0.072	—	0.177
J	—	0.229	0.135	—

TABLE 7: Net thickness (ft) for all reservoir sands.

Reservoir	Apete 05	Apete 06	Apete 15	Apete 16
A	—	78.50	—	57.00
B	12.00	45.50	—	31.50
C	34.00	42.00	—	—
D	—	89.00	102.00	62.00
E	78.50	40.00	—	13.00
F	169.50	95.00	156.00	67.50
G	—	126.50	—	45.00
H	27.50	22.00	—	11.00
I	—	81.00	—	70.00
J	—	20.00	12.00	—

is that the product of irreducible water saturation and porosity is constant. Empirical relationships were also established for porosity-depth, water saturation-depth, water saturation-porosity, and permeability-depth to check the trends.

4. Results and Discussions

Petrophysical attributes as porosity (effective and total), reservoir thickness (net and gross), water saturation (Archie and dual water model), and volume of shale (Tables 1, 2, 3, 4, 5, 6, and 7) were delineated in this research work. Results from petrophysical analysis revealed reservoir Sand F to be the most viable reservoir with net thickness as high as 126.50 ft. All the ten reservoirs exhibited good petrophysical attributes with high porosity and hydrocarbon saturation. The sands are shaly sands with high volume of shale resulting in overestimated water saturation in the reservoirs; dual water model was used for estimating the water saturation for better appraisal of the reservoirs due to their shaly nature.

4.1. Crossplot. Correlation analysis was performed to determine whether the petrophysical attributes (water saturation and porosity) are interdependent. Generally, the effective porosity decreases with depth (Figure 4) with high correlation coefficient except in Apete 15 where there was increase in porosity with depth. The observed reduction in depth would likely be due to the effect of compaction resulting from overburden pressure. Water saturation generally increases with depth (Figure 5) except in Apete 16. This implies that reservoirs in Apete field occur in shallow depth hence the unavailability of reservoirs as we go deeper into the subsurface. Efforts made to delineate trends for water saturation and effective porosity were marginally efficient. From crossplot results (Figure 6), effective porosity reduces with an increase in water saturation in Apete 05 and 06. Specifically, in Apete 15 and 16, there was no correlation at all,

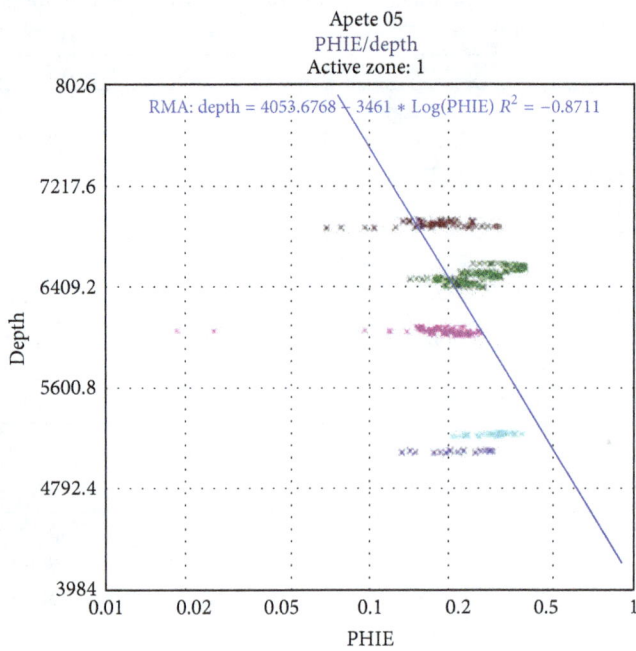

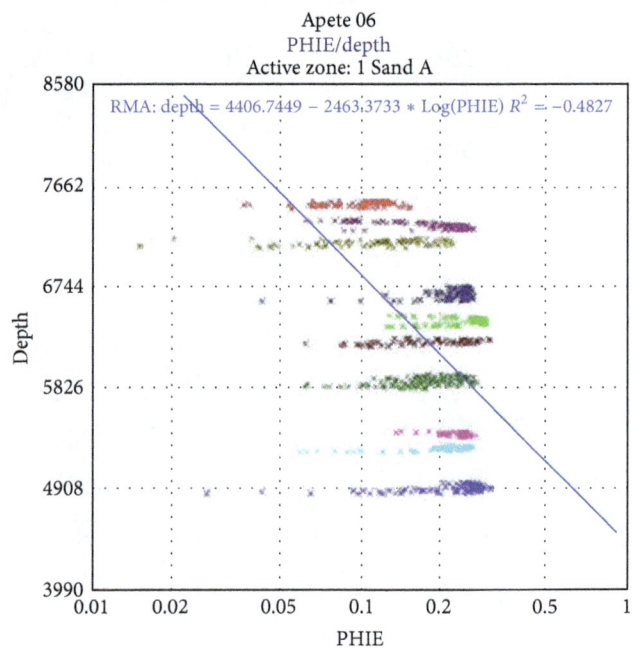

376 points plotted out of 376

Zone	Depths		Zone	Depths
× (1)	5097 F–5111 F		× (4)	6413 F–6601 F
× (2)	5224 F–5258 F		× (5)	6877 F–6940 F
× (3)	6024 F–6096 F			

869 points plotted out of 875

Zone	Depths		Zone	Depths
× (1) Sand A	4860 F–4959 F		× (6) Sand F	6380 F–6482 F
× (2) Sand B	5241 F–5300 F		× (7) Sand G	6617 F–6746 F
× (3) Sand C	5385 F–5430 F		× (8) Sand H	7100 F–7185 F
× (4) Sand D	5818 F–5943 F		× (9) Sand I	7257 F–7352 F
× (5) Sand E	6206 F–6271 F		× (10) Sand J	7462 F–7523 F

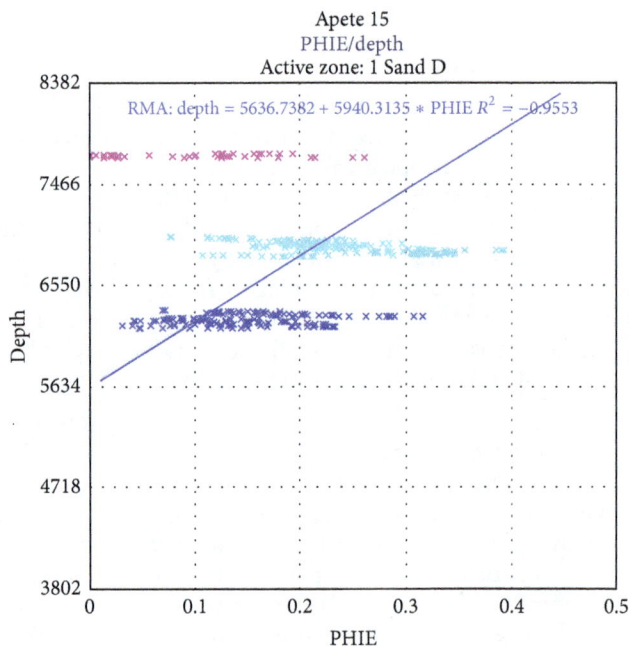

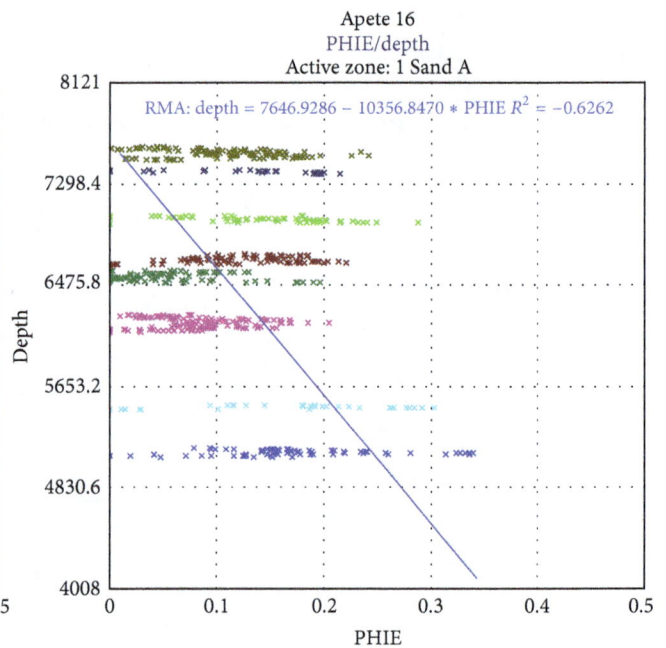

386 points plotted out of 386

Zone	Depths
× (1) Sand D	6160 F–6324 F
× (2)	6810 F–6986 F
× (3)	7706 F–7749 F

652 points plotted out of 652

Zone	Depths		Zone	Depths
× (1) Sand A	5080 F–5149 F		× (5)	6642 F–6728 F
× (2) Sand B	5469 F–5506 F		× (6)	6980 F–7045 F
× (3) Sand D	6098 F–6236 F		× (7)	7378 F–7416 F
× (4) Sand E	6494 F–6598 F		× (8)	7492 F–7599 F

FIGURE 4: Porosity-depth plot for Apete 05, 06, 15, and 16.

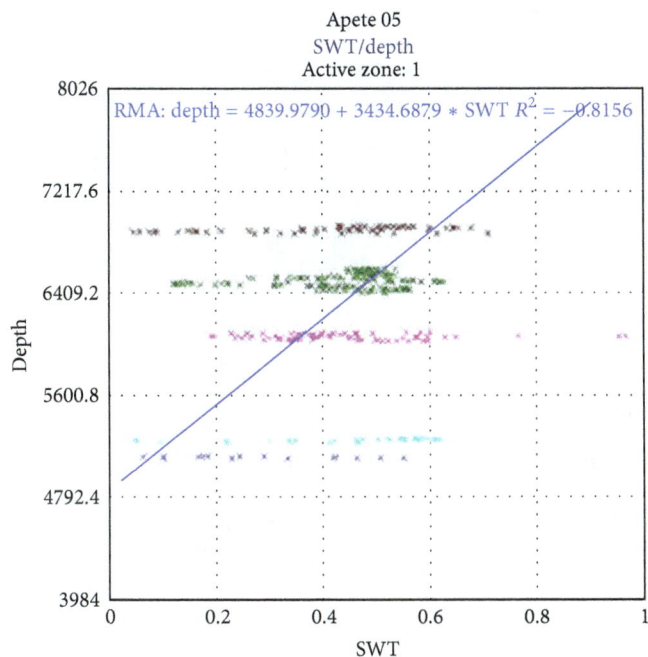

Apete 05
SWT/depth
Active zone: 1

RMA: depth = 4839.9790 + 3434.6879 ∗ SWT R^2 = −0.8156

376 points plotted out of 376

Zone	Depths	Zone	Depths
× (1)	5097 F–5111 F	× (4)	6413 F–6601 F
× (2)	5224 F–5258 F	× (5)	6877 F–6940 F
× (3)	6024 F–6096 F		

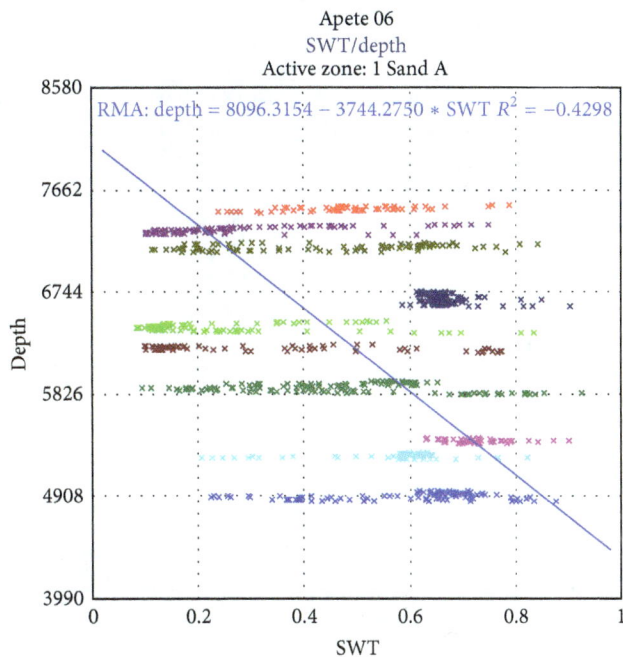

Apete 06
SWT/depth
Active zone: 1 Sand A

RMA: depth = 8096.3154 − 3744.2750 ∗ SWT R^2 = −0.4298

875 points plotted out of 875

Zone	Depths	Zone	Depths
× (1) Sand A	4860 F–4959 F	× (6)	6380 F–6482 F
× (2) Sand B	5241 F–5300 F	× (7)	6617 F–6746 F
× (3) Sand C	5385 F–5430 F	× (8)	7100 F–7185 F
× (4) Sand D	5818 F–5943 F	× (9)	7257 F–7352 F
× (5) Sand E	6206 F–6271 F	× (10)	7462 F–7523 F

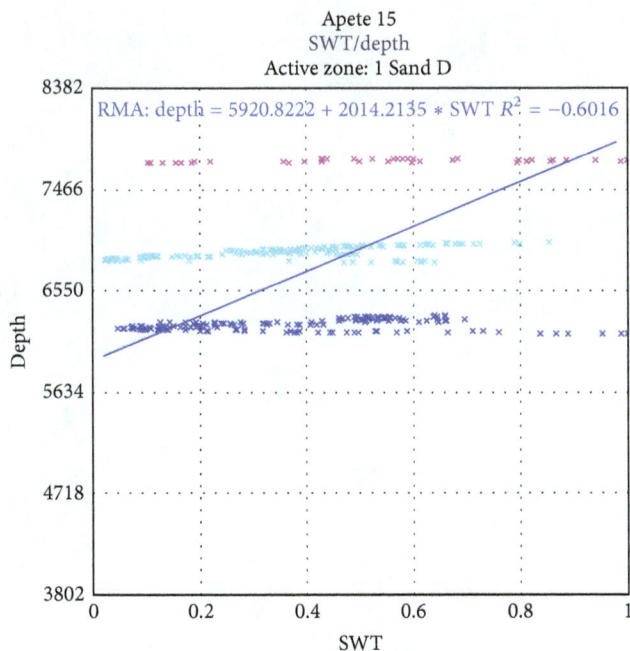

Apete 15
SWT/depth
Active zone: 1 Sand D

RMA: depth = 5920.8222 + 2014.2135 ∗ SWT R^2 = −0.6016

386 points plotted out of 386

Zone	Depths
× (1) Sand D	6160 F–6324 F
× (2)	6810 F–6986 F
× (3)	7706 F–7749 F

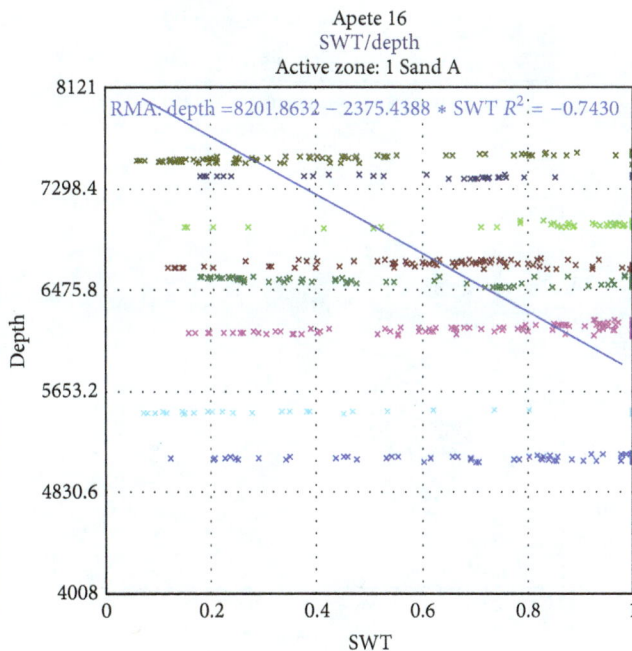

Apete 16
SWT/depth
Active zone: 1 Sand A

RMA: depth = 8201.8632 − 2375.4388 ∗ SWT R^2 = −0.7430

652 points plotted out of 652

Zone	Depths	Zone	Depths
× (1) Sand A	5080 F–5149 F	× (5)	6642 F–6728 F
× (2) Sand B	5469 F–5506 F	× (6)	6980 F–7045 F
× (3) Sand D	6098 F–6236 F	× (7)	7378 F–7416 F
× (4) Sand E	6494 F–6598 F	× (8)	7492 F–7599 F

FIGURE 5: Water saturation-depth plot for Apete 05, 06, 15, and 16.

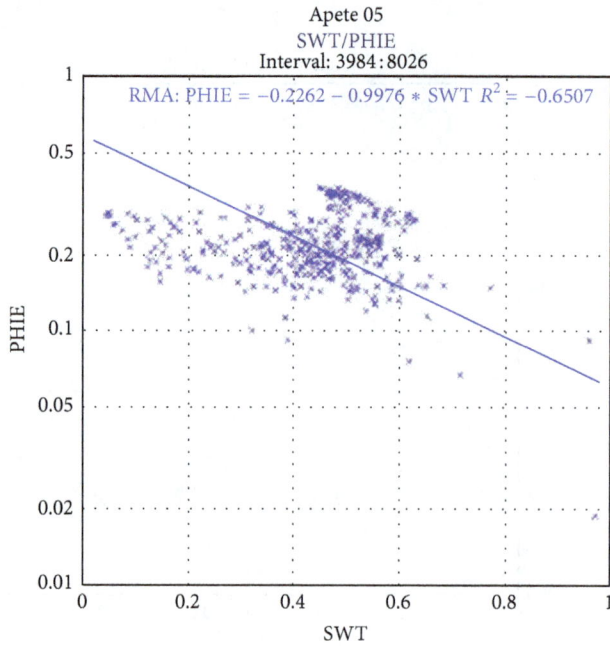

424 points plotted out of 4043

Well	Depths
× (1) Apete 05	3984 F–8026 F

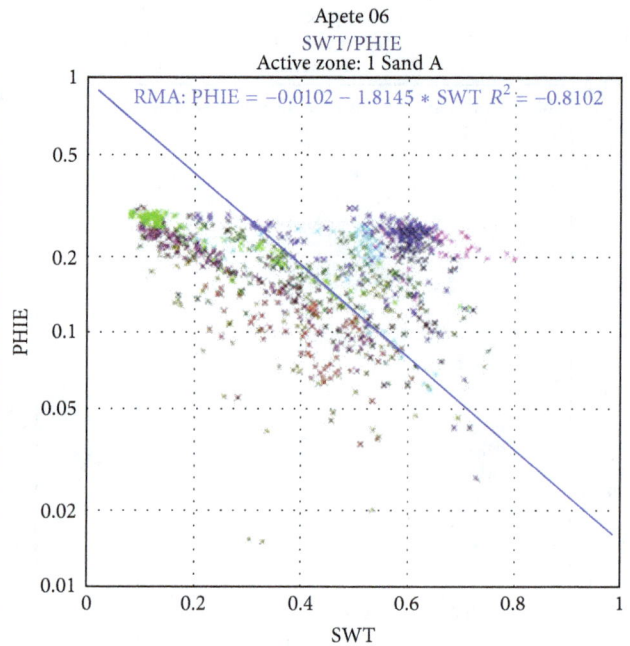

869 points plotted out of 875

Zone	Depths		Zone	Depths
× (1) Sand A	4860 F–4959 F		× (6) Sand F	6380 F–6482 F
× (2) Sand B	5241 F–5300 F		× (7) Sand G	6617 F–6746 F
× (3) Sand C	5385 F–5430 F		× (8) Sand H	7100 F–7185 F
× (4) Sand D	5818 F–5943 F		× (9) Sand I	7257 F–7352 F
× (5) Sand E	6206 F–6482 F		× (10) Sand J	7462 F–7523 F

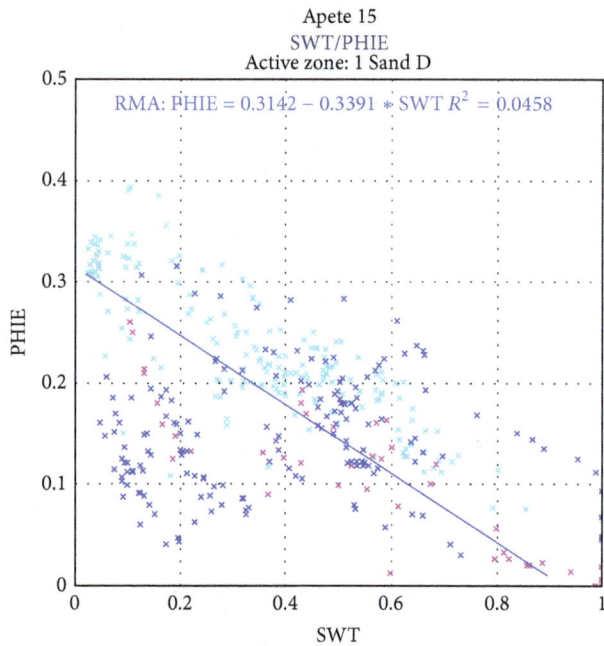

386 points plotted out of 386

Zone	Depths
× (1) Sand D	6160 F–6324 F
× (2)	6810 F–6986 F
× (3)	7706 F–7749 F

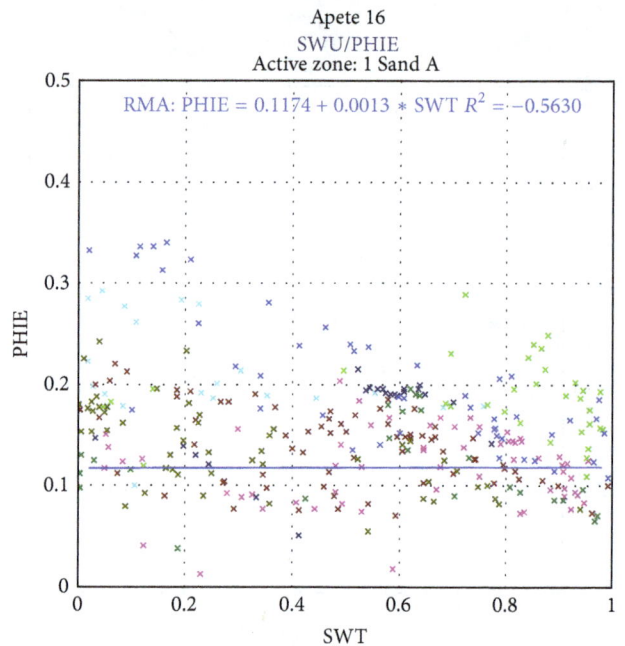

328 points plotted out of 652

Zone	Depths		Zone	Depths
× (1) Sand A	5080 F–5149 F		× (5)	6642 F–6728 F
× (2) Sand B	5469 F–5506 F		× (6)	6980 F–7045 F
× (3) Sand D	6098 F–6236 F		× (7)	7378 F–7416 F
× (4) Sand E	6494 F–6598 F		× (8)	7492 F–7599 F

FIGURE 6: Effective porosity—water saturation trend for Apete 05, 06, 15, and 16.

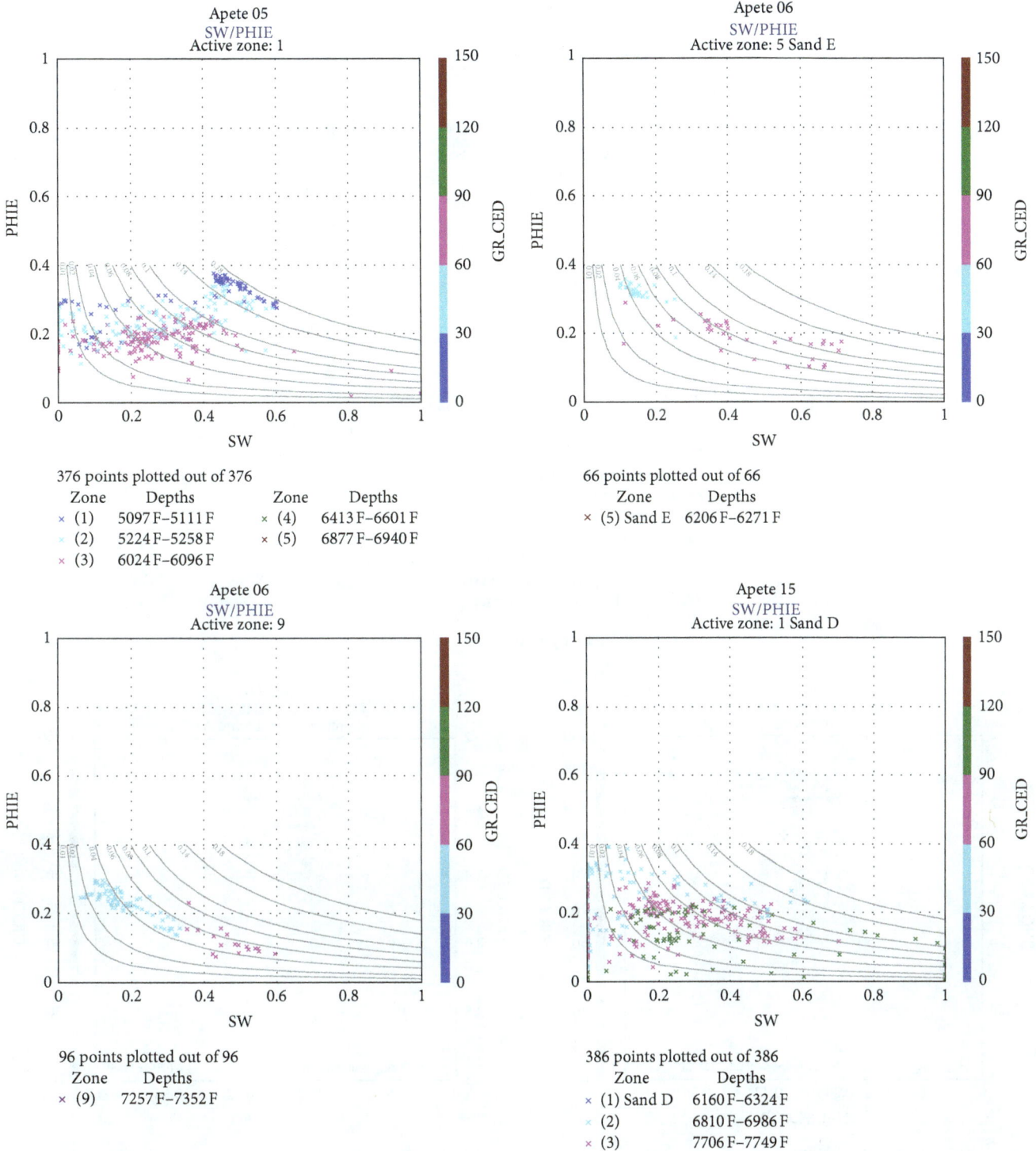

FIGURE 7: Buckles plot for Apete 05, 06, and 15.

as the correlation coefficient was extremely low which suggests that no relationship exists between the two petrophysical attributes. Buckles plot for the reservoirs in the wells were generated (Figure 7); the results from these plots reveal only Apete 06 to be at irreducible water saturation as the data points align along the bulk volume of water (BVW) trend line due to the consistency of the data points. The reservoirs

zones in this well are considered to be homogenous; therefore, hydrocarbon production from Apete 06 should be water free [14]; that is, the reservoirs would have a low water cut.

Pickett plot (Figure 8) reveals the reservoirs to be somewhat shaly which is observed with saturation exponent being less than 2 in the best porosity type. This is further confirmed from the Neutron-Density crossplot (Figure 9). Pickett plot

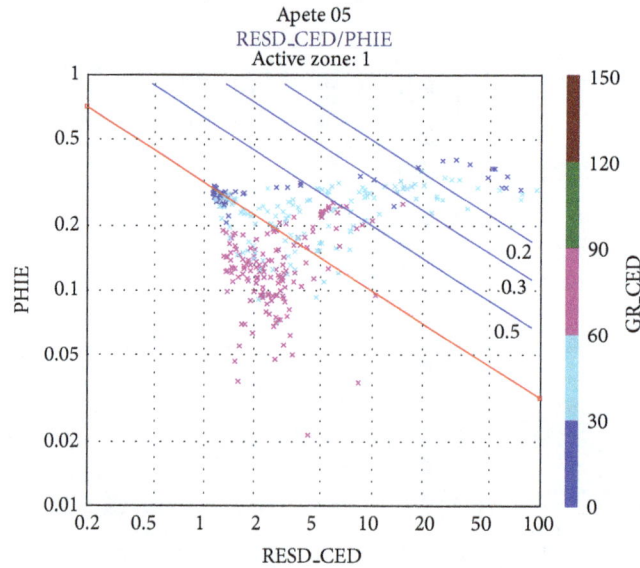

369 points plotted out of 376

Zone	Depths	
× (1)	5097 F–5111 F	Parameter: *Rw*: 0.1
× (2)	5224 F–5258 F	Parameter: *Rw* form temp: 0.1
× (3)	6024 F–6096 F	Parameter: *m* exponent: 2
× (4)	6413 F–6601 F	Parameter: *n* exponent: 2
× (5)	6877 F–6940 F	Parameter: *a* factor: 1

FIGURE 8: Pickett plot of Sand C in Apete 05.

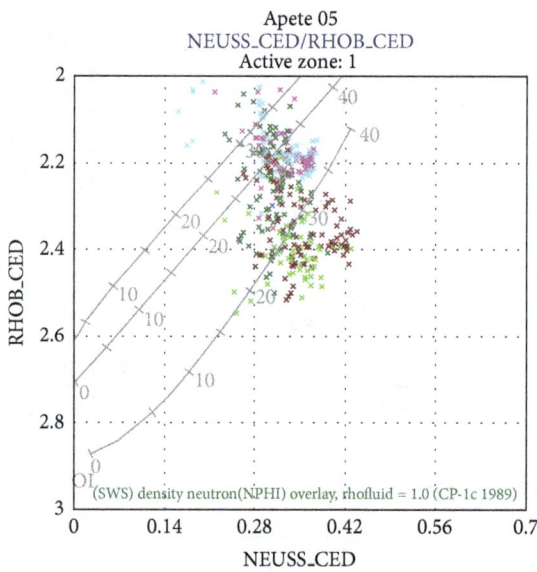

374 points plotted out of 376

Zone	Depths	Zone	Depths
× (1)	5097 F–5111 F	× (4)	6413 F–6601 F
× (2)	5224 F–5258 F	× (5)	6877 F–6940 F
× (3)	6024 F–6096 F		

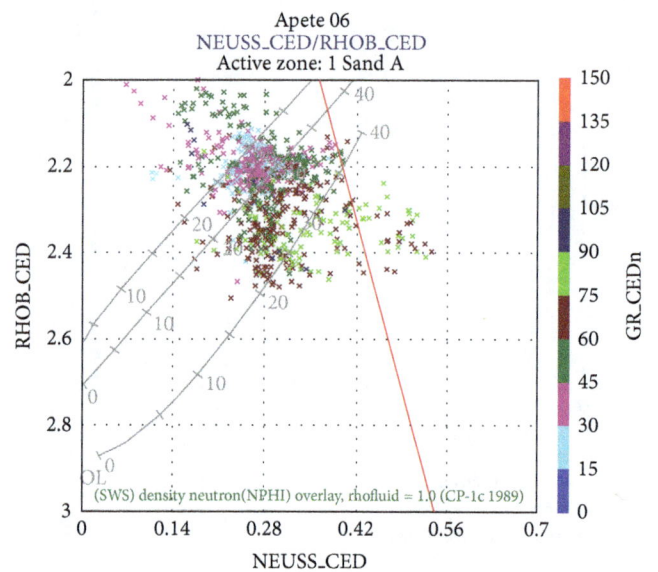

885 points plotted out of 886

Zone	Depths	Zone	Depths
× (1) Sand A	4860 F–4959 F	× (6) Sand F	6380 F–6482 F
× (2) Sand B	5241 F–5300 F	× (7) Sand G	6617 F–6746 F
× (3) Sand C	5385 F–5430 F	× (8) Sand H	7100 F–7185 F
× (4) Sand D	5818 F–5943 F	× (9) Sand I	7246 F–7352 F
× (5) Sand E	6206 F–6271 F	× (10) Sand J	7462 F–7523 F

Parameter : GR clean : 20.6   Parameter : GR method : 0
Parameter : GR clay : 114   Parameter : Neu clay : 0
Parameter : Neu clean : 0   Parameter : Neu use : 0

FIGURE 9: Neutron-density crossplot for Apete 05 and 06.

was used to determine Archie parameters, tortuosity (a), and cementation exponent (m) which is approximately 1 and 2, respectively. Neutron density crossplot reveals more of laminated clay in the reservoirs; this should be taken into consideration during well planning. The shale morphology generally changed from laminated to dispersed which affects saturation mixing function hence the need to use another saturation model (Schlumberger's dual water model) which is designed specifically for shaly sands, rather than the conventional Archie's water saturation model. Results from both water saturation models used show a wide disparity which could not have been noticed if only the conventional Archie's model was used which could have led to bypassing some reservoirs as well as undervaluating the reserves in this field.

5. Conclusion

In the study of well logs from the Apete field, Niger Delta, it was observed that Apete 06 is the most economic well drilled in this field; apart from having the most presence of reservoirs it also has the highest net thickness of 639.50 ft (Table 7). Generally, water saturation increases with depth and porosity reduces with depth as a result of compaction. Water saturation-porosity trends cannot be emphatically established except in Apete 05 and 06 where porosity reduces with increase in water saturation. The reservoirs are shaly sands with the shales mostly occurring as laminated clays which could act as impediment to flow during production and therefore causing reservoir compartmentalization. Shale morphology changes from laminated to dispersed, thereby affecting saturation mixing functions.

Glossary

Buckle's plot: A plot of water saturation (S_w) against porosity (Φ) generated to depict whether or not the sands are at irreducible water saturation (Φ on y-axis and S_w on the x-axis)

Bulk volume of water (BVW): Percentage of the total rock volume occupied by water

Core data: Set of data derived from analysis of core (rock) samples

Eustatic sea level: Sea level change which occurs on a global scale

Fluviatile: Pertaining or relating to rivers, found in or near rivers

Fossiliferous: Means containing fossils

Laminated: Composed of layers bonded together

Marine flooding surface: A surface of deposition at the time the shoreline is at its highest landward position

Morphology: Means this refers to the description of the shape of geologic features

Offlap: The arrangement of strata deposited on the sea floor during the progressive withdrawal of the sea from land

Pickett plot: Plot of Archie's saturation parameters against resistivity of water so as to estimate the water saturation in such a reservoir

Transgression: Progressive movement of the sea towards land

Transitional environment: Environment situated between the continental realm and the marine

Water cut: Amount of water produced with oil.

References

[1] D. Toby, *Well Logging and Formation Evaluation*, Elsevier, San Francisco, Calif, USA, 2005.

[2] Next, Seismic Reservoir Analysis Course Notes, 2003.

[3] N. G. Obaje, "Fairways and reservoir prospects of plocene—recent sands in the shallows offshore Niger delta," *Journal of Mining and Geology*, vol. 40, pp. 25–38, 2005.

[4] K. J. Weber, "Sedimentological aspects of oil fields in the Niger delta," *Geologie en Mijnbouw*, vol. 50, pp. 559–576, 1972.

[5] J. E. Ejedawe, "The eastern Niger delta: geological evolution and hydrocarbon occurrences," SPDC Internal Report Exploration Note 89. 002, 1989.

[6] K. C. Short and J. Stauble, "Outline geology of the Niger delta," *AAPG Bulletin*, vol. 51, no. 5, pp. 761–779, 1967.

[7] P. Shannon and D. Naylor, *Petroleum Basin Studies*, Graham & Trotman, London, UK, 1989.

[8] H. Doust and E. Omatsola, "Niger delta," in *Divergent/Passive Margin Basins*, J. D. Edwards and P. A. Santogrossi, Eds., vol. 48, pp. 239–248, American Association of Petroleum Geologists, Tulsa, Okla, USA.

[9] T. A. Reijers, C. S. Nwajide, and A. A. Adesida, "Sedimentology and lithostratigraphy of the Niger delta," in *Proceedings of the 15th International Conference of the Nigerian Association of Petroleum Explorationist (NAPE)*, Lagos, Nigeria, 1997.

[10] A. A. Adesida, T. J. A. Reijers, and C. S. Nwajide, "Sequence stratigraphic framework of the Niger delta," in *Proceedings of the AAPG International Conference and Exhibition*, Vienna, Austria, 1997.

[11] W. E. Galloway, "Genetic stratigraphic sequences in basin analysis I: architecture and genesis of flooding-surface bounded depositional units," *American Association of Petroleum Geologists Bulletin*, vol. 73, no. 2, pp. 125–142, 1989.

[12] *Log Interpretation, Principles and Application*, Schlumberger Wireline and Testing, Houston, Tex, USA, pp. 21–89, 1989.

[13] V. V. Larionov, *Borehole Radiometry*, National Electric Drag Racing Association, Moscow, Soviet Union, 1969.

[14] R. L. Morris and W. P. Biggs, "Using log-derived values of water saturation and porosity," in *Proceedings of the South West Powerlifting Association Annual Logging Symposium*, vol. 10, p. 26.

Estimating the Thickness of Sedimentation within Lower Benue Basin and Upper Anambra Basin, Nigeria, Using Both Spectral Depth Determination and Source Parameter Imaging

Adetona A. Abbass[1] and Abu Mallam[1,2]

[1] *Department of Physics, Federal University of Technology P.M.B. 65 Minna, Niger State, Nigeria*
[2] *Department of Physics, University of Abuja, P.M.B. 117Abuja, Nigeria*

Correspondence should be addressed to Abu Mallam; mallamabu@yahoo.com

Academic Editors: A. Donnellan and S. Pullammanappallil

The Total Aeromagnetic Data covering the study area was subjected to First Vertical Derivative, Spectral Depth Analysis, and Source Parameter Imaging (SPI). The result from the First Vertical Derivative shows that the Northern part of the area is covered by the young biotite granite of Precambrian origin, and the western edge is covered by the old granite, gneisses, and migmatite of Western Nigeria, while the remaining area is covered by the cretaceous sedimentary deposits. The entire area was divided into forty-eight sections. Spectral Depth Analysis was run for each of these forty-eight sections; the result shows that a maximum depth above 7 km was obtained within the cretaceous sediments of Idah, Ankpa, and below Udegi at the middle of the study area. Minimum depth estimates between 188.0 and 452 meters were observed around the basement regions. Results from Source Parameter Imaging show a minimum depth of 76.983 meters and a maximum thickness of sedimentation of 9.847 km, which also occur within Idah, Ankpa, and Udegi axis. The disparity observed in depth obtained by each method is discussed based on the merit and demerit of each method, and the depths obtained were compared with results from previous researchers. Geophysical implication of the result to oil and gas exploration in the area is briefly discussed.

1. Introduction

Of all the magnetic minerals that occur in nature, magnetite is the most abundant. Aeromagnetic surveys reflect almost exclusively the distribution of magnetite and pyrrhotite in rocks. On a global basis, the others can probably be ignored [1, 2]. Thus aeromagnetic surveys, in particular terms, map the magnetite in the rocks below the aircraft. While aeromagnetic surveys are extensively used as reconnaissance tools, there has been an increasing recognition of their value for evaluating prospective areas by virtue of the unique information they provide. Outline of the roles of aeromagnetic survey is as follows [3].

(i) Delineation of volcano-sedimentary belts under sand or other recent cover, or in strongly metamorphosed terrains when recent lithologies are otherwise unrecognizable.

(ii) Recognition and interpretation of faulting, shearing, and fracturing not only as potential hosts for a variety of minerals, but also an indirect guide to epigenetic, stress related mineralization in the surrounding rocks.

(iii) Identification and delineation of post-tectonic intrusive. Typical of such targets are zoned syenite or carbonatite complexes, kinerlites, tin-bearing granites, and mafic intrusions.

(iv) Direct detection of deposits of certain iron ores.

(v) In prospecting for oil, aeromagnetic data can give information from which one can determine depths to basement rocks and thus locate and define the extent of sedimentary basins. Sedimentary rocks however exert such a small magnetic effect compared with

Estimating the Thickness of Sedimentation within Lower Benue Basin and Upper Anambra Basin, Nigeria, Using Both
Spectral Depth Determination and Source Parameter Imaging

13

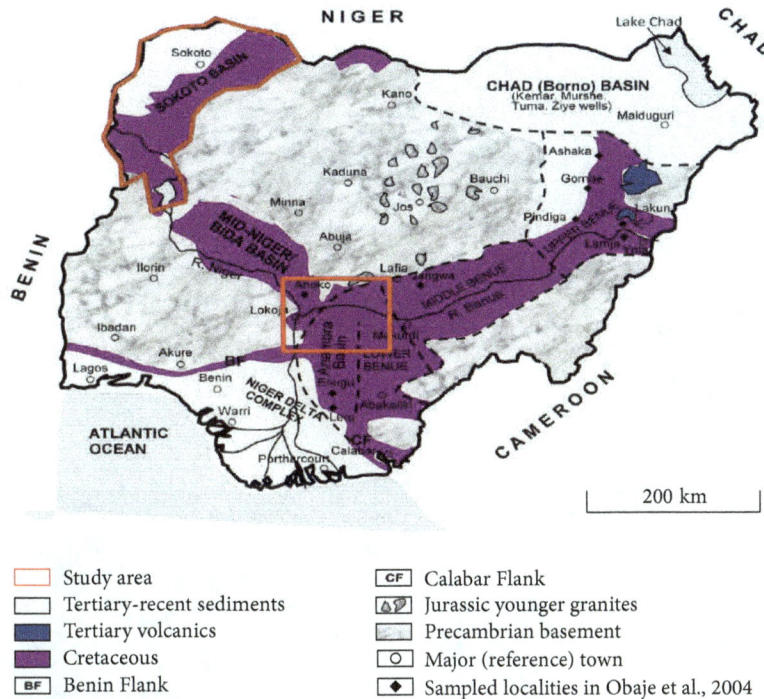

FIGURE 1: Geology map of Nigerian showing the location of the study area.

igneous rocks that virtually all variations in magnetic intensity measurable at the surface result from topographic or lithologic changes associated with the basement or from igneous intrusions [4].

In this paper, a combination of source parameter imaging and Euler deconvolution were employed to evaluate the depth to source magnetic rocks within lower parts of Benue Basin and upper parts of Anambra basin.

2. Location and Extent of the Study Area

The study area covers the Lower Benue Trough, the Upper part of Anambra Basin, and the basement complexes bounding it at the West and Northern edges (Figure 1). The area is bounded by Latitude 7.0°N to 8.5°N and Longitude 6.5°E to 8.5°E. The physiological features recognized in the area are the river Benue, river Anambra, and river Okulu. Twelve aeromagnetic maps covered the study area and are numbered (227, 228, 229, 230, 247, 248, 249, 250, 267, 268, 269, and 270), a total area of 36,300 square kilometers. The study area touches four states majorly, which are Nassarawa at the upper part, Kogi, Enugu, and Benue States at the lower part (Figure 2).

2.1. Geology of Lower Benue and Upper Anambra Basin. Sedimentation in the Lower Benue Trough commenced with the marine Albian Asu River Group, although some pyroclastics of Aptian-Early Albian ages have been sparingly reported [5]. The Asu River Group in the Lower Benue Trough comprises the shales, limestones, and sandstone lenses of the Abakaliki

Formation in the Abakaliki area and the Mfamosing Limestone in the Calabar Flank [6]. The marine Cenomanian-Turonian Nkalagu Formation (black shales, limestones, and siltsones) and the interfingering regressive sandstones of the Agala and Agbani Formations rest on the Asu River Group. Mid-Santonian deformation in the Benue Trough displaced the major depositional axis westward which led to the formation of the Anambra Basin. Post-deformational sedimentation in the Lower Benue Trough, therefore, constitutes the Anambra Basin. Sedimentation in the Anambra Basin thus commenced with the Campanian-Maastrichtian marine and paralic shales of the Enugu and Nkporo Formations, overlaid by the coal measures of the Mamu Formation. The fluviodeltaic sandstones of the Ajali and Owelli Formations lie on the Mamu Formation and constitute its lateral equivalents in most places. In the Paleocene, the marine shales of the Imo and Nsukka Formations were deposited, overlain by the tidal Nanka Sandstone of Eocene age. Downdip, towards the Niger Delta, the Akata Shale and the Agbada Formation constitute the Paleogene equivalents of the Anambra Basin (Figure 2). The Basin Formation and the Imo Shale mark the onset of another transgression in the Anambra during the aleocene. The shales contain significant amount of organic matter and may be potential source for the hydrocarbons in the northern part of the Niger Delta [7]. In the Anambra Basin, they are only locally expected to reach maturity levels for hydrocarbon expulsion.

The Enugu and the Nkporo Shales represent the brackish marsh and fossiliferous pro-delta facies of the Late Campanian-Early Maastrichtian depositional cycle [7]. Deposition of the sediments of the Nkporo/Enugu

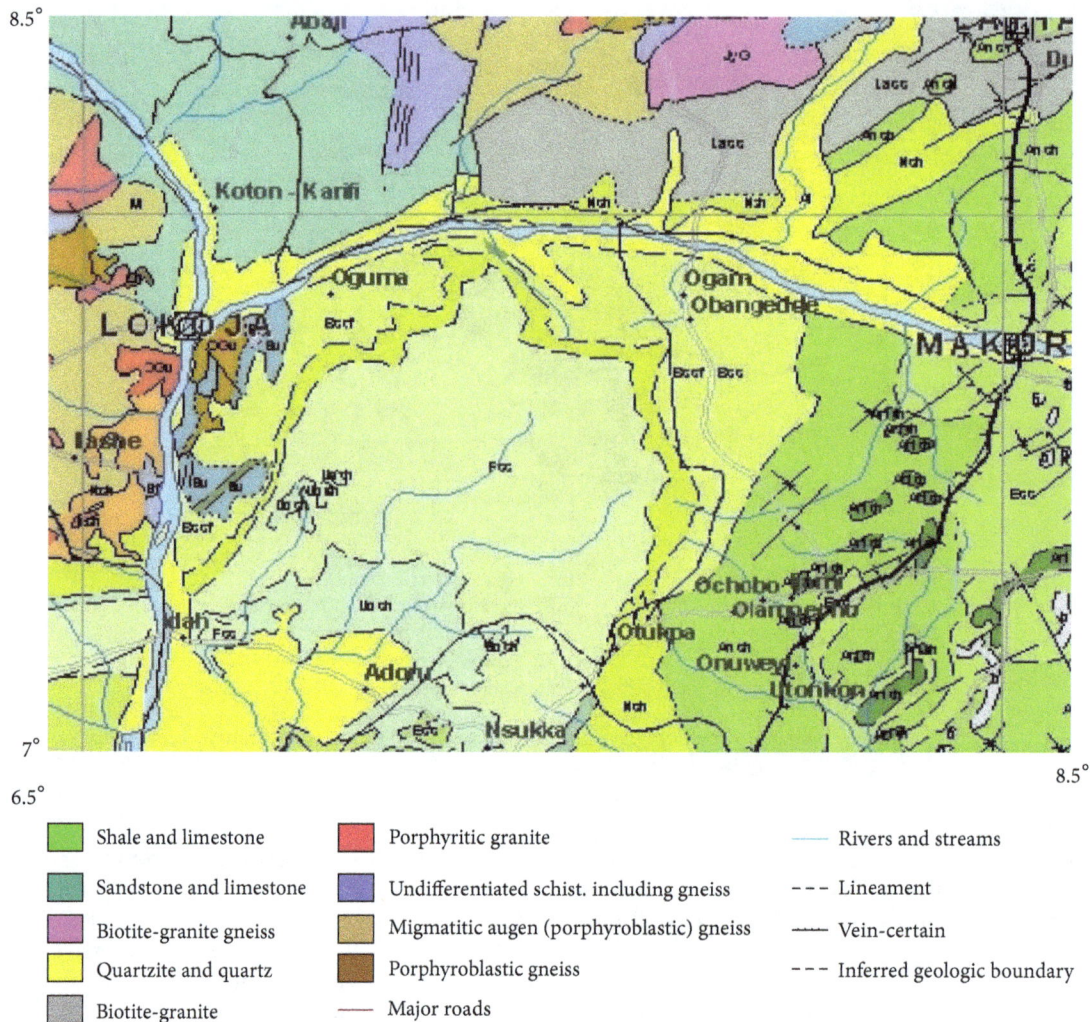

FIGURE 2: Geology map of the study area (adapted from the Geological and Mineral Map of Nigeria, 2009, Nigerian Geological Survey Agency).

Formations reflects a funnel-shaped shallow marine setting that graded into channeled low-energy marshes. The coal-bearing Mamu Formation and the Ajali Sandstone accumulated during this epoch of overall regression of the Nkporo cycle. The Mamu Formation occurs as a narrow strip trending north-south from the Calabar Flank, swinging west around the Ankpa plateau and terminating at Idah near the River Niger. The best exposure of the Nkporo Shale is at the village of Leru (Lopauku). The Ajali Sandstone marks the height of the regression at a time when the coastline was still concave. The converging littoral drift cells governed the sedimentation and are reflected in the tidal sand waves which are characteristic for the 72 km south of Enugu on the Enugu-Portharcourt express road, while that of Enugu Shale is at Enugu, near the Onitsha-Road flyover. The Mamu Formation is best exposed at the Miliken Hills in Enugu, with well-preserved secions along the road cuts from the King Petrol Station up the Miliken Hills and at the left bank of River Ekulu near the bridge to Onyeama mine.

3. Materials and Methods

The procedures employed in this research include the following.

(1) Production of Total Magnetic Intensity (TMI) map of the study area using MONTAJ software.

(2) Perform vertical derivative of the TMI data to enhance shallow geological features and horizontal derivative to identify geology boundaries in the profile data.

(3) Production of the Geomagnetic Map of the Study Area from Magnetic Signatures and Geology Map.

(4) Spectral depth determination to buried magnetic rocks within the study area.

(5) Depth evaluation using Source Parameter Imaging.

3.1. Source of Aeromagnetic Data. A new dataset has been generated from the largest airborne geophysical survey ever undertaking in Nigeria, which is helping to position the country as an exciting destination for explorers. This survey which was conducted in three phases between 2005 and 2010 was partly financed the Nigerian Federal Government and the World Bank as part of a major project known as the Sustainable Management for Mineral Resources Project. All of the airborne geophysical work, data acquisition processing and compilation, was carried out by Fugro Airborne Surveys; the survey acquired both magnetic and radiometric data compilation. The recent survey has a Tie-line spacing of 500 m, flight line spacing of 100 m, and Terrain clearance of 100 m using TEMPEST system. Compared with the 1970s survey which has a Tie-line spacing of 20 km, flight line spacing of 2 km, and flying altitude of 200 m, these levels of survey are intensive and detailed for the objectives of this research. Data covering the twelve aeromagnetic sheets numbered (227, 228, 229, 230, 247, 248, 249, 250, 267, 268, 269, and 270) was acquired from The Nigerian Geological Survey Agency, 31, Shetima Mangono Crescent Utako District, Garki, Abuja.

3.2. Spectral Depth-Determination Methods. The Fourier transform of the potential filed due to a prismatic body has a broad spectrum whose peak location is a function of the depth to the top and bottom surfaces and whose amplitude is determined by its density or magnetization. You can relate the peak wavenumber (ω') to the geometry of the body according to the following expression [8]:

$$\omega' = \frac{\lin (h_b/h_t)}{h_b - h_t}. \tag{1}$$

ω' is the peak wavenumber in radian or ground-unit, h_t is the depth to the top, and h_b is the depth to the bottom. For a bottomless prism, the spectrum peak at the zero wavenumber is according to the expression

$$f(\omega) = e^{-h\omega}, \tag{2}$$

where ω is the angular wavenumber in radians/ground-unit and h is depth to the top of the prism [9].

When considering a line that is long enough to include many sources, you can use the log spectrum of these data to determine the depth to the top of a statistical ensemble of sources using the relationship

$$\Log E(k) = 4\pi hk, \tag{3}$$

where h is the depth in ground-units and k is the wavenumber in cycles/ground-unit.

You can determine the depth of an "ensemble" of source by measuring the slope of the energy (power) spectrum and dividing by 4π. A typical energy spectrum for magnetic data may exhibit three parts—a deep source component, a shallow source component, and a noise component.

Figure 9 illustrates the interpretation of an energy spectrum into these three components [8].

3.3. Theory of Source Parameter Imaging. The basics are that for vertical contacts, the peaks of the local wave number define the inverse of depth. In other words,

$$\text{Depth} = \frac{1}{K_{\max}} = \frac{1}{\left(\sqrt{(\partial \text{Tilt}/\partial x)^2 + (\partial \text{Tilt}/\partial y)^2} \right)_{\max}}, \tag{4}$$

where the Tilt is given as

$$\text{Tilt} = \arctan \left(\frac{(\partial T/\partial z)}{\left(\sqrt{(\partial T/\partial x)^2 + (\partial T/\partial y)^2} \right)} \right) \tag{5}$$

$$= \arctan \left(\frac{\partial T/\partial z}{\text{HGRAD}} \right).$$

The Source Parameter Imaging (SPI) method calculates source parameters from gridded magnetic data. The method assumes either a 2D sloping contact or a 2D dipping thin-sheet model and is based on the complex analytic signal. Solution grids show the edge locations, depths, dips, and susceptibility contrasts. The estimate of the depth is independent of the magnetic inclination, declination, dip, strike, and any remanent magnetization. Image processing of the source-parameter grids enhances detail and provides maps that facilitate interpretation by nonspecialists [10].

Estimation of source parameters can be performed on gridded magnetic data. This has two advantages. First, this eliminates errors caused by survey lines that are not oriented perpendicular to strike. Second, there is no dependence on a user-selected window or operator size, which other techniques like the Naudy [11] and Euler methods require. In addition, grids of the output quantities can be generated, and subsequently image processed to enhance detail and provide structural information that otherwise may not be evident.

4. Analysis and Results

4.1. The Total Magnetic Intensity (TMI) Map of the Study Area. The total magnetic intensity map of the study area bounded by 7.0°–8.5°N latitude and 6.50°–8.0°E longitude is produced into maps (Figure 4) which is in color aggregate. The magnetic intensity of the area ranges from −2415.97 minimum to 1264.72 maximum with an average value of 33.87 nT. The total number of data points is 3,667,251. The area is marked by both high and low magnetic signatures, which could be attributed to several factors such as (1) variation in depth, (2) difference in magnetic susceptibility, (3) difference in lithology, and (4) degree of strike. Subsequent interpretation under qualitative analysis will reveal more information of each. The northern edge and the western edge of the study area are inhibited by short wavelength (high frequency in occurrence) signatures, which are generally attributed to basement areas. These are mostly prominent on sheets (228) Katakwa, (229) Udegi, and (247) Lokoja (Figure 3).

The left part of the northern edge comprised of Koton Karfi, Katakwa, Udegi sheets and part of the western edge of Lokoja sheet shows a lot of activity, as they are dotted

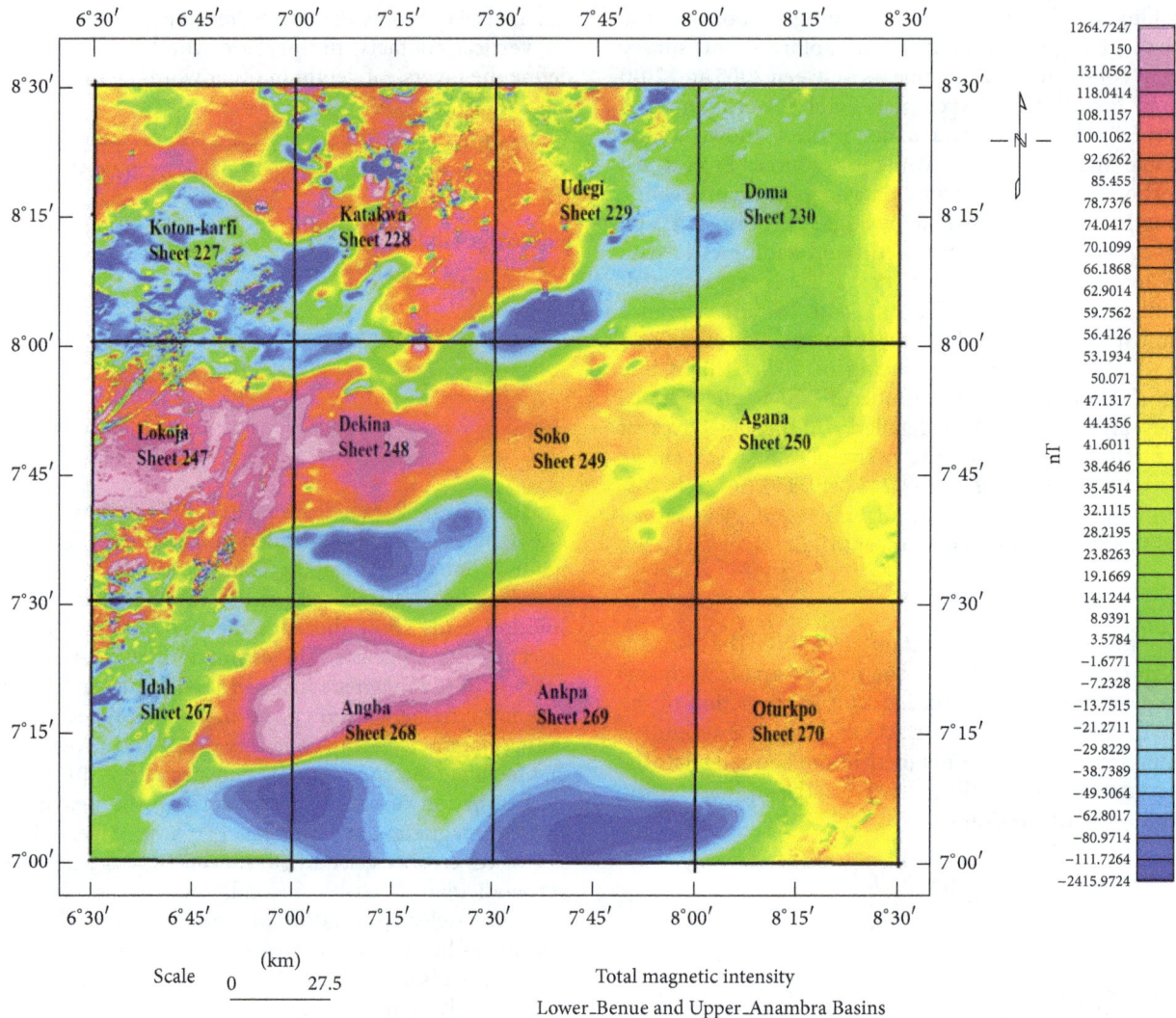

FIGURE 3: Total Magnetic Intensity Map of the study area showing major towns flown over.

by mixtures of both high and low magnetic structures with features that are characteristics of surface to near surface structures such as outcrops.

Structural trends within the study area are N-E and NE-SW (First Vertical Derivative of the TMI and First Vertical Derivative in Figure 4). High Frequency (short wavelength) signatures observed at the Northern and Western portion of the study area revealed a shallow depth to magnetic source typical of Basement Complex. Long wavelength signatures observed at the major part of the study area are as a result of deep magnetic source typical of Sedimentary basin.

4.2. Interpretation of Geo-Magnetic Map. From the aeromagnetic anomaly map, Figure 3, a geo-magnetic sketch map over the study area, is compiled, which provides a new insight on the lithology of the study area. The geo-magnetic sketch Figure 5 inferred from geology Figure 2 and the magnetic susceptibility shows that the basin is covered with the Chhattisgarh sediments with intrusions of granite

and greenstone. The striking feature with NW-SE trending maximum negative magnetic anomaly zone in the northern part of basin is associated with Sonakhan greenstone belt Undifferentiated Older granite, mainly porphyritic buried under the sediments. Granite, granitized gneiss, and porphyroblastic granite. Rock type at the Northern portion is identified as Biotite gneiss. False bedded sandstone, coal, sandstone and shale.

Rock type at the western portion of the study area is identified from geology as Undifferentiated Older granite, mainly porphyritic granite granitized gneiss with porphyroblastic granite. Rock type at the Northern portion is identified as Biotite gneiss. False bedded sandstone, coal, sandstone, and shale are the lithologic units at the surface within the sedimentary basin. River Alluvium deposition was identified along the river channel.

4.3. Application of Spectral Depth and Results. The entire study area was divided into forty eight (48) sections, Figure 6,

Estimating the Thickness of Sedimentation within Lower Benue Basin and Upper Anambra Basin, Nigeria, Using Both
Spectral Depth Determination and Source Parameter Imaging

17

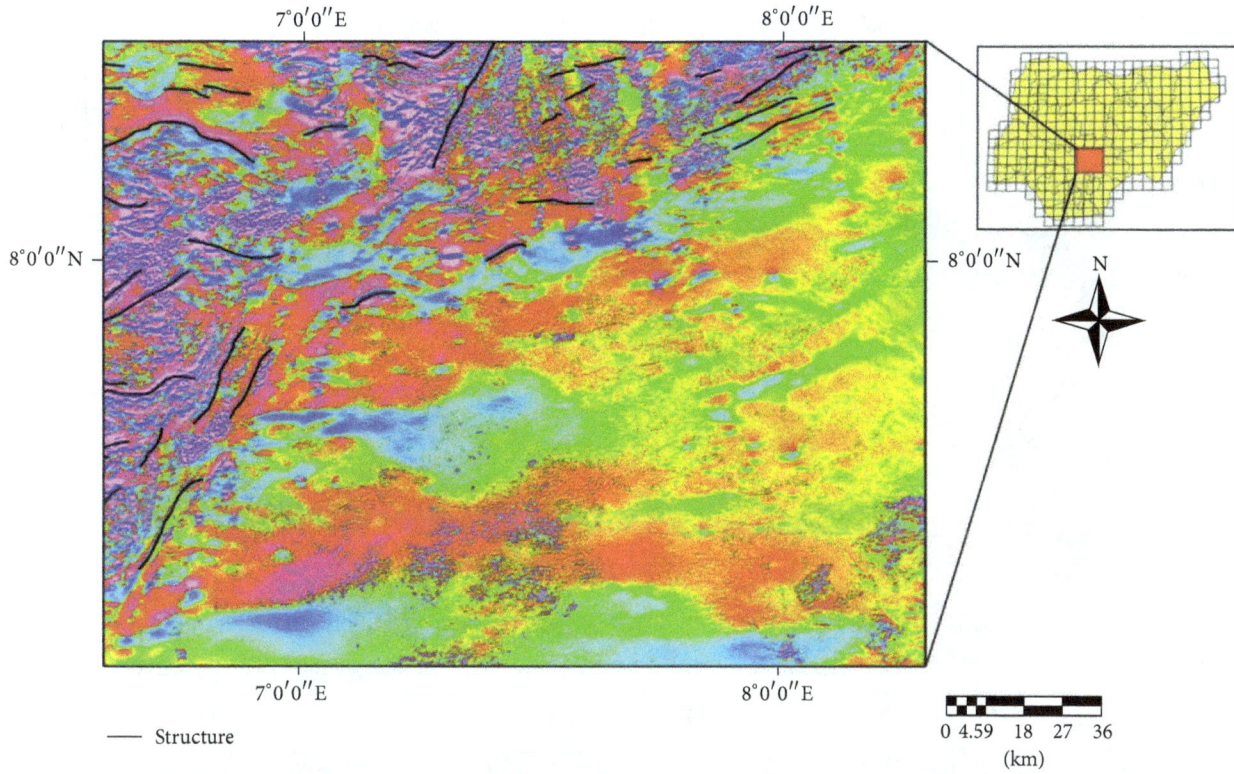

FIGURE 4: First Vertical Derivative Map showing identified structures.

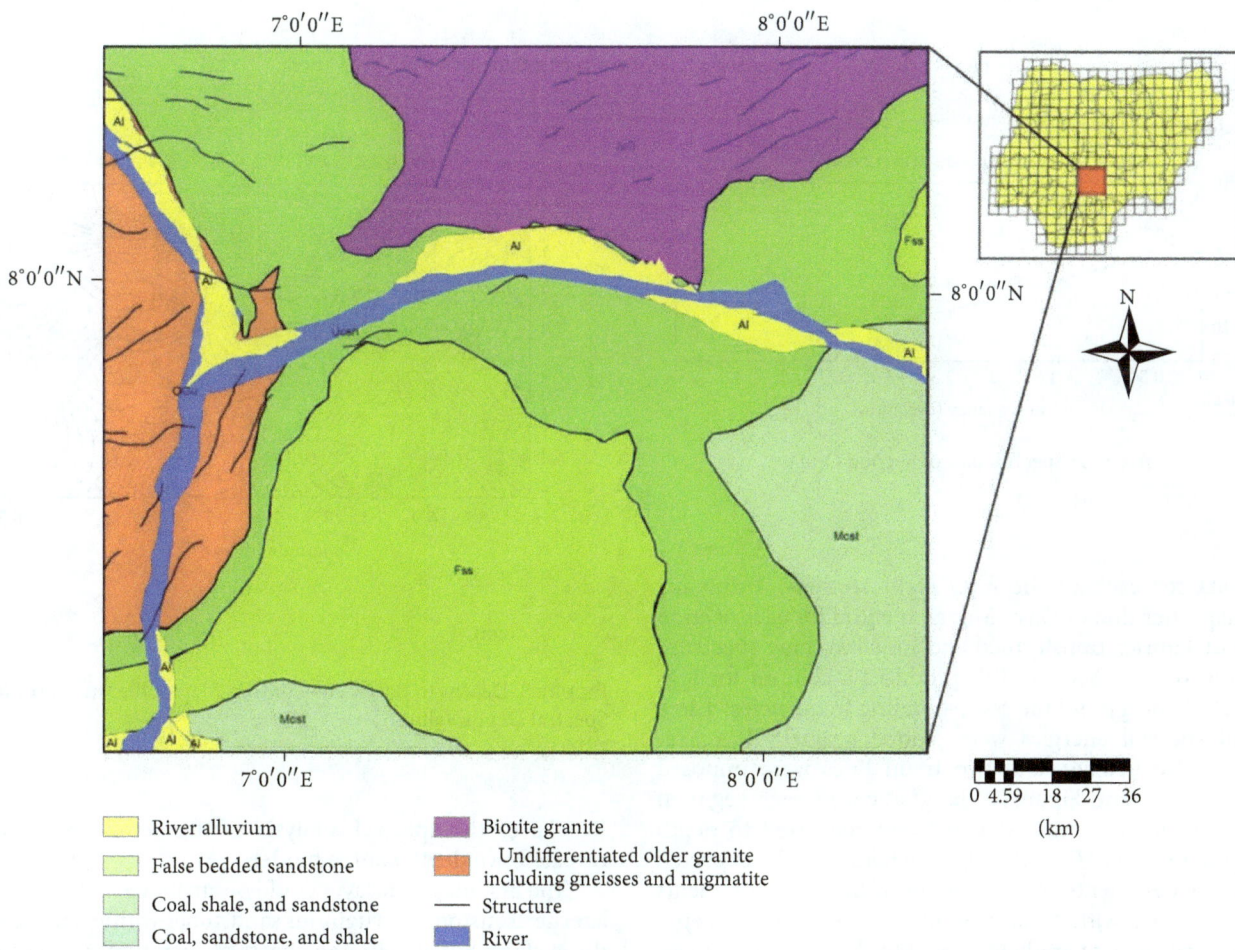

FIGURE 5: Geo-magnetic sketch of the study area produced from geology and magnetic susceptibility.

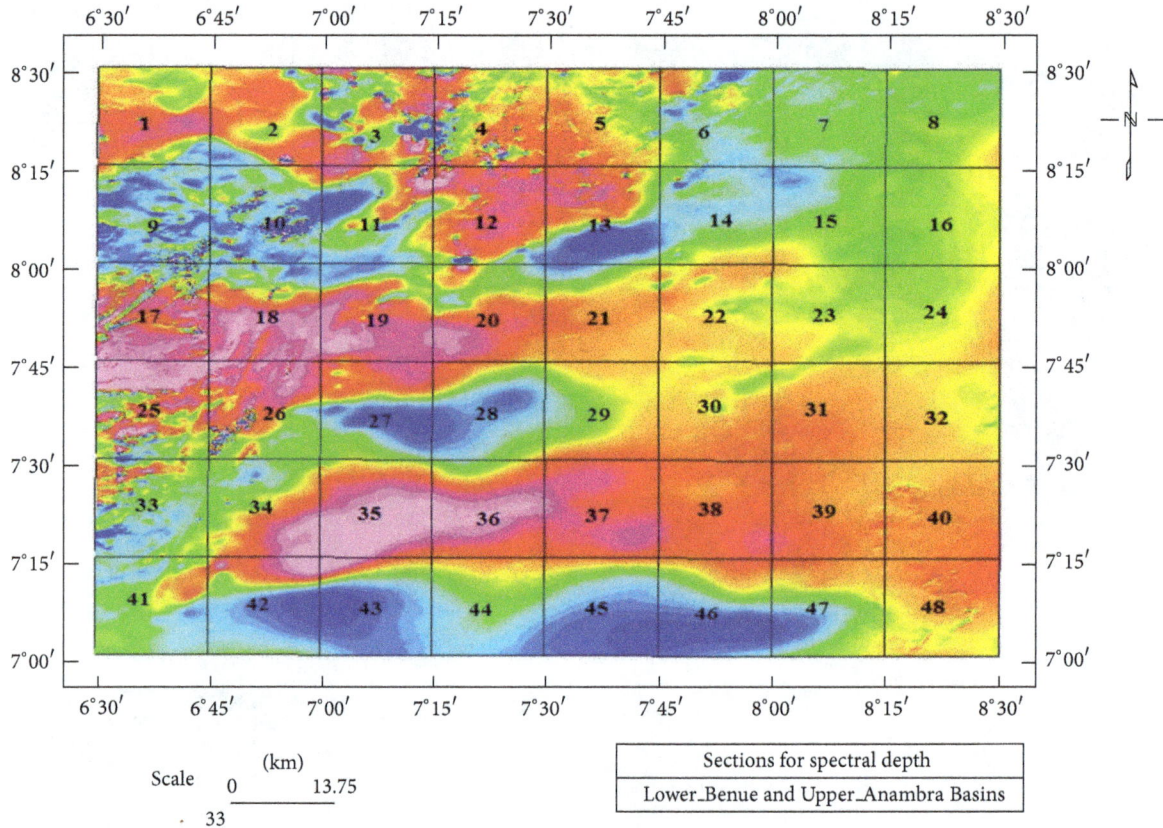

FIGURE 6: TMI Map showing sections for spectral depths of the study area.

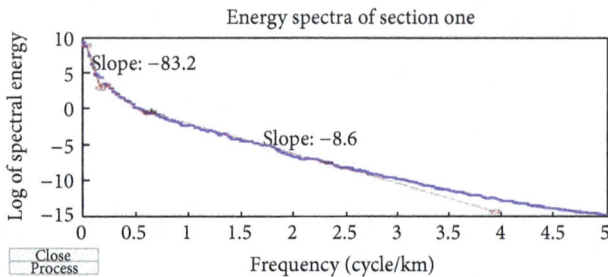

FIGURE 7: Spectral plot of Section One.

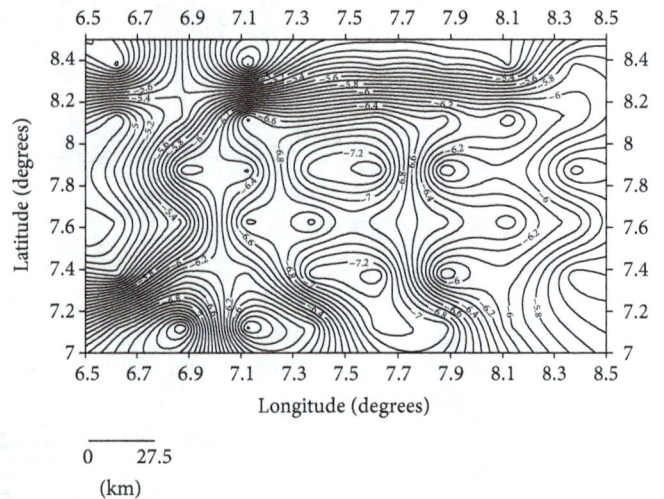

FIGURE 8: Depths to Basement Contour Map of the study area form Spectral Depth Values.

and data for each of the forty eight sections. Using the Magmap extension of Oasis Montaj the grid for each section was Fast Fourier transformed and radial average spectrum was run for each section; this produces a column for logs of spectral energy and the corresponding frequencies. These logs of spectral energies were plotted against the corresponding frequencies, and two trend lines were imposed on linear segment, Figure 7. The gradient of each segment of the straight line was evaluated and converted to depth using the formula $H = GRAD/4\pi$, where H is the expected depth. Two gradients corresponding to the linear segments were evacuated, with the steep gradient related to the deeper sources and the low gradient related to the shallow sources.

Results of the spectral analysis of the aeromagnetic data revealed two depth source models.

The first magnetic layer could be attributed to the effect of laterite, ironstone, ferruginous sandstones within and close to the surface. Equally are the effects of surrounding basement

Estimating the Thickness of Sedimentation within Lower Benue Basin and Upper Anambra Basin, Nigeria, Using Both
Spectral Depth Determination and Source Parameter Imaging

19

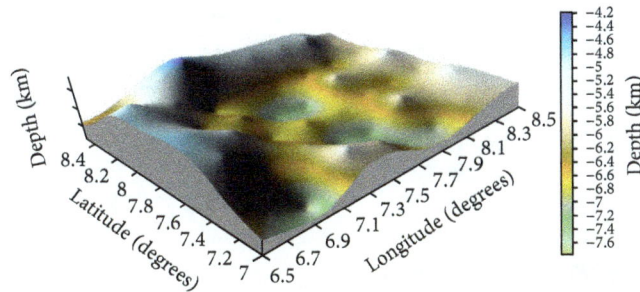

FIGURE 9: 3D construction of the subsurface using Deeper Magnetic Sources Spectral Depth Values.

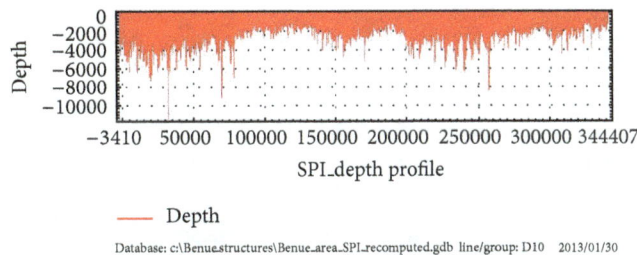

—— Depth

Database: c:\Benue_structures\Benue_area_SPI_recomputed.gdb line/group: D10   2013/01/30

FIGURE 10: Source Parameter Imaging Depth Profile Map of the study area.

rocks at the Northern and Western flanks of the study area. The shallow magnetic sources depth ranges in depth from 188.0 meters to 452.796 meters with an average value of 313.440 meters (Table 1).

The second layer could be attributed to magnetic rocks intrusion into the basement surface, lateral discontinuities in basement susceptibilities, and inter-basement features like faults and fractures. The second layer depth thus represents the depth to basement in the area and this depth has a minimum value of 4.186 km and maximum of 7.369 km with average value of 6.013 km.

This represents the average thickness of the sedimentary formation that overlay the basement complex within the lower portion of Benue Basin and the upper portion of Anambra Basin. The deepest parts of the basins are the lower southern edge, these are around Idah, Angba, and Ankpa and the mid-portion of the study area, below Udegi, which record sedimentation above 7 km.

Two maps were produced from the results of the spectral analysis for visual interpretation; these are Figure 8, Depths to Basement Contour Map and Figure 9, 3D construction of the subsurface.

4.4. Application of Source Parameter Imaging and Results. The Source Parameter Imaging (SPI) module from Oasis Montaj software was applied to the TMI data of the study area; the SPI statistics show a minimum depth of 76.983 meters and a maximum depth of 9847.4 meters. Depths to Basement Profile Map (Figure 10) and Depths to Basement Map (in color aggregate) of the study area (Figure 11) show that the deepest part of the basin coincided with those obtained from spectral analysis.

5. Conclusion

Spectral Depth Analysis which uses the radial average energy spectrum of an assemblage within a square or rectangular shape area to determine the depth to magnetic source bodies was employed on 48 sections in the study area. With this, spectral depths to magnetic rocks were determined at every 13.75 km distance across the entire field. Summary of the results shows a maximum depth of 7.36 km around Idah, Angba, and Ankpa on sheets 267, 268, and 269, respectively; other areas with depth in the range of 6 kilometers are located on sheet 229 Udegi and on sheet 268 Angba.

Source Parameter Imaging was equally applied to evaluate the thickness of sedimentation within the same area and a maximum depth of 9,847 km was obtained.

This depth obtained is higher than that which was arrived at while using Average Radial Spectral Analysis (7.3 km). The spectral method is not devoid of human error because position of the trend line on the spectral plots which is manually done affects the final result, and most essentially, it has been established that errors in depth estimation increase with depth of source [10].

The result obtained above can be compared to those obtained by other researchers, who have worked around this area; for example, Depth of Sedimentation that ranged from 0.5 to 7 km was obtained by Ofoegbu in the lower and middle Benue Basin [12]. Likkasson, O. K., equally worked around lower and middle Benue Basin and obtained spectral matching yielding three dipoles equivalent source layers at 0.89 km to 4.33 km and 18.22 km, where 4.33 km corresponds to maximum thickness within middle Benue Basin; research has shown that thickness of sedimentation increases southward within lower Benue Basin [13].

TABLE 1: First and second layer depth estimates.

SEC_NO	Long	LAT	Grad. of deeper sources	Grad. of shallow sources	Depth to deeper sources in km	Depth to shallow sources in meters
1	6.375	8.225	−83.2	−8.36	−6.621	−332.6
2	6.525	8.225	−67.8	−6.9	−5.395	−274.5
3	7.075	8.225	−52.6	−8.72	−4.186	−347.0
4	7.225	8.225	−66	−8.9	−5.252	−354.1
5	7.375	8.225	−66.4	−9.34	−5.284	−371.6
6	7.52	8.225	−65.4	−8.44	−5.204	−335.8
7	8.075	8.225	−62.6	−7.54	−4.982	−300.0
8	8.225	8.225	−74.4	−8.84	−5.921	−351.7
9	6.375	8.075	−59.2	−6.54	−5.395	−368.4
10	6.525	8.075	−67.8	−4.44	−5.828	−238.7
11	7.075	8.075	−85.8	−7.18	−6.557	−410.6
12	7.225	8.075	−82.4	−10.38	−6.732	−432.9
13	7.375	8.075	−84.6	−10.68	−6.541	−452.8
14	7.52	8.075	−82.2	−8.56	−6.462	−293.6
15	8.075	8.075	−81.2	−8.56	−6.096	−292.1
16	8.225	8.075	−76.6	−6.8	−5.045	−441.7
17	6.375	7.52	−63.4	−6.12	−5.066	−252.3
18	6.525	7.52	−80	−6.3	−6.175	−320.7
19	7.075	7.52	−77.6	−8.80	−6.273	−237.9
20	7.225	7.52	−91.4	−8.22	−6.369	−309.6
21	7.375	7.52	−92.6	−7.24	−5.793	−357.3
22	7.52	7.52	−72.8	−8.68	−6.064	−305.6
23	8.075	7.52	−76.2	−7.88	−5.443	−323.1
24	8.225	7.52	−68.4	−8.42	−5.100	−188.0
25	6.375	7.375	−57.2	−7.24	−5.311	−288.1
26	6.525	7.375	−59.8	−5.7	−5.395	−226.8
27	7.075	7.375	−82.6	−7.22	−6.828	−287.3
28	7.225	7.375	−71.8	−7.64	−6.557	−304.0
29	7.375	7.375	−67.6	−7.08	−6.732	−281.7
30	7.52	7.375	−74.4	−5.64	−6.541	−224.5
31	8.075	7.375	−79.6	−8.76	−6.462	−348.6
32	8.225	7.375	−74.2	−8.36	−6.096	−332.6
33	6.375	7.225	−54	−7.08	−5.045	−281.7
34	6.525	7.225	−83.6	−9	−6.366	−358.1
35	7.075	7.225	−81.8	−6.62	−6.175	−263.4
36	7.225	7.225	−75.6	−6.34	−7.273	−252.3
37	7.375	7.225	−68.4	−7.86	−7.369	−312.7
38	7.52	7.225	−90.4	−8.04	−6.693	−319.9
39	8.075	7.225	−67.2	−8.62	−6.064	−343.0
40	8.225	7.225	−71.4	−8.74	−5.443	−347.8
41	6.375	7.075	−87.80	−8.04	−6.987	−319.9
42	6.525	7.075	−98.20	−7.42	−7.215	−295.2
43	7.075	7.075	69.35	−7.46	−5.470	−270.3
44	7.225	7.075	−74.60	−6.16	−5.936	−245.1
45	7.375	7.075	−90.00	−8.40	−7.162	−334.2
46	7.52	7.075	−89.00	−7.14	−7.082	−284.1
47	8.075	7.075	−75.60	−6.48	−6.016	−257.8
48	8.225	7.075	−71.20	−8.36	−5.666	−332.6

Estimating the Thickness of Sedimentation within Lower Benue Basin and Upper Anambra Basin, Nigeria, Using Both
Spectral Depth Determination and Source Parameter Imaging

21

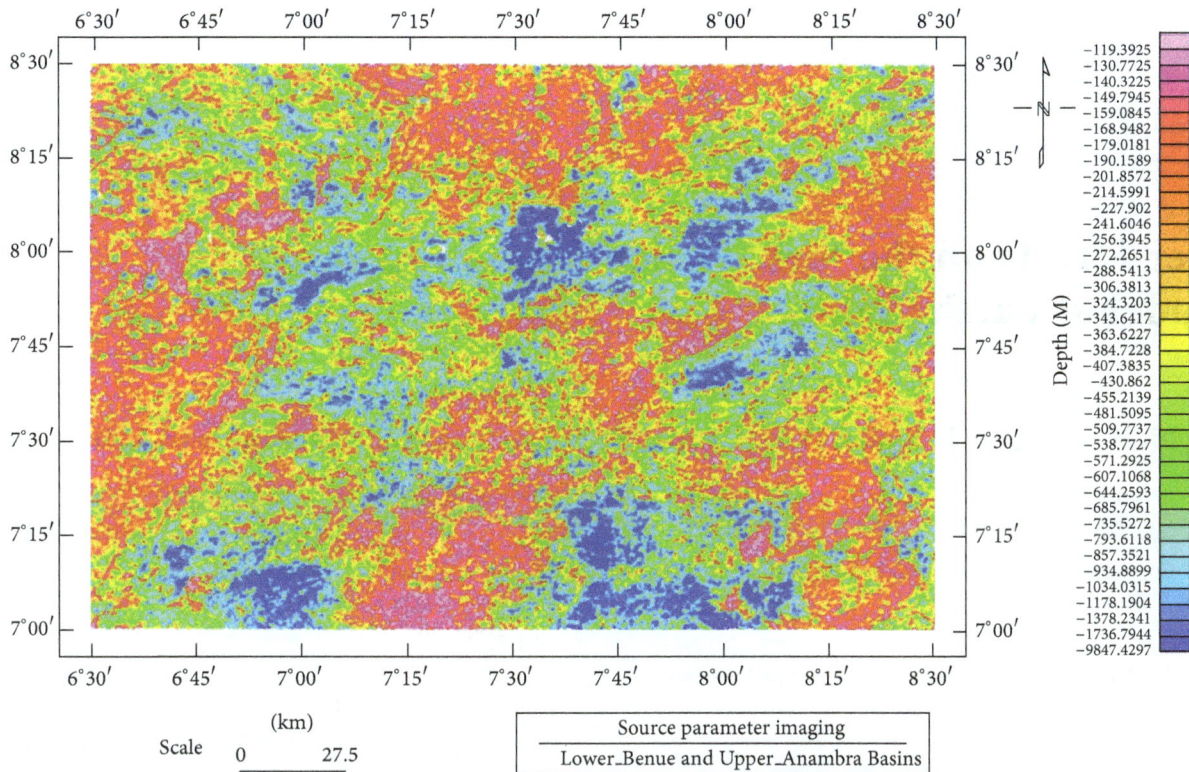

FIGURE 11: Source Parameter Imaging Depth Map of the study area.

The outcome of this research shows that the maximum depth of sedimentation within the study area is approximately 10 km.

It is understood by earth researchers that the condition for hydrocarbon generation and accumulation is guided by source rock lithology, thickness of sediments, and geothermal history. The result of depth estimates obtained by this research especially around the Anambra basin below Idah and Angba in Kogi state is of interest because it is high enough for the attainment of temperatures of approximately 60°C and higher than is required for thermal degradation of kerogen yielding hydrocarbons.

References

[1] C. V. Reeves, "The Kalahari Desert, central southern Africa: a case history of regional gravity and magnetic exploration," in *The Utility of Gravity and Magnetic Surveys, Society of Exploration Geophysicists, Special Volume*, W. J. Hinze, Ed., pp. 144–156, 1985.

[2] F. S. Grant, "Aeromagnetics, geology and ore environments, I. Magnetite in igneous, sedimentary and metamorphic rocks: an overview," *Geoexploration*, vol. 23, no. 3, pp. 303–333, 1985.

[3] C. V. Reeves, *Aeromagnetic Surveys Principles, Practice and Interpretation*, 2005.

[4] M. B. Dobrin, *Introduction To Geophysical Prospecting*, 2nd edition, 1960.

[5] A. Spector and F. S. Grant, "Statistical methods for interpreting aeromagnetic data," *Geophysics*, vol. 35, pp. 293–302, 1970.

[6] S. W. Petters, "Mid Cretaceous Paleoenvironments and biostratigraphy of the BenueTrough, Nigeria," *Geological Society of America Bulletin*, vol. 89, pp. 151–154, 1977.

[7] T. J. A. Reijers and C. S. Nwajide, "Geology of the Southern Anambra Basin," Field Course Note 66, Chevron Nigeria Limited, 1998.

[8] J. B. Thurston and R. S. Smith, "Automatic conversion of magnetic data to depth,dip, and susceptibility contrast using the SPI method," *Geophysics*, vol. 62, no. 3, pp. 807–813, 1997.

[9] B. K. Bhattacharyya, "Continuous spectrum of the total magnetic field anomaly due to a rectangular prismatic body," *Geophysics*, vol. 31, pp. 91–121, 1996.

[10] K. A. Ojoh, "The Southern part of the Benue Trough (Nigeria) Cretaceous stratigraphy, basin analysis, paleo-oceanography and geodynamic evolution in the equatorial domain of the south Atlantic," *NAPE Bulletin*, vol. 7, pp. 131–152, 1992.

[11] H. Naudy, "Method for analyzing aeromagnetic profiles," *Geophysical Prospecting*, vol. 18, no. 1, pp. 56–63, 1970.

[12] C. O. Ofoegbu, "A model for the tectonic evolution of the Benue Trough of Nigeria," *Geologische Rundschau*, vol. 73, no. 3, pp. 1007–1018, 1984.

[13] O. K. Likkason, C. O. Ajayi, E. M. Shemang, and E. F. C. Dike, "Directional filtering and spectral analysis of aeromagnetic Data over the Middle Benue Trough, Nigeria European," *Journal of Scientific Research*, vol. 2, no. 2, pp. 76–112, 2005.

3

Water Effects on the First-Order Transition in a Model of Earthquakes

M. W. Dongmo, L. Y. Kagho, F. B. Pelap, G. B. Tanekou, Y. L. Makenne, and A. Fomethe

Laboratory of Mechanics and Modelling of Physical Systems, Department of Physics, University of Dschang, P.O. Box 69, Dschang, Cameroon

Correspondence should be addressed to M. W. Dongmo; mathurin.wamba@yahoo.fr

Academic Editors: E. Del Pezzo, E. Liu, and A. Stovas

The study of 1D spring-block model of earthquake dynamics with consideration of water effects in preexisting fault deals with new forms of frictional force. An analytical study of the equation of motion enables us to establish that motion of geological fault is accelerated by water pressure. In the same setting the critical value of frictional velocity for which appears the discontinuous (first-order) transition from a stick-slip behavior to a creep motion strongly depends on water pressure. The investigation also displays the magnitude and probability of events as a function of water pressure; these two quantities decrease and increase, respectively, with the variation of water pressure.

1. Introduction

Despite significant advances made in the study of geological fault structures and plate tectonics, our understanding of the physical mechanisms responsible for the initiation, propagation, and termination of earthquake rupture remains unfinished. Burridge and Knopoff [1], in 1967, introduced a one-dimensional chain block and spring-discrete model, aiming at explaining the earthquake mechanism [2]. In 1996, Vasconcelos simplified Knopoff's model with a single spring-block model and attempted to facilitate the understanding of earthquake using this model. The mechanism of slip instabilities in laboratory experiments has been proposed to be dependent on several factors including reduced frictional force during sliding (slip weakening) and a decrease in slip velocity [3], which is in concordance with the observation of the geologist Rick Sibson [4], who in 1981 maintained the idea affirming that water pressure in the fault was opposite to the rocks' pressure, which intensified friction between the fault's sides. The mathematical model which takes into consideration this idea was not yet established; however, some studies have been done in dry faults with the help of well-known mathematical models [5, 6]. Moreover, Vasconcelos [6] had investigated the phase transition in this single block

model and demonstrated that it occurs (from stick-slip to creep motion) only when the characteristic velocity is equal to 0.5. This result was obtained because he considered dry fault in his study. In other cases there are several values of characteristic speed (less than 0.5) which are able to lead to the transition.

The aim of this paper is to study Giovanni's modified single block model, by considering the new shape of frictional forces (that contains water pressure), and investigate the effects of water pressure on the earthquake dynamics. The content of this paper is organized as follows.

Section 2: frictional model and water effects.

Section 3: equation of motion and temporal evolution of displacements.

Section 4: effects of pressure on the first-order transition.

Section 5: effects of pressure on the magnitude and the probability of occurrence of an event.

The last section is devoted to discussion and conclusion.

2. Description of Water Effects in the Fault and Equation of Motion

Figure 1(a) displays a preexisting fault broken on one side and impervious at the other side. Water gets into the fault

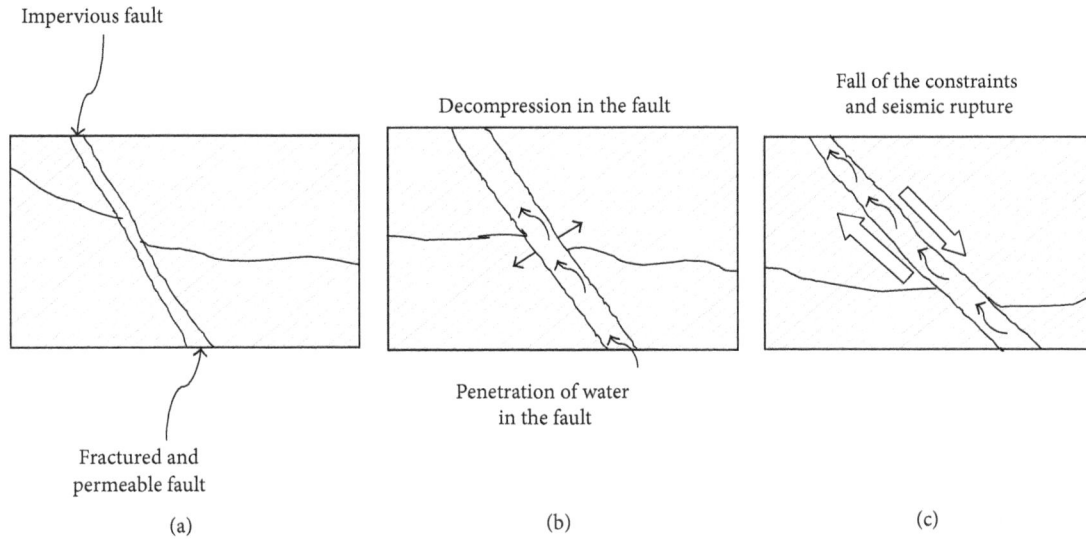

FIGURE 1: (a) Fault with permeable side and impervious side. (b) Very high pressure of water. (c) Flux of water in the fault.

through the fractured side; it thwarts the pressure of the rocks which intensifies friction between the two sides of the fault (Figure 1(b)). This will provoke a seismic rupture when water pressure becomes increasingly intensified [4] (Figure 1(c)).

Since our investigation focused on the role of water on the fault's dynamics, we hereby propose a new form of frictional force that takes into consideration water characteristic in the process. The model below has a great advantage of being analytically tractable, so that one hopes that a thorough understanding of such a simple model might in turn shed further light on the basic principles governing real earthquakes. We will for convenience choose the frictional force such as $f(dX/dt) = F_0\Phi((1/V_f)(dX/dt)) - \Gamma_0\Phi((1/V_f)(dX/dt))$, where F_0 corresponds to the maximum frictional force (greater than Γ_0).

$\Gamma_0 = \rho S V_e^2/2$ is the strength of pressure, ρ is the density of water, S is the surface of the fault, and V_e is the speed of water in the fault assumed to vary weakly, such as a constant in our equation. The quantities (dX/dt) and V_f represent the velocity of the fault and the characteristic velocity for the friction, respectively.

$\Phi(x)$ is assumed to be a continuous function for $x \geq 0$ satisfying the initial conditions $\Phi(0) = 1$ and $(d\Phi/dx)_{x=0} = -1$; the second conditions simply express the velocity-weakening effect of the friction, since it implies that $f(dX/dt)$ will be a decreasing function of the block velocity dX/dt, at least in a neighborhood of origin. In this model (Figure 1), a block of mass m is connected by spring of constant K (corresponding to the linear elastic properties of the medium surrounding the fault) to a rigid pulling rod that moves at a small constant velocity V (Figure 2). The block rests upon a stationary surface, which provides a velocity dependent on frictional force f, which impedes the motion of the block. It is important to note that water pressure acts between the block and the stationary surface, such that it reduces the frictional force f.

When the force due to the spring exceeds the threshold force denoted by $F_0 - \Gamma_0$, the block is set into motion and the corresponding equation of motion is

$$m\frac{d^2X}{dt^2} = K(Vt - X) - f\left(\frac{dX}{dt}\right), \qquad (1)$$

where $X(t)$ is the position of the block. Owing to the velocity-weakening effect of friction, the block undergoes a rapid motion (earthquakes), during which most of the accumulated stress is released. Before proceeding with the analysis, it is convenient to introduce dimensionless variables:

$$U = \left(\frac{K}{F_0}\right)X, \qquad \tau = \left[\frac{K}{m}\right]^{1/2}t. \qquad (2)$$

So that the equation of motion (1) takes the dimensionless form

$$\frac{d^2U}{d\tau^2} = -U + v\tau + \left(-1 + \frac{\Gamma_0}{F_0}\right)\Phi\left(\frac{1}{v_f}\frac{dU}{d\tau}\right) \qquad (3)$$

with $v = V/V_0$, $v_f = V_f/V_0$, and $v_e = V_e/V_0$, $\mu_e = \rho S(V_0^2/F_0)$, $p_e = \mu_e v_e^2/2$, and $V_0 = F_0/\sqrt{mK}$. The velocity scale V_0 corresponds to the maximum velocity attained by a block that experiences no (kinematic) friction as it moves. The dimensionless parameters v and v_f are, respectively, the pulling speed and friction characteristic velocity measured in this scale. In the same manner μ_e, p_e, v_e are, respectively, the density, the pressure, and the velocity of water in the same scale.

For this model to be relevant for real earthquakes, dimensionless water pressure must be less than one (i.e., $p_e < 1$; we recall that $\Gamma_0 < F_0$) and v must be taken to be very small. Indeed, during an earthquake the relative velocity between the two sides of the fault is of the order m/s, while the typical relative plate velocity is of the order cm/yr.

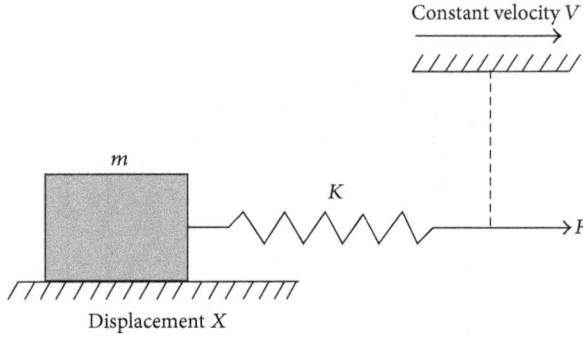

FIGURE 2: Spring-block model for earthquakes.

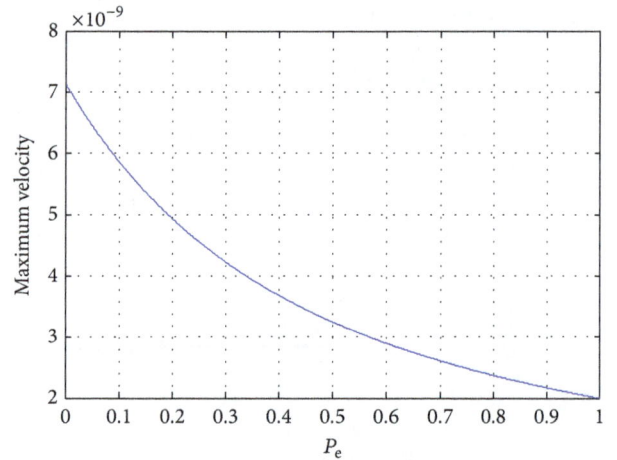

FIGURE 3: Effects of water pressure on the magnitude of maximum velocity ($\nu = 10^{-9}$ and $\alpha = 0.5$).

We start the analysis by considering first the linearized version of the motion equation. In view of (1) and (2) the linearization of (3) yields

$$\frac{d^2U}{d\tau^2} - 2\alpha\left(1 - p_e\right)\frac{dU}{d\tau} + U = \nu\tau. \tag{4}$$

For convenience we have introduced the parameter $\alpha = 1/2\nu_f$ and $\gamma = \alpha(1 - p_e)$, with $0 \leq p_e < 1$. After this simplification and redefinition of the origin of displacement so as to eliminate the constant that would otherwise appear on the right-hand side of (4), we have

$$\frac{d^2U}{d\tau^2} - 2\gamma\frac{dU}{d\tau} + U = \nu\tau. \tag{5}$$

The linear approximation above will be valid only if the block velocity is small compared to the friction characteristic velocity (i.e., $(dU/d\tau) \ll \nu_f$). The initial conditions are given by $(d^2U(0)/d\tau^2) = (dU(0)/d\tau) = U(0) = 0$.

Equation (5) presents two cases to be considered: (i) $\nu_f > (1 - p_e)/2$ and (ii) $\nu_f \leq (1 - p_e)/2$.

3. Oscillatory Solutions

Now, our attention is focused on the creep motion of the block; for this we consider the case (i) and then the solution of (5) is

$$U(\tau) = \nu\left[\exp\left(\gamma\tau\right)\left(\frac{2\gamma^2 - 1}{\omega}\sin\omega\tau - 2\gamma\cos\omega\tau\right)\right.$$
$$\left. + \tau + 2\gamma\right], \tag{6}$$

where $\omega = \sqrt{1 - \gamma^2}$. The velocity of block is obtained by time derivative of displacement of the block. From (6) we have

$$\dot{U}(\tau) = \nu\left[\left(\frac{\gamma}{\omega}\sin\omega\tau - \cos\omega\tau\right)\exp\left(\gamma\tau\right) + 1\right], \tag{7}$$

where the dots indicate time derivatives; the maximum velocity attained by the block is

$$\dot{U}_{\max} = \nu\left[1 + \exp\left(\frac{\gamma\pi}{\omega}\right)\right]. \tag{8}$$

It appears from Figure 3 that water pressure strongly affects the magnitude of maximum velocity attained by the block; at high pressures, the block rapidly reaches its cruise velocity and thus it can be easily understood that the fault's motion is accelerated by water pressure.

During the description of earthquake, it is important to define the position of the block at the end of the slip. But, when the block ceases to move, it occupies an unspecified position. However, determination of this position requires the knowledge of the time τ_0 at which the block ceased to slip. Concretely, when the block stops, its speed vanishes. Therefore, the corresponding time τ_0 is the solution of the following equation:

$$\frac{\gamma}{\omega}\sin\omega\tau_0 - \cos\omega\tau_0 + \exp\left(-\gamma\tau_0\right) = 0. \tag{9}$$

Substituting (9) into (6) we notice that the block's displacement $\Delta = U(\tau_0)$ after such slip event is given by

$$\Delta = U\left(\tau_0\right) = \nu\left[\tau_0 + \gamma + \sqrt{\exp\left(2\gamma\tau_0\right) - \omega^2}\right]. \tag{10}$$

Since τ_0 does not depend on ν, it then follows that as $\nu \to 0$ the displacement Δ vanishes whenever $\nu_f > (1 - p_e)/2$.

Next we investigate the situation when $\nu_f \leq (1 - p_e)/2$.

4. A Periodic Solution of the Motion Equation

The main objective of this section is to investigate the behavior of the block in stick-slip motion. Considering the case (ii): $\nu_f \leq (1 - p_e)/2$, the solution of (5) is given by

$$U(\tau) = \nu\left[\left(\frac{2\gamma^2 - 1}{\omega}\sinh\omega\tau - 2\gamma\cosh\omega\tau\right)\right.$$
$$\left. \times \exp\left(\gamma\tau\right) + \tau + 2\gamma\right] \tag{11}$$

with $\omega = \sqrt{\gamma^2 - 1}$.

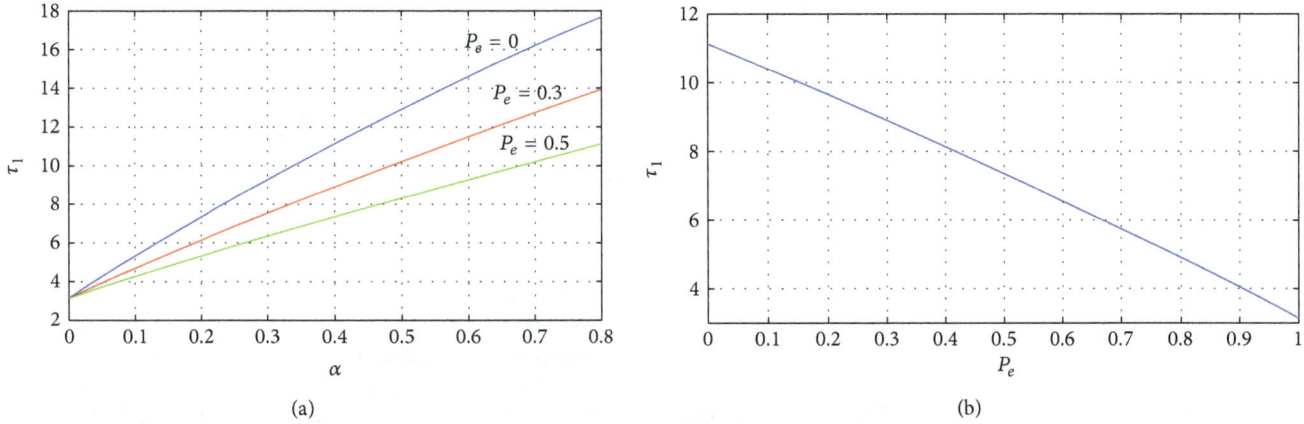

FIGURE 4: (a) Effects of water pressure on the closed time of linear motion part for ($p_e = 0$, $p_e = 0.3$, $p_e = 0.5$). (b) Closed time of linear part of motion versus water pressure (with $\alpha = 0.4$).

Since the velocity $dU/d\tau$ is now a monotonously increasing function of time, the block eventually reaches a velocity comparable to the characteristic velocity v_f of the friction, independently of the smallness of v. Around such point, the linear approximation is useless since the motion of the block becomes nonlinear. It is thus necessary to consider another model of the friction law which will not present such behavior.

Several friction models have been recently considered in the literature [7, 8]. Hereafter, we consider the simple model proposed by Langer and Tang [7] to examine the behavior of our system; that is,

$$\Phi(x) = \begin{cases} 1 - x, & \text{if } 0 \leq x \leq 1, \\ 0, & \text{if } x \geq 1. \end{cases} \tag{12}$$

With this model, the motion equation (5) remains valid and presents two distinguished cases which correspond, respectively, to (i) and (ii). We recall that if $v_f > (1 - p_e)/2$, then $\Delta \to 0$ as $v \to 0$, regardless of the nonlinear features of the friction law. In this case, the solution for the block motion can be divided into three parts [6] as follows.

Initially, when $(dU/d\tau) < v_f$ the motion of the block is confined to the linear part of the friction and hence the solution of (5) is given by (11). This solution is valid until the time τ_1, where $(dU(\tau_1)/d\tau) = v_f$, after which the block enters the nonlinear regime of the friction law. In the limit $v \to 0$, this time τ_1 diverges logarithmically with the term v:

$$\tau_1 = (\gamma - \omega) \ln \left[\frac{\omega(\omega + \gamma)}{\gamma v} \right]. \tag{13}$$

We examine here the influence of the pressure on the time τ_1 marking the onset of nonlinear motion.

Figures 4(a) and 4(b) show, respectively, the reduction of τ_1 (as function of α) for the greater values of the pressures and decreasing of τ_1 as a function of pressure for the fixed value of α.

From here it is easy to understand that water can considerably reduce the time of transition (from a stick-slip

behavior to a creep motion). Although the block spends a very long time in this linear regime, considerable motion will occur only for times close to the instant τ_1. Therefore, it is convenient to introduce the renormalized time $t = \tau - \tau_1$. The block position in terms of t can now be obtained by substituting (13) into (11) and taking the limit $v \to 0$. After some simplifications, we find that

$$U(t) = v_f (\gamma - \omega) \exp(\gamma + \omega) t. \tag{14}$$

For time $t > 0$, the motion consists of two parts [6]. First, the block will swing frictonlessly until the time t_1 (defined below) at which its velocity is again equal to v_f. Afterwards, the block experiences once more a nonzero (linear) friction until it finally stops later at the time t_2. Since we are interested in the linear behavior of the system in the limit $v \to 0$, we set $v = 0$ in (5) and exploit the Langer and Tang model [7] to examine the dynamics of the block in these two regions. Computations lead to the complete solution of the motion equation that defines the position of the block for time > 0:

$U(t)$

$$= \begin{cases} v_f \left[\sin t - (\gamma + \omega) \cos t \right], & \text{if } 0 \leq t \leq t_1, \\[2ex] v_f e^{\gamma(t-t_1)} \left[(3\gamma + \omega) \cosh \omega (t - t_1) \right. \\[1ex] \qquad - \left(\gamma + 3\omega + \left(\dfrac{2}{\omega} \right) \right) \\[1ex] \qquad \left. \times \sinh \omega (t - t_1) \right], & \text{if } t_1 \leq t \leq t_2, \end{cases} \tag{15}$$

in which explicit expressions of the times t_1 and t_2 are

$$t_1 = 2 \arctan(\gamma + \omega); \qquad t_2 = t_1 + \frac{1}{2\omega} \ln \left(1 + \frac{\omega}{\gamma} \right). \tag{16}$$

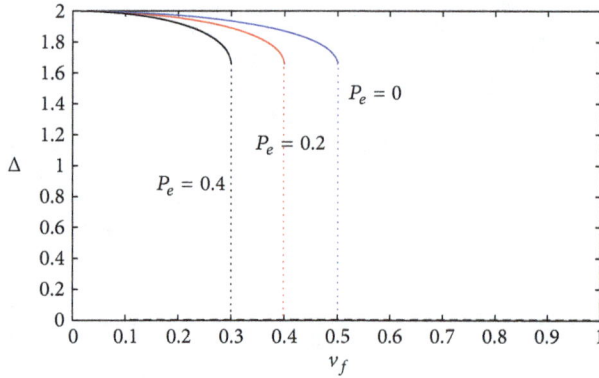

FIGURE 5: The block displacement Δ versus the friction characteristic ν_f for three values of water pressure in model (12).

After the earthquake, the position of the block (fault's displacement) is obtained by evaluating the quantity $\Delta = U(t_2)$; that is,

$$\Delta = \begin{cases} \left(1 + \dfrac{\omega}{\gamma}\right)^{(1/2)(1+(\gamma/\omega))}, & \text{if } \nu_f \leq \dfrac{(1-p_e)}{2}, \\ 0, & \text{if } \nu_f > \dfrac{(1-p_e)}{2}. \end{cases} \quad (17)$$

Here, we also collect the aforementioned results that Δ vanish for $\nu_f > (1-p_e)/2$ (with respect to $\nu \to 0$). Then, we note that at $\nu_f = (1 - p_e)/2$, the system undergoes a phase transition in the sense that Δ vanishes for $\nu_f > (1 - p_e)/2$ while it takes finite values for $\nu_f \leq (1 - p_e)/2$.

A curve given the evolution of Δ versus ν_f for several values of p_e is plotted in Figure 5.

The curve of Figure 5 exhibits that water pressure has a remarkable effect on Δ (fault's displacement) at the end of sliding, whereof it is easy to note that the first-order transition depends on the value of water pressure. The transition occurs increasingly with increase in pressure; thus a fast transition allows for short-period earthquake. Worthy of note is the fact that $p_e = 0$ corresponds to the dry fault studied in [6]. Actually the system undergoes a discontinuous (first-order) transition from stick-slip behavior to creep motion as the friction parameter is varied.

Following the study of water effects on the first-order transition, it is worthy to investigate the influence of water on energy release during an earthquake.

5. Magnitude of Earthquake and Probability of Occurrence

The objective of this section is to investigate the effects of water pressure on the earthquake magnitude. Several magnitude expressions have been investigated. Some of these take into consideration the parameters characterizing the seismic wave [9–11], and others consider the physical parameter of the rocks [12, 13].

Hereafter, we consider the magnitude expression formulated by Kanamori et al. [14, 15]:

$$M = \frac{2}{3}\left[\log_{10}\left(\frac{M_0}{1N \cdot m}\right) - 9.1\right], \quad (18)$$

where the seismic moment is

$$M_0 = \mu S \Delta. \quad (19)$$

See [16, 17].

Wherein μ defines the rocks rigidity, S represents the fractured surface and Δ the fault's displacement. Owing to the fact that the surface S is proportional to the length of the fault which is itself proportional to the position Δ, the seismic moment (19) takes the form [4]

$$M_0 = K_0 \Delta^3. \quad (20)$$

Substituting (20) and (19) into (18), then (18) becomes

$$M$$
$$= \begin{cases} \dfrac{3}{2}\left[\log_{10}\left(K_0\left(1 + \dfrac{\omega}{\gamma}\right)^{(3/2)(1+(\gamma/\omega))}\right) - 9.1\right], & \text{if } \nu_f \leq \dfrac{(1-p_e)}{2}, \\ \text{not defined}, & \text{if } \nu_f > \dfrac{(1-p_e)}{2}. \end{cases} \quad (21)$$

In (20), K_0 is a constant which can be determined by exploiting the experimental curve giving the magnitude as a function of displacement in [4], ($K_0 \sim 1.2589254.10^{18}$).

The effect of water on the magnitude is investigated by plotting the magnitude M as function of pressure p_e (Figure 6).

From Figure 6 we notice that the magnitude decreases logarithmically with the pressure until it reaches its smallest value, at $p_e = 0.75$ (corresponding to the transition point for $\nu_f = 0.125$). Note equally in Figure 6 that when $p_e \to 0$ the magnitude has a maximum value. When $p_e \geq 0.75$, the magnitude is not defined because the logarithm of zero does not exist. Recall that $p_e = 0.75$ corresponds to the value for which the phase transition occurs (i.e., when Δ vanishes). It is important to remark that the critical value of p_e from which the magnitude is not defined changes with fixed values of ν_f. It arises from this curve (Figure 6) that the magnitude of the earthquake decreases when the pressure of water in geological fault increases; from here it is important to note that when water gets into the fault, there is an impact on the energy release during its sliding.

Now let us examine the influence of pressure on the probability of earthquakes occurrence; as defined by Gutenberg and Richter in 1944 [18] it is

$$W = 10^{(c - d \times \log M_0)}. \quad (22)$$

In this previous quantity, M_0 is defined by (20); after substitution of (20) into (22) we have

$$W = 10^{c - d \times \log(K_0 \cdot \Delta^3)}, \quad (23)$$

where c and d are arbitrary constants.

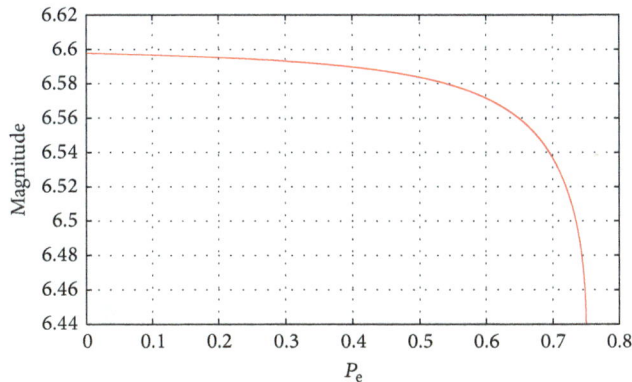

FIGURE 6: Magnitude M versus the pressure of water p_e (with $\nu_f = 0.125$).

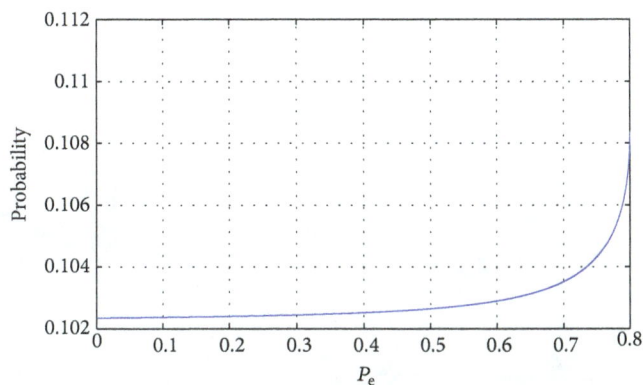

FIGURE 7: Probability of events in terms of water pressure with $\nu_f = 0.1$; $c = 0.91$; and $d = 0.1$.

It appears from Figure 7 that the probability of occurrence of an event increases with water pressure. So this result is in accordance with the results obtained in Figures 4 and 5, because the closed time of linear motion decays with pressure, and the transition occurs rapidly with greater values of pressure. This means that water in the fault can cause more earthquakes with smaller magnitude (precisely for $p_e < 1$).

It is important to remark that the value of frictional velocity which leads to the transition at $p_e = 0.8$ is $\nu_f = 0.1$.

6. Conclusion

In this paper, we have studied water effects on the dynamics of a 1D spring-block model for earthquakes by performing the Vasconcelos frictional force [6]. We have shown that the system undergoes a discontinuous phase transition from a stick-slip motion to a creep motion as the frictional parameter varies. We established that this transition is not dependent only on the greatness of the driving plate's velocity (great improvement of the Vasconcelos' results [6]) but depends also on the value of water pressure in the fault; contrary to Vasconcelos we revealed that the value of characteristic frictional velocity which leads to the transition is not unique (0.5) but that it varies according to the value of water pressure

in the fault. Through our investigation, it is established that for the greater values (less than one) of water pressure the transition occurs rapidly (i.e., short-period earthquake) as compared to the smallest or zero pressure. These high pressures allow for weaker magnitude earthquakes. Thus the propagation of water in the fault could be of an advantage as it permits the occurrence of earthquakes with lower magnitude. In general, more events will be recorded (with weak magnitude) in faults containing water than in dry ones.

Conflict of Interests

The authors do not have any conflict of interests regarding the publication of this paper.

Acknowledgment

Professor F. B. Pelap is grateful to Professor Giovani L. Vasconcelos for valuable discussions during his last visit in the University of Pernembuco, Brazil.

References

[1] R. Burridge and L. Knopoff, "Model and theoretical seismicity," *Bulletin of the Seismological Society of America*, vol. 57, pp. 341–371, 1967.

[2] J. M. Carlson and J. S. Langer, "Properties of earthquakes generated by fault dynamics," *Physical Review Letters*, vol. 62, no. 22, pp. 2631–2635, 1989.

[3] A. Ruina, "Slip instability and state variable friction laws," *Journal of Geophysical Research*, vol. 88, no. 12, pp. 10359–10370, 1983.

[4] P. Bernard, "Qu'est-ce qui fait trembler la terre?" EDP Science, 2003.

[5] R. Montagne and G. L. Vasconcelos, "Complex dynamics in a one-block model for earthquakes," *Physica A*, vol. 342, no. 1-2, pp. 178–185, 2004.

[6] G. L. Vasconcelos, "First-order transition in a model of earthquakes," *Physical Review Letters*, vol. 76, p. 25, 1996.

[7] J. S. Langer and C. Tang, "Rupture propagation in a model of an earthquake fault," *Physical Review Letters*, vol. 67, no. 8, pp. 1043–1046, 1991.

[8] J. M. Carlson and J. S. Langer, "Mechanical model of an earthquake fault," *Physical Review Letters*, vol. 62, p. 2632, 1989.

[9] C. F. Richter, "An instrumental earthquake magnitude scale," *Bulletin of the Seismological Society of America*, vol. 25, pp. 1–32, 1935.

[10] B. Gutenberg, "Amplitudes of surface waves and magnitude of shallow earthquakes," *Bulletin of the Seismological Society of America*, vol. 35, pp. 3–12, 1945.

[11] B. Gutenberg and C. F. Richter, "Earthquake magnitude, intensity, energy, and acceleration," *Bulletin of the Seismological Society of America*, vol. 46, pp. 105–146, 1956.

[12] H. Kanamori, "The energy release in great earthquakes," *Journal of Geophysical Research*, pp. 2981–2987, 1977.

[13] T. C. Hanks and H. Kanamori, "A moment magnitude scale," *Journal of Geophysical Research B*, vol. 84, no. 5, pp. 2348–2350, 1979.

[14] H. Kanamori and D. Hadley, "Crustal structure and temporal velocity change in southern California," *Pure and Applied Geophysics*, vol. 113, no. 1, pp. 257–280, 1975.

[15] H. Kanamori, H.-K. Thio, D. Dreger, E. Hauksson, and T. Heaton, "Initial investigation of the Landers, California, earthquake of 28 June 1992 using TERRAscope," *Geophysical Research Letters*, vol. 19, no. 22, pp. 2267–2270, 1992.

[16] K. Aki, "Generation and propagation of G waves from the Niigata Earthquake of June 16, 1964: Part 1. A statistical analysis," *Bulletin of the Earthquake Research Institute*, vol. 44, pp. 23–72, 1966.

[17] J. Vaneck, A. Zapotek, V. Karnik et al., "Standardization of magnitude scales," *Izvestiya Akademii Nauk SSSR Seriya Geofizicheskii*, pp. 153–158, 1962.

[18] B. Gutenberg and C. F. Richter, "Frequency of earthquakes in California," *Bulletin of the Seismological Society of America*, vol. 34, pp. 185–188, 1944.

4

Using Microseismicity to Estimate Formation Permeability for Geological Storage of CO_2

D. A. Angus[1] and J. P. Verdon[2]

[1] CiPEG, University of Leeds, Leeds LS2 9JT, UK
[2] School of Earth Sciences, University of Bristol, Bristol BS8 1RJ, UK

Correspondence should be addressed to D. A. Angus; d.angus@leeds.ac.uk

Academic Editors: E. Del Pezzo and A. Donnellan

We investigate two approaches for estimating formation permeability based on microseismic data. The two approaches differ in terms of the mechanism that triggers the seismicity: pore-pressure triggering mechanism and the so-called seepage-force (or effective stress) triggering mechanism. Based on microseismic data from a hydraulic fracture experiment using water and supercritical CO_2 injection, we estimate permeability using the two different approaches. The microseismic data comes from two hydraulic stimulation treatments that were performed on two formation intervals having similar geological, geomechanical, and *in situ* stress conditions, yet different injection fluid was used. Both approaches (pore-pressure triggering, and the seepage-force triggering) provide estimates of permeability within the same order of magnitude. However, the seepage-force mechanism (i.e., effective stress perturbation) provides more consistent estimates of permeability between the two different injection fluids. The results show that permeability estimates using microseismic monitoring have strong potential to constrain formation permeability limitations for large-scale CO_2 injection.

1. Introduction

Fracture stimulation has been applied for the past 60 years to enhance recovery from hydrocarbon reservoirs, with an estimated 70% of wells being fracture stimulated, and hence is a key factor in the economic exploitation of unconventional reserves, such as tight-gas and shale-gas reservoirs [1]. Over the past 20 years, microseismic monitoring has developed into one of the most effective methods of monitoring fracture stimulation and hence is routinely applied to monitor fracture stimulation programs.

The spatial and temporal variations in microseismicity can be used to monitor changes in the stress field and hence potentially be used to monitor perturbations in fluid pathways as well as top-seal and well-bore integrity. Furthermore, microseismicity has been used also to characterise spatial and temporal variations within the reservoir and surrounding rock mass by monitoring changes in seismic attributes between the source and receiver (e.g., shear-wave splitting analysis to characterise fracture-induced anisotropy [2–4]).

Additional information can be gained by evaluating microseismic failure mechanisms to characterise the rock mass at the source and provide a measure of the strength, orientation, and type of elastic failure to potentially quantify damage (e.g., [5–7]).

Although microseismicity can provide fairly accurate temporal and spatial locations of brittle failure, how the measured microseismicity relates to the evolution of the induced pressure front and effective stress field as well as creation and enhancement of cracks and fractures is still not well constrained. Examination of the distribution of microseismic events can help characterize the flow and mechanical properties of the stimulated reservoir. In particular, by assuming that seismicity is triggered by the diffusion of pore-pressure from the injection point, Shapiro [8] has shown that the permeability of a formation can be estimated from the rate of increase in distance between injection well and event hypocenter distance through time. This spatiotemporal behaviour is commonly visualized on the so-called r-t plot, (where r is injection-well-to-event distance and t is time).

This method has shown potential for predicting apparent formation permeability and hydrocarbon production for various fracture stimulation case studies (e.g., [9]).

Although this pore-pressure diffusion approach has shown some promise in estimating reservoir permeability, there are some nonphysical aspects to the theory, such as weak pore-pressure perturbation triggering seismicity [10]. As an alternative to the pore-pressure diffusion approach, Rozhko [11] introduces the concept of seepage-force triggering to predict the r-t response of seismicity, which considers diffusion of effective stress perturbations as the driving force of microseismicity.

In this paper we compare both methods, predicting formation permeability by modelling observed r-t behaviour during hydraulic fracturing, where water and supercritical CO_2 have been used as the injected fluids. By estimating permeability using microseismic monitoring, we hope to explore the potential of using microseismic monitoring to constrain formation permeability limitations for large-scale CO_2 injection sites.

2. Models Describing Spatiotemporal Evolution of Seismicity

2.1. Pore-Pressure Triggering. The r-t pore-pressure triggering approach is based on the concept that the spatial and temporal evolution of microseismicity is hydraulically induced and characterised in terms of a low frequency pore-pressure relaxation mechanism described by Biot [12]. The key assumption for application of this approach to hydraulic fracture-induced microseismic data is that the tectonic stress in the subsurface is close to the critical stress needed for brittle failure (e.g., [13]). As such, increasing fluid pressure (i.e., injecting fluid) within the reservoir results in a transient increase of the reservoir pore-pressure and a decrease in effective stress. If the decrease in effective stress is sufficient, it can lead to relaxation of normal stresses along preexisting fractures and hence slip along the fracture and associated microseismicity. The following derivations are from Shapiro [8] and so the reader is referred to this paper for a more detailed presentation of the approach.

Assuming a point-source injector and a homogeneous and isotropic medium, the triggering front (i.e., the distance between outer enveloe of the microseismic "cloud" and the fluid injection point) is described by

$$r(t) = \sqrt{4\pi D (t - t_o)}, \quad (1)$$

where t is observation time, t_0 is injection start time, and D is the scalar apparent hydraulic diffusivity. By plotting the microseismic events on a time-distance plot and matching the best-fitting r-t curve to the triggering front an estimate of the hydraulic diffusivity can be obtained. The calculated apparent diffusivity can then be used to estimate other reservoir and flow parameters, such as formation permeability.

Assuming that the injected fluid is incompressible, the fluid volume balance is such that the total injected fluid is equal to the sum of the fluid volume within the fracture and lost to the surrounding formation. Further, assuming that the

induced fracture is straight and of fixed height (i.e., the PKN model; see [14]), then the fracture half-length is approximated by

$$L(t) \approx \frac{q_i t}{4h_f C_l \sqrt{2t} + 2h_f w}, \quad (2)$$

where q_i is the average injection rate, h_f is the fracture height (either estimated from perforation interval or vertical extent of microseismicity), C_l is the fluid-loss coefficient, and w is the average fracture width. The fluid-loss coefficient C_l is given by

$$C_l \approx \frac{q_i}{8h_f \sqrt{2\pi D}}. \quad (3)$$

Another surface, the back front, characterises the seismically quiet zone after injection stops and tracks the propagation of maximum pore-pressure perturbation. The back front is given by

$$r_b(t) = \sqrt{2dDt \left(\frac{t}{t_s} - 1\right) \ln\left(\frac{t}{t - t_s}\right)}, \quad (4)$$

where d is the dimension of the pressure diffusion (1D, 2D, or 3D) and t_s is the injection shut-off time.

Neglecting induced fracture surface effects such as filter cake permeability damage and effects on pore space and fractures within the vicinity of the fracture treatment, the permeability of the reservoir can be estimated:

$$\kappa \approx \frac{q_i^2 \mu_f}{128 h_f^2 \Delta p^2 \phi c_f D}, \quad (5)$$

where c_f and μ_f are the compressibility and viscosity of the reservoir fluid, respectively, Δp is the difference in the average injection pressure and the initial (or far-field) reservoir pressure, and ϕ is the reservoir porosity.

2.2. Seepage-Force Triggering. The nonlinear diffusion approach of Shapiro [8] and the so-called Coulomb failure stress (CFS) criteria (e.g., [15]) applied to microseismicity make the assumption that seismicity is triggered by the propagation of a fluid pressure perturbation front. Based on the previous mechanism, it is implied that small changes in pore-pressure are sufficient to trigger seismicity, and this is often explained by assuming that most of the faults are critically stressed [13, 15]. Rozhko [11] argues that induced microseismicity is explained and predicted better by linear diffusion coupled to linear poroelastic deformation rather than the highly nonlinear fluid diffusion mechanism [8] or CFS and critically stressed faults [15].

In the Rozhko [11] approach, the seismicity is triggered by the propagation of an effective stress perturbation front. The role of fluid pressure in rock strength is significant, yet comes about through the Terzaghi [16] effective stress law

$$\underline{\sigma}_{ij} = \sigma_{ij} + \delta_{ij} P, \quad (6)$$

where $\underline{\sigma}_{ij}$ is the effective stress tensor, σ_{ij} is the stress tensor, P is pressure, and δ_{ij} is the Kronecker delta function. Rozhko [11] refers to this as the so-called seepage-force and makes use of the Coulomb yielding criteria (CYS) written as

$$\text{CYS} = \frac{\sigma_1 - \sigma_3}{2} + \sin\phi_y \left(\frac{\sigma_1 + \sigma_3}{2} + P \right) + C_y \cos\phi_y, \quad (7)$$

where CYS is the Coulomb yielding stress and σ_1 and σ_3 are the maximum and minimum principal stresses (positive in tension). The parameters ϕ_y and C_y are the friction angle and cohesion during dilatancy and can be determined from geomechanical triaxial laboratory measurements. Application of CYS stems from laboratory measurements, where observed acoustic emissions during loading have been shown to correlate with the onset of dilatancy. CYS not only describes the onset of dilatancy and hence microseismicity, but also incorporates the Kaiser effect [17]; during unloading deformation is elastic with no additional fracturing, and during reloading no additional fracturing and seismicity develop until overcoming the previous loading maximum.

The seismicity-triggering front (the CYS equivalent to (1)) is given by

$$\Delta\text{CYS} = \frac{1}{r} P_{c0} H \left(t - t_0 \right)$$
$$\times \left[\eta f_D \left(R_0 \right) + \left(\eta f_M \left(R_0 \right) + f_F \left(R_0 \right) \right) \sin\phi_y \right]$$
$$+ \cdots + \frac{1}{r} P_{c1} H \left(t - t_1 \right)$$
$$\times \left[\eta f_D \left(R_1 \right) + \left(\eta f_M \left(R_1 \right) + f_F \left(R_1 \right) \right) \sin\phi_y \right], \quad (8)$$

where r is radial distance, η is the poroelastic stress coefficient, $f_F(R)$, $f_D(R)$, and $f_M(R)$ are nondimensional functions given by Rozhko [11, equations 7, 10, and 11], $R = r/(4Dt)^{1/2}$, P_{c0} and P_{c1} are the pressure perturbations for times t_0 and t_1, $R_0 = r/[4D_0(t - t_0)]^{1/2}$ and $R_1 = r/[4D_1(t - t_1)]^{1/2}$, and D is the pressure diffusivity constant. The seismicity-suppression front (the CYS equivalent to (4)) is written as

$$\Delta\text{CYS}_* = \frac{1}{r} P_{c0} H \left(t - t_0 \right)$$
$$\times \left[\eta f_D \left(R_0 \right) + \left(\eta f_M \left(R_0 \right) + f_F \left(R_0 \right) \right) \sin\phi_y \right]$$
$$+ \cdots + \frac{1}{r} P_{c1} H \left(t - t_1 \right)$$
$$\times \left[-\eta f_D \left(R_1 \right) + \left(\eta f_M \left(R_1 \right) + f_F \left(R_1 \right) \right) \sin\phi_y \right]. \quad (9)$$

By fitting the seismicity-triggering and seismicity-suppression fronts to the induced seismicity, the formation diffusivity can be estimated. Apparent permeability can then be estimated from

$$\kappa = D\mu_f \varphi \left(c_m + c_f \right), \quad (10)$$

where c_m is the pore volume compressibility.

3. Microseismic Data

Verdon et al. [20] compared the microseismicity produced when first water and then CO_2 (in a supercritical state) were used as the injection fluids for hydraulic fracture of a tight gas reservoir, with the purpose of identifying any characteristic differences in event locations and/or magnitudes induced by the different fluids. A total of 9 injection stages were performed in a vertical well, with each stage at a slightly shallower depth than the previous stage (see Maxwell et al. [21]). The first 7 stages used water, while the final 2 used supercritical CO_2. No major lithologic differences have been identified between the stages. The fracture stimulations were monitored with a downhole array of 12 three-component geophones installed in a nearby vertical well. Verdon et al. [20] presented data from stages 4 (water) and 8 (CO_2). For both fluids, microseismic event locations indicated the formation of fracture networks parallel to the maximum horizontal stress (Figure 1). Event magnitudes showed a weak correlation with injection pressure, while the influence of the differing fluids was found to be minimal.

4. Permeability Estimates from Pore-Pressure and Seepage-Force Triggering

4.1. Pore-Pressure Triggering

4.1.1. Water-Gel Injection. Figure 1(a) displays a map view of the recorded microseismicity during the water-gel fracture treatment. The microseismicity follows an approximately linear trend with an absolute correlation coefficient of 0.83 using simple linear regression. The length and width of the microseismic cloud are approximately 220 m and 70 m. In Figure 2, the microseismic events are plotted with respect to distance from injection well and injection time (i.e., r-t space). The vertical error bars represent the estimated location errors based on the residuals between the predicted and observed travel times. The events were located using the in-house location algorithm of Pinnacle Technologies (see Zimmer et al. [22]) using an isotropic one-dimensional velocity model (see Figure 3). Also shown are horizontal error bars that serve as a qualitative (and not quantitative) measure of event measurement confidence and so by no means reflect error in time (see Zimmer et al. [22] for description of uncertainty characterization). In other words, they attempt to present additional information about microseismic event quality to help further scrutinize the r-t plot. Three microseismic trigger (or forward front) r-t curves are shown for apparent diffusivities of $1.25\,\text{m}^2/\text{s}$, $0.60\,\text{m}^2/\text{s}$, and $0.40\,\text{m}^2/\text{s}$ using (1). These curves represent subjective end-member r-t curves for the forward front microseismicity. Also shown is an r-t curve for the back front for an apparent diffusivity of $0.10\,\text{m}^2/\text{s}$ using (4).

4.1.2. Supercritical CO_2 Injection. Figure 1(b) displays a map view of the recorded microseismicity during the supercritical CO_2 fracture treatment. The microseismicity follows a more diffuse trend compared with the water-gel treatment with an absolute correlation coefficient of 0.65. The length and

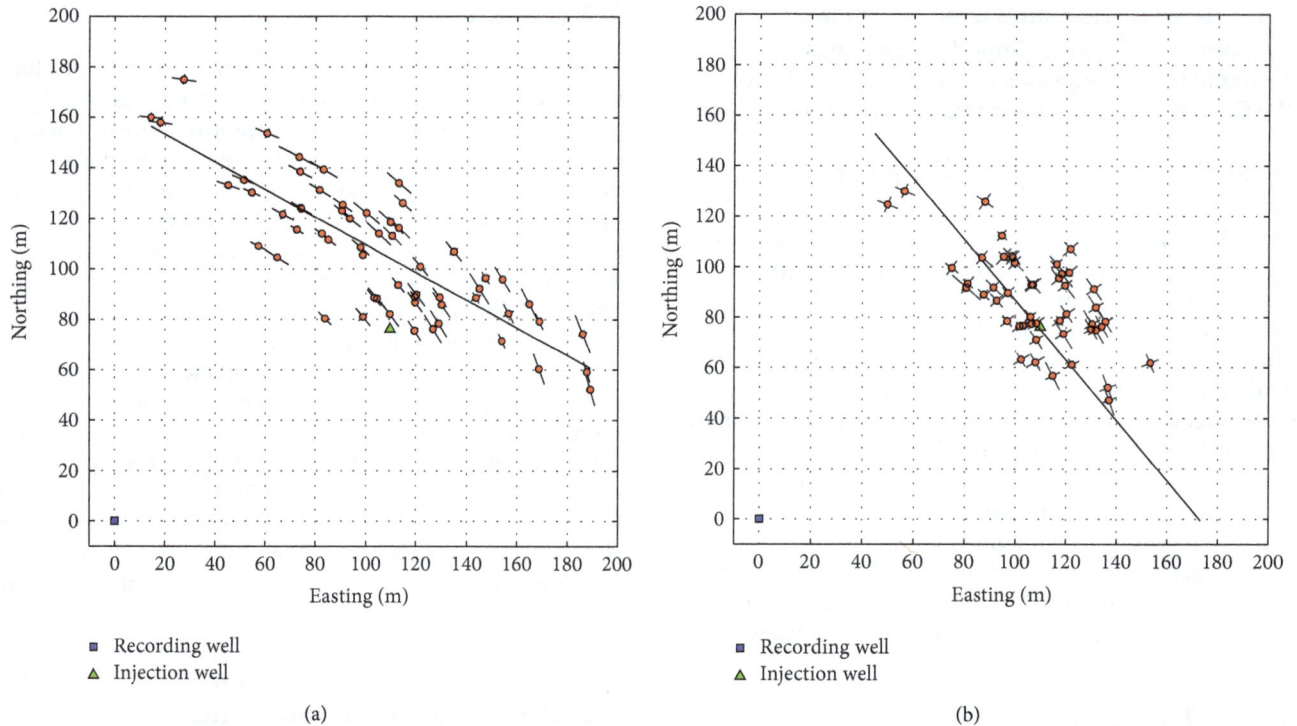

FIGURE 1: Map views of event locations during hydraulic fracture stimulation for water (a) and CO_2 (b) fluid injection. The locations of the injection well and monitoring array are also marked. Error bars represent one-standard-deviation errors based on arrival time residuals and particle motion analysis. As such, they do not account for the additional errors introduced by velocity model discrepancies (e.g., [18, 19]) and so should be considered a lower bound of the true location error.

width of the microseismic cloud are approximately 120 m and 50 m. In Figure 4, the microseismic events are plotted in r-t space. Three forward front r-t curves are shown for apparent diffusivities of $1.20\,\text{m}^2/\text{s}$, $0.80\,\text{m}^2/\text{s}$, and $0.30\,\text{m}^2/\text{s}$ and a back front r-t curve for an apparent diffusivity of $0.90\,\text{m}^2/\text{s}$.

4.2. Seepage-Force Triggering. Figures 5 and 6 show the same observed r-t data with the predicted seismicity-triggering and seismicity-suppression fronts based on seepage-force modelling. In these figures, we assume $\phi_y = 30°$ and $\eta = 0.30$. For the water treatment (Figure 5), we use an average differential injection pressure of 14 MPa to define $P_{c0} = 14\,\text{MPa}$ and $P_{c1} = -14\,\text{MPa}$. The best fitting seismicity-triggering and seismicity-suppression fronts were obtained using the following values: perforation interval of 20 m, $\Delta\text{CYS}_h = 0.1\,\text{MPa}$, $D_0 = 2.75 \times 10^3\,\text{m}^2/\text{h}$ and $D_1 = 3.85 \times 10^3\,\text{m}^2/\text{h}$. For the supercritical CO_2 treatment (Figure 6), we use an average differential injection pressure of 15 MPa to define $P_{c0} = 15\,\text{MPa}$ and $P_{c1} = -15\,\text{MPa}$. The best fitting seismicity-triggering and seismicity-suppression fronts were obtained using the following values: perforation interval of 20 m, $\Delta\text{CYS}_h = 0.1\,\text{MPa}$, $D_0 = 2.15 \times 10^3\,\text{m}^2/\text{h}$, and $D_1 = 3.75 \times 10^3\,\text{m}^2/\text{h}$.

4.3. Permeability Estimates

4.3.1. Pore-Pressure Triggering. The average reservoir porosity is assumed to be 10% (an upper end for tight sand

reservoirs). The reservoir fluid viscosity and compressibility are estimated to be $1.00 \times 10^{-3}\,\text{Pa·s}$ and $1.45 \times 10^{-11}\,\text{Pa}^{-1}$, respectively, based on typical values for oil given by Dake [23]. For the water-gel treatment, the average injection rate is $0.09\,\text{m}^3/\text{s}$, the pressure difference 15 MPa, and fracture height 70 m. For the supercritical CO_2 treatment, the average injection rate is $0.08\,\text{m}^3/\text{s}$, the pressure difference 15 MPa, and fracture height 100 m. We assume a fracture width of 0.01 m for both fracture stimulations. Table 1 compiles the results for the estimated fluid-loss coefficient (3), fracture half-length (2), and reservoir permeability (5).

4.3.2. Seepage-Force Triggering. Assuming that the pore volume compressibility is negligible with respect to the reservoir fluid compressibility (i.e., $C_m \ll C_f$), the permeability estimates based on (10) are summarized in Table 2. The assumption of negligible pore volume compressibility suggests that our estimates of formation permeability are conservative (i.e., an underestimate).

Estimates of apparent permeability from both approaches are within the same order of magnitude between 10 mD and 100 mD. Note that we use the term "apparent" permeability for two reasons. First, the permeability estimates are typically higher than the true formation permeability through enhancement via hydraulic stimulation. Second, the apparent permeability is indirectly calculated from microseismic data and depends on assumed values of Biot's constant, Poisson's ratio, and the internal angle of friction.

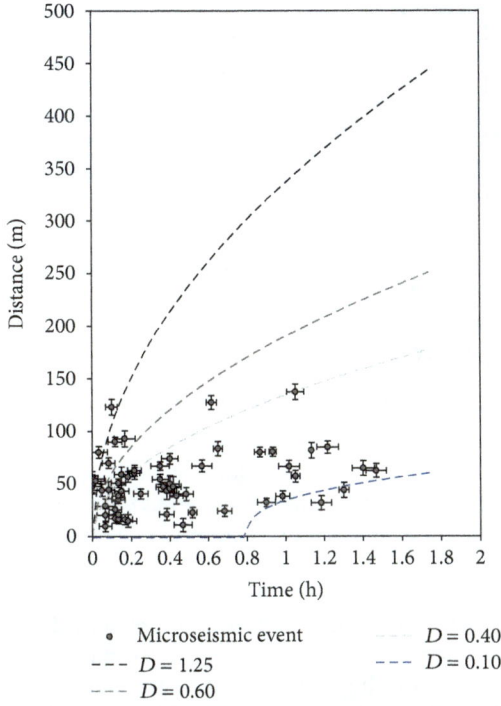

FIGURE 2: r-t pore-pressure triggering plot for water-gel injection treatment. The circles represent the spatiotemporal location of each microseismic event, with the vertical error bar being the total estimated location error and the horizontal error bar representing a scaled event confidence term. The black, the grey, and the light-grey curves are the triggering front r-t curves for diffusivities of 1.25, 0.60, and 0.40 m^2/s, respectively. The blue curve is the back front r-t curve with diffusivity of 0.10 m^2/s.

TABLE 1: Estimated fluid loss, fracture half-length, and formation permeability based on pore-pressure triggering.

Fracture treatment fluid	Diffusivity (m^2/s)	C_l (m/s$^{1/2}$)	$L(t)$ (m)	κ (mD)
	1.25	5.91×10^{-5}	108.80	34.07
Water-gel	0.60	8.53×10^{-5}	93.19	70.99
	0.40	1.04×10^{-5}	84.33	106.48
	1.20	3.68×10^{-5}	79.59	14.54
CO$_2$	0.80	4.73×10^{-5}	73.41	21.81
	0.30	7.72×10^{-5}	57.91	58.15

TABLE 2: Estimated formation permeability based on seepage-force triggering.

Fracture treatment fluid	Diffusivity (10^3 m^2/h)	κ (mD)
Water-gel	$D_0 = 2.75$	67
	$D_1 = 3.85$	94
CO$_2$	$D_0 = 2.15$	53
	$D_1 = 3.75$	92

For the water-gel fluid injection, the formation permeability estimates based on pore-pressure triggering range

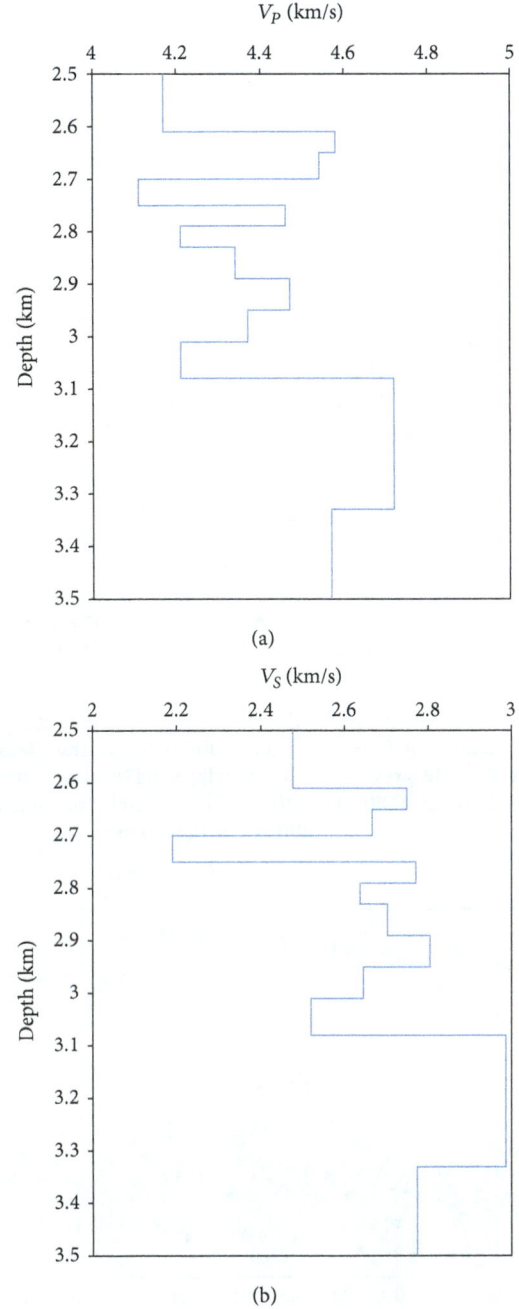

(a)

(b)

FIGURE 3: One-dimensional isotropic P-wave velocity profile (a) and S-wave velocity profile (b) used for locating the microseismic events (the S-wave velocity model shows similar structure).

between 34 mD and 106 mD, whereas those based on seepage-force triggering range between 67 mD and 94 mD. For seepage-force triggering, there are two estimates for formation permeability. This is because two values of diffusivity were needed to fit the microseismic data: D_0 for the diffusivity during hydraulic stimulation and D_1 for the diffusivity due to negative pore-pressure perturbation in fractured rock. Thus, the estimate of $\kappa = 67$ mD during hydraulic stimulation is more representative of formation permeability during fluid injection, whereas the estimate of $\kappa = 94$ mD is more representative of the formation after fracture damage. For the

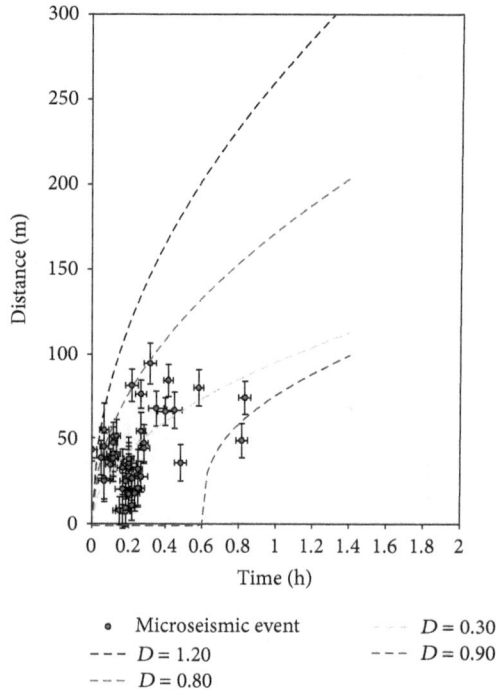

FIGURE 4: r-t pore-pressure triggering plot for supercritical CO_2 injection treatment (refer to Figure 2 for details). The black, the grey, and the light-grey curves are the triggering front r-t curves for diffusivities of 1.20, 0.80, and 0.30 m^2/s, respectively. The blue curve is the back front r-t curve for diffusivity of 0.90 m^2/s.

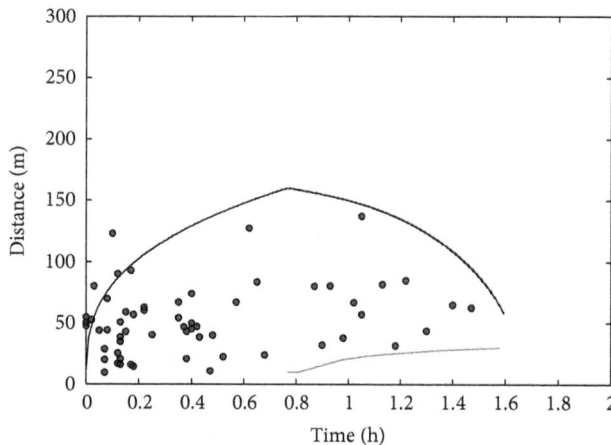

FIGURE 5: r-t seepage-force predictions for the water treatment injection. The black curve is the seismicity-triggering front and the grey curve is the seismicity-suppression front.

supercritical CO_2 fluid injection, the formation permeability estimates based on pore-pressure triggering are lower and range between 14 mD and 58 mD, whereas those based on seepage-force triggering range between 53 mD and 92 mD. Permeability estimates based on seepage-force triggering are more consistent between the water-gel and supercritical CO_2 injection. The permeability estimates for the fractured formation are nearly equal as would be expected for similar geological formations. As per the conclusions of Verdon et al.

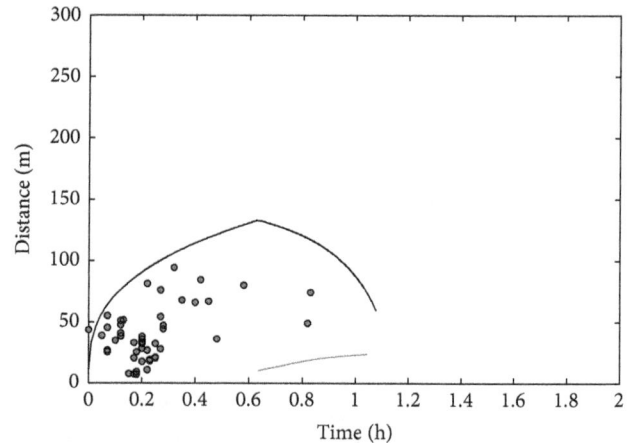

FIGURE 6: r-t seepage-force predictions for the supercritical CO_2 injection treatment. The black curve is the seismicity-triggering front and the grey curve is the seismicity-suppression front.

[20], we see little evidence for a different seismic response when CO_2 rather than water is the injected fluid.

Besides examining event locations and magnitudes, Verdon et al. [20] used shear-wave splitting to image the induced fracture networks. Although not robustly constrained, inversions based on the SWS measurements appeared to show that the fracture network created during water injection was slightly more intense. This may account for what differences are there in formation permeability between supercritical CO_2 and water-gel injection, as estimated by the pore-pressure method. However, the permeability estimates for the fractured formation (i.e., based on diffusivity estimate D_1) would suggest that both fluids generate similar fracture density. One possible explanation for the difference might be in terms of the size of fractures generated. The water-gel treatment may generate large fractures that are effectively constant in dimension (e.g., displaying a Gaussian distribution), whereas the supercritical CO_2 treatment may generate fewer large fractures yet many smaller fractures (i.e., skewed distribution). Based on the geometry of the microseismic monitoring array, the detectability limitations would be biased towards larger fractures. This would certainly explain the lower number of events recorded from the supercritical CO_2 injection yet similar fractured formation permeability.

5. Discussion and Conclusions

We examined two approaches of estimating formation permeability using microseismic data: the pore-pressure triggering and the seepage-force triggering mechanisms. Based on microseismic data from a hydraulic fracture experiment injecting water and supercritical CO_2 during different stages, we compared permeability estimates between the two approaches. The two hydraulic stimulation treatments were performed separately on two formation intervals having similar geological, geomechanical, and *in situ* stress conditions and only differed in terms of the injection fluid used. Both approaches (pore-pressure triggering and the seepage-force triggering) provided estimates of permeability within the same order of magnitude. However, the seepage-force mechanism (i.e., effective stress perturbation) provided more consistent estimates of permeability between the two different injection fluids.

Urbancic et al. [24] monitored significant microseismicity during a field test injecting 10,000 tons CO_2 over a period of one month. They observed that microseismicity can be used to identify the position of the CO_2 plume, and, although microseismicity was significant, there was no evidence of reduced cap rock integrity. Verdon et al. [25] demonstrated the added benefit of microseismic monitoring for the geological storage of CO_2 during injection, where, for example, it was shown that microseismic activity is a natural consequence of fluid production and injection and does not necessarily imply leakage from the storage formation. Our results show that permeability estimates based on the seepage-force triggering mechanism technique using microseismic monitoring have strong potential to constrain formation permeability limitations for large-scale CO_2 injection.

Acknowledgments

The authors would like to thank Pinnacle Technologies Ltd. for making the microseismic data available and Alexander Rozhko for providing the Matlab scripts to compute the seepage-force curves. This research was partially funded by the BUMPS consortium, UK Energy Research Center (UKERC), and Research Councils UK (RCUK). J. P. Verdon is a Natural Environment Research Council (NERC) Early-Career Research Fellow (Grant no. NE/I021497/1).

References

[1] J. R. Jones and L. K. Britt, *Design and Appraisal of Hydraulic Fractures*, SPE, Richardson, Tex, USA, 1997.

[2] N. Teanby, J. M. Kendall, R. H. Jones, and O. Barkved, "Stress-induced temporal variations in seismic anisotropy observed in microseismic data," *Geophysical Journal International*, vol. 156, no. 3, pp. 459–466, 2004.

[3] J. P. Verdon, J. M. Kendall, and A. Wüstefeld, "Imaging fractures and sedimentary fabrics using shear wave splitting measurements made on passive seismic data," *Geophysical Journal International*, vol. 179, no. 2, pp. 1245–1254, 2009.

[4] A. Wuestefeld, O. Al-Harrasi, J. P. Verdon, J. Wookey, and J. M. Kendall, "A strategy for automated analysis of passive microseismic data to image seismic anisotropy and fracture characteristics," *Geophysical Prospecting*, vol. 58, no. 5, pp. 755–773, 2010.

[5] A. McGarr, "Violent deformation of rock near deep-level tabular excavations-seismic events," *Bulletin of the Seismological Society of America*, vol. 61, pp. 1453–1466, 1971.

[6] C. I. Trifu, D. Angus, and V. Shumila, "A fast evaluation of the Seismic moment tensor for Induced Seismicity," *Bulletin of the Seismological Society of America*, vol. 90, no. 6, pp. 1521–1527, 2000.

[7] A. Al-Anboori, M. Kendall, D. Raymer, and R. Jones, "Spatial variations in microseismic focal mechanisms, Yibal Field, Oman," in *Proceedings of the 68th EAGE Conference & Exhibition*, June 2006.

[8] S. A. Shapiro, *Microseismicity: A Tool for Reservoir Characterization*, EAGE Publications, Houten, The Netherlands, 2008.

[9] V. Grechka, P. Mazumdar, and S. A. Shapiro, "Predicting permeability and gas production of hydraulically fractured tight sands from microseismic data," *Geophysics*, vol. 75, no. 1, pp. B1–B10, 2010.

[10] S. A. Shapiro, J. Kummerow, C. Dinske et al., "Fluid induced seismicity guided by a continental fault: injection experiment of 2004/2005 at the German Deep Drilling Site (KTB)," *Geophysical Research Letters*, vol. 33, no. 1, Article ID L01309, 2006.

[11] A. Y. Rozhko, "Role of seepage forces on seismicity triggering," *Journal of Geophysical Research B*, vol. 115, no. 11, Article ID B11314, 2010.

[12] M. A. Biot, "Mechanics of deformation and acoustic propagation in porous media," *Journal of Applied Physics*, vol. 33, no. 4, pp. 1482–1498, 1962.

[13] S. Crampin and S. Peacock, "A review of shear-wave splitting in the compliant crack-critical anisotropic Earth," *Wave Motion*, vol. 41, no. 1, pp. 59–77, 2005.

[14] W. Narr, D. W. Schechter, and L. B. Thompson, *Naturally Fractured Reservoir Characterization*, SPE, Richardson, Tex, USA, 1997.

[15] M. D. Zoback and H. P. Harjes, "Injection-induced earthquakes and crustal stress at 9 km depth at the KTB deep drilling site, Germany," *Journal of Geophysical Research B*, vol. 102, no. 8, pp. 18477–18491, 1997.

[16] K. Terzaghi, *Theoretical Soil Mechanics*, John Wiley & Sons, New York, NY, USA, 1943.

[17] E. J. Kaiser, *A study of acoustic phenomena in tensile test [Doctoral thesis]*, Technische Hochschule München, Munich, Germany, 1959.

[18] L. Eisner, P. M. Duncan, W. M. Heigl, and W. R. Keller, "Uncertainties in passive seismic monitoring," *Leading Edge*, vol. 28, no. 6, pp. 648–655, 2009.

[19] P. J. Usher, D. A. Angus, and J. P. Verdon, "Influence of a velocity model and source frequency on microseismic waveforms: some implications for microseismic locations," *Geophysical Prospecting*. In press.

[20] J. P. Verdon, J. M. Kendall, and S. C. Maxwell, "A comparison of passive seismic monitoring of fracture stimulation from water and CO_2 injection," *Geophysics*, vol. 75, no. 3, pp. MA1–MA7, 2010.

[21] S. C. Maxwell, J. Shemata, E. Campbell, and D. Quirk, "Microseismic deformation rate monitoring," in *Proceedings of the SPE Annual Technical Conference and Exhibition (ATCE '07)*, pp. 21–24, Denver, Colo, USA, September 2008.

[22] U. Zimmer, S. Maxwell, C. Waltman, and N. Warpinski, "Microseismic monitoring quality-control (QC) reports as an interpretative tool for nonspecialists," in *Proceedings of the SPE Annual Technical Conference and Exhibition (ATCE '07)*, pp. 3074–3080, Anaheim, Calif, USA, November 2007.

[23] L. P. Dake, *The Practice of Reservoir Engineering*, Elsevier, London, UK, 2001.

[24] T. I. Urbancic, J. Daugherty, S. Bowman, and M. Prince, "Microseismic monitoring of a carbon sequestration field test," in *Proceedings of the 2nd EAGE Passive Seismic Workshop*, Limassol, Cyprus, 2009.

[25] J. P. Verdon, J. M. Kendall, D. J. White, and D. A. Angus, "Linking microseismic event observations with geomechanical models to minimise the risks of storing CO_2 in geological formations," *Earth and Planetary Science Letters*, vol. 305, no. 1-2, pp. 143–152, 2011.

5

A Simulation Study of the Formation of Large-Scale Cyclonic and Anticyclonic Vortices in the Vicinity of the Intertropical Convergence Zone

Igor V. Mingalev,[1] Natalia M. Astafieva,[2] Konstantin G. Orlov,[1] Victor S. Mingalev,[1] Oleg V. Mingalev,[1] and Valery M. Chechetkin[3]

[1] Polar Geophysical Institute, Kola Scientific Center of the Russian Academy of Sciences, Murmansk Region, Apatity 184209, Russia
[2] Space Research Institute of the Russian Academy of Sciences, Moscow 117997, Russia
[3] Keldysh Institute of Applied Mathematics of the Russian Academy of Sciences, Moscow 125047, Russia

Correspondence should be addressed to Victor S. Mingalev; mingalev@pgia.ru

Academic Editors: A. De Santis, M. Ernesto, and A. Streltsov

A regional nonhydrostatic mathematical model of the wind system of the lower atmosphere, developed recently in the Polar Geophysical Institute, is utilized to investigate the initial stage of the origin of large-scale vortices at tropical latitudes. The model produces three-dimensional distributions of the atmospheric parameters in the height range from 0 to 15 km over a limited region of the Earth's surface. Time-dependent modeling is performed for the cases when, at the initial moment, the simulation domain is intersected by the intertropical convergence zone (ITCZ). Calculations are made for various cases in which the initial forms of the intertropical convergence zone are different and contained convexities with distinct shapes, which are consistent with the results of satellite microwave monitoring of the Earth's atmosphere. The results of modeling indicate that the origin of convexities in the form of the intertropical convergence zone, having distinct configurations, can lead to the formation of different large-scale vortices, in particular, a cyclonic vortex, a pair of cyclonic-anticyclonic vortices, and a pair of cyclonic vortices, during a period not longer than three days. The radii of these large-scale vortices are about 400–600 km. The horizontal wind velocity in these vortices can achieve values of 15–20 m/s in the course of time.

1. Introduction

It is known that severe tropical cyclonic storms and hurricanes can cause tremendous damage and numerous fatalities. Therefore, prediction of tropical cyclone formation is a very important problem. Despite considerable efforts, the physical theory of tropical cyclone formation is still far from completion, even if some aspects of tropical cyclogenesis are commonly understood, in particular, in the late stages of formation as well as in a fully developed stage (see [1–4] and references therein). For tropical storm forecasting, it is necessary to investigate the pregenesis evolution of tropical cyclone. This investigation assumes that physical aspects of tropical cyclone formation must to be studied.

To investigate physical mechanisms responsible for the tropical cyclone formation, mathematical models may be utilized. Most of numerical studies of tropical cyclogenesis explore how a tropical cyclone forms from vortices, those are precursors to tropical cyclones, or from other preexisting large-scale disturbances of the troposphere [5–12].

In the Polar Geophysical Institute (PGI), the nonhydrostatic model of the global wind system in the Earth's atmosphere has been developed not long ago [13, 14]. This model enables to calculate three-dimensional global distributions of the zonal, meridional, and vertical components of the wind at levels of the troposphere, stratosphere, mesosphere, and lower thermosphere, with whatever restrictions on the vertical transport of the atmospheric gas being absent. This

A Simulation Study of the Formation of Large-Scale Cyclonic and Anticyclonic Vortices in the Vicinity of the
Intertropical Convergence Zone

37

model has been utilized in order to investigate numerically how the horizontal nonuniformity of the atmospheric gas temperature affects the formation of the middle atmosphere circulation for conditions corresponding to different seasons [13–16] and how solar activity affects the formation of the large-scale global circulation of the mesosphere and lower thermosphere [17].

Recently, a regional mathematical model of the wind system of the lower atmosphere has been developed in the Polar Geophysical Institute [18]. This model was applied to investigate the formation mechanisms of a large-scale vortex over a warm water band on the ocean surface. The results of modeling have allowed the authors to distinguish one of the formation mechanisms of moderate cyclones over the ocean [18].

Another formation mechanism of a cyclone was investigated, using this mathematical model, in the study by Belotserkovskii et al. [19]. In this study, it was shown that cyclones can appear in horizontally stratified shear flows of warm and wet air masses with a meridional direction of gradients of the wind velocity components as a result of small disturbances of pressure which can be produced by Rossby waves.

Also, this mathematical model has been applied to verify the hypothesis of the influence of the shape of the intertropical convergence zone (ITCZ) on the process of the formation of tropical cyclones. It was shown that the origin of a convexity in the configuration of the intertropical convergence zone can lead to the formation of a cyclonic vortex during the period of about one day. Its center is close to the southern edge of the initial intertropical convergence zone [20]. The results of mathematical modeling have indicated that the origin of a convexity of the intertropical convergence zone, having the specific forms, can lead to the formation of not only a single cyclonic vortex but also a pair of large-scale vortices [21–23].

The purpose of the present work is to continue these studies and to investigate numerically, applying the regional mathematical model of the wind system of the lower atmosphere developed in the PGI, the initial stage of the origin of large-scale vortices in the vicinity of the intertropical convergence zone. The applied model produces distributions of the lower atmosphere parameters in the limited three-dimensional simulation domain. Time-dependent modeling is performed for various cases when, at the initial moment, the simulation domain is intersected by the intertropical convergence zones with different configurations. Calculations are made for some cases in which the initial forms of the intertropical convergence zone are different and contained convexities with distinct shapes.

2. Mathematical Model

In this study, the regional nonhydrostatic mathematical model of the wind system of the lower atmosphere, developed not long ago at the Polar Geophysical Institute, is applied. In this model, the atmospheric gas is considered as a mixture of air and water vapor, in which two types of precipitating

water (namely, water microdrops and ice microparticles) can exist. The model is based on the numerical solution of the system of transport equations containing the equations of continuity for air and for the total water content in all phase states, momentum equations for the zonal, meridional, and vertical components of the air velocity, and energy equation. The characteristic feature of the model is that the vertical component of the air velocity is calculated without using the hydrostatic equation. Instead, the vertical component of the air velocity is obtained by means of a numerical solution of the appropriate momentum equation, with whatever simplifications of this equation being absent. In the momentum equations for all components of the air velocity, the effect of the turbulence on the mean flow is taken into account by using an empirical subgrid-scale parameterization similar to the global circulation model of the Earth's atmosphere developed earlier in the PGI [13, 14].

Thus, the utilized mathematical model is based on numerical solving of nonsimplified gas dynamic equations and produces three-dimensional time-dependent distributions of the wind components, temperature, air density, water vapor density, concentration of microdrops of water, and concentration of ice particles. The model takes into account heating/cooling of the air due to absorption/emission of infrared radiation, as well as due to phase transitions of water vapor to microdrops of water and ice particles, which play an important role in energetic balance. The finite-difference method and an explicit scheme are applied for solving the system of governing equations.

In the model calculations, the following variables are computed at each grid node: the temperature of the mixture of air and water vapor, T; densities of air and water vapor, ρ_a and ρ_v, respectively; hydrodynamic velocity of the mixture (a 3D vector), $\vec{v}$; and the total mass of water microdrops and ice microparticles in a unit volume, ρ_w and ρ_i, respectively. The governing equations in vectorial form can be written as follows:

$$\frac{\partial \rho_a}{\partial t} + \operatorname{div}(\rho_a \vec{v}) = 0,$$

$$\frac{\partial (\rho_v + \rho_w + \rho_i)}{\partial t} + \operatorname{div}\left[\rho_v \vec{v} + \rho_w \left(\vec{v} + \vec{v}_w^{\text{prec}}\right)\right.$$
$$\left. + \rho_i \left(\vec{v} + \vec{v}_i^{\text{prec}}\right)\right] = 0, \quad (1)$$

$$\frac{\partial (\rho_{\text{mix}} \vec{v})}{\partial t} + \operatorname{div}(\rho_{\text{mix}} \vec{v} \otimes \vec{v})$$
$$= -\nabla p + \operatorname{div}\hat{\tau} + (\rho_{\text{mix}} + \rho_w + \rho_i)\vec{F},$$

$$\frac{\partial W}{\partial t} + \operatorname{div}\left[W_{\text{mix}} \vec{v} + W_w \left(\vec{v} + \vec{v}_w^{\text{prec}}\right) + W_i \left(\vec{v} + \vec{v}_i^{\text{prec}}\right)\right]$$
$$= \left[\rho_{\text{mix}} \vec{v} + \rho_w \left(\vec{v} + \vec{v}_w^{\text{prec}}\right) + \rho_i \left(\vec{v} + \vec{v}_i^{\text{prec}}\right), \vec{F}\right] \quad (2)$$
$$+ \operatorname{div}\left(\hat{\tau} \cdot \vec{v} - p\vec{v} - \vec{j}\right) + Q,$$

where $\vec{v}_w^{\text{prec}}$ and $\vec{v}_i^{\text{prec}}$ are the precipitation velocities of water microdrops and ice microparticles, respectively, conditioned by the presence of an external force field and determined from the Stokes relation with Cunningham's correction; $\rho_{\text{mix}} = \rho_a + \rho_v$; p is the pressure of the mixture defined as $p = (\rho_a R_a + \rho_v R_v)T$, where R_a and R_v are the gas constants of air and water vapour, respectively; $\hat{\tau}$ is the extra stress tensor whose components are given by the rheological equation of state or the law of viscous friction which is the same as in the global circulation model of the Earth's atmosphere developed earlier [13, 14], with the effect of a small-scale turbulence, having the scales equal and less than the steps of finite-difference approximations, on the mean flow having been taken into account; $\vec{F}$ is the acceleration due to an external force field, which consists of the gravity acceleration, Coriolis acceleration, and acceleration of translation, that can be written in the form:

$$\vec{F} = \vec{g} - 2\vec{\Omega} \times \vec{v} - \vec{\Omega} \times \left(\vec{\Omega} \times \vec{r} \right), \qquad (3)$$

where $\vec{g}$ is the acceleration due to gravity, $\vec{\Omega}$ is the Earth's angular velocity, and $\vec{r}$ is a radius vector from the center of the Earth to the point where the equation is applied. The following notations are used in (2):

$$W_i = \rho_i \left[\frac{1}{2}(\vec{v} + \vec{v}_i^{\text{prec}})^2 + C_i T \right],$$

$$W_w = \rho_w \left[\frac{1}{2}(\vec{v} + \vec{v}_w^{\text{prec}})^2 + C_w (T - T_0) + q_{\text{mel}} + C_i T_0 \right], \qquad (4)$$

$$W = W_{\text{mix}} + W_w + W_i,$$

where C_i and C_w are the specific heat capacities of ice and water, respectively, which are assumed constant; T_0 is the freezing temperature of water; q_{mel} is the specific heat of ice melting at $T = T_0$; q_{ev}^0 is the specific heat of water evaporation at $T = T_0$; also, the vector of heat flux, $\vec{j}$, is given by the well-known formula, $\vec{j} = -\hat{\lambda} \nabla T$, where $\hat{\lambda}$ is the symmetric tensor of thermal conductivity coefficients; and Q is the rate of change of energy in a unit volume due to absorption/emission of infrared radiation. Concrete expressions of the model parameters, those appear in (1)-(2), may be found in the studies of Belotserkovskii et al. [18, 19].

It can be noticed that the model assumes that the water microdrops can exist only in the presence of saturated water vapor on condition that $T \geq T_0$, while the ice microparticles can exist only in the presence of saturated water vapor on condition that $T \leq T_0$. At $T = T_0$, the temperature of the matter cannot increase until all ice microparticles melt, and it cannot decrease until all water microdrops freeze.

The three-dimensional simulation domain of the model is a part of a spherical layer stretching from land and ocean surface up to the altitude of 15 km over a limited region of the Earth's surface. The dimensions of this region in longitudinal and latitudinal directions are 32° and 25°, respectively. The finite-difference method and explicit scheme are applied for solving the system of governing equations. The calculated parameters are determined on a uniform grid. The latitude and longitude steps are equal to 0.08°, and height step is equal to 200 m. Complete details of the utilized finite-difference method and numerical schemes have been presented in the paper of Mingalev et al. [24]. More complete details of the

applied regional mathematical model may be found in the studies of Belotserkovskii et al. [18, 19] and Mingalev et al. [23].

3. Presentation and Discussion of Results

Modern scientific facility does not allow somebody to measure detailed three-dimensional fields of thermodynamical and gas dynamical parameters of the lower atmosphere with sufficient accuracy to understand the intrinsic nature of tropical cyclone genesis. However, for a better conceptual understanding of tropical cyclogenesis, satellite microwave observations may be applied, in particular, the results of satellite microwave monitoring of the Earth's atmosphere, collected at the Space Research Institute of the Russian Academy of Sciences (RAS) and included in the electronic data base "GLOBAL-Field" (http://www.iki.rssi.ru/asp/). This data base contains global radiothermal fields of the Earth at the frequencies containing the information about a moisture and water integral content distribution in the troposphere. The data were obtained from the spacecraft mission, DMSP (Defense Meteorological Satellite Program), with the help of the instrument, SSM/I (Special Sensor Microwave/Imager).

As examples of data from the electronic collection "GLOBAL-Field," fragments of global radiothermal fields over the Atlantic Ocean are presented in Figures 1 and 2. In these Figures, the various colors of the image show the distribution of radiobrightness temperature of surface of ocean and land. It is essential that this temperature is consistent with the moisture and water integral content distribution in the troposphere. From Figures 1 and 2, one can see that the Atlantic Ocean is intersected by a red band stretched from Africa to South America. This band represents an intertropical convergence zone. It can be noticed that an intertropical convergence zone is similar to a band, in which zonal westward flow of air predominates, with the air velocity being enhanced. The width of the intertropical convergence zone can achieve a value of some hundreds of kilometers.

Studying numerous results of satellite microwave monitoring of the Earth's atmosphere, included in the electronic data base "GLOBAL-Field," we noticed that the form of an intertropical convergence zone may be different and, sometimes, can contain convexities with distinct shapes. Just such examples are presented in Figures 1 and 2, where the convexities are indicated by black pointers. As a consequence of our studies of results of satellite microwave monitoring of the Earth's atmosphere, we have advanced a hypothesis of the important role of the shape of the intertropical convergence zone in the process of the formation of tropical cyclones.

To verify the advanced hypothesis, we have decided to make calculations with the help of the regional nonhydrostatic mathematical model of the wind system of the lower atmosphere, described above, with the south boundary of the simulation domain having been disposed in the vicinity of the equator. It was supposed that the simulation domain is located between 5°S and 20°N and situated over the Atlantic Ocean. Simulations were performed for the cases

A Simulation Study of the Formation of Large-Scale Cyclonic and Anticyclonic Vortices in the Vicinity of the
Intertropical Convergence Zone

39

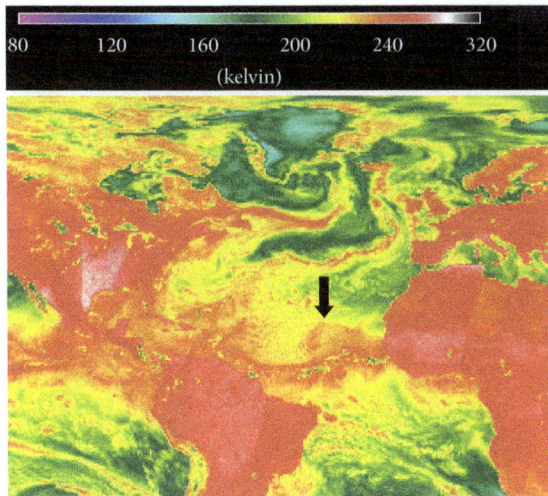

FIGURE 1: The fragment of global radiothermal field, derived from the results of satellite microwave monitoring of the Earth's atmosphere, obtained on June 29, 2005.

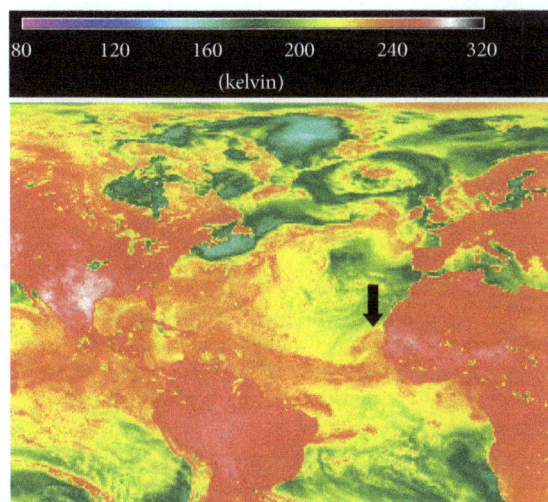

FIGURE 2: The same as in Figure 1, but obtained on July 3, 2005.

when the simulation domain is intersected by an intertropical convergence zone in the westeast direction.

From observation, it is known that, in an intertropical convergence zone, a zonal flow of air is westward, with the horizontal wind velocity being enhanced inside an intertropical convergence zone. A meridional wind velocity directs towards the centerline of an intertropical convergence zone at levels less than approximately 3 km and directs from the centerline of an intertropical convergence zone at levels higher than approximately 3 km. A vertical wind velocity in an intertropical convergence zone is upward. Therefore, an intertropical convergence zone may be considered as a fluid stream, having enhanced zonal velocities, in the ambient atmospheric gas. In our calculations, we define the initial and boundary conditions as consistent with the situation when the intertropical convergence zone intersects the simulation domain in the westeast direction. Calculations are made for

various cases in which the initial forms of the intertropical convergence zone are different and contained convexities with distinct shapes, which are consistent with observational data, included in the electronic collection "GLOBAL-Field." Since the obtained results are different, it is convenient to present them separately.

3.1. The Formation of a Cyclonic Vortex. Initially, let us consider the first case when, at the initial moment, the intertropical convergence zone contains a convexity in the north direction, with the deviation achieving a value of a few hundreds of kilometers. The initial form of the intertropical convergence zone may be easy seen from Figure 3(a), where it is like a light curved band. It is essential to note that, in the considered first case, the west crook of the convexity is sharp while the east crook of the convexity is gently sloping, with the west and east ends of the convexity being at the same latitudes.

The time evolution of model parameters was numerically simulated using the mathematical model during the period of about one day. The results of time-dependent modeling are partly shown in Figure 3. As can be seen from this figure, in the course of time, the initial distribution of horizontal component of the air velocity was considerably transformed. A cyclonic vortex flow arose whose center is close to the southern edge of the initial intertropical convergence zone. The horizontal wind velocity in this cyclonic vortex achieved a value of 20 m/s during the period of twenty-seven hours. The radius of this large-scale cyclonic vortex is about 600 km.

In addition, we made simulations for the second and third cases when, at the initial moment, the intertropical convergence zone has different configurations. For both cases, the initial forms of the intertropical convergence zone contained the convexities analogous to the convexity of the first case. However, for the second case, the east end of the convexity is situated at more northern latitudes than the west end of the convexity (see Figure 4(a)). On the contrary, for the third case, the east end of the convexity is situated at more southern latitudes than the west end of the convexity (see Figure 5(a)). The results of time-dependent modeling for the second and third cases of the initial configurations of the intertropical convergence zone are partly shown in Figures 4 and 5. As can be seen from these figures, in the course of time, cyclonic vortex flows arose whose centers are close to the southern edge of the initial intertropical convergence zone. These vortices are analogous to those obtained for the first case.

The simulation results indicate that physical reason of the formation of the calculated tropical cyclonic vortices is the origin of a convexity in the configuration of the intertropical convergence zone. As a rule, such convexities are observed during the periods of rebuilding of the global circulation of the atmosphere. The origin of a convexity of the intertropical convergence zone leads to the beginning of an instability of stream air flow. As a consequence, a large-scale vortex flow arises in the lower atmosphere, with its center being close to the southern edge of the initial intertropical convergence zone. When the mixture of air and water vapor moves to

(a)

(b)

(c)

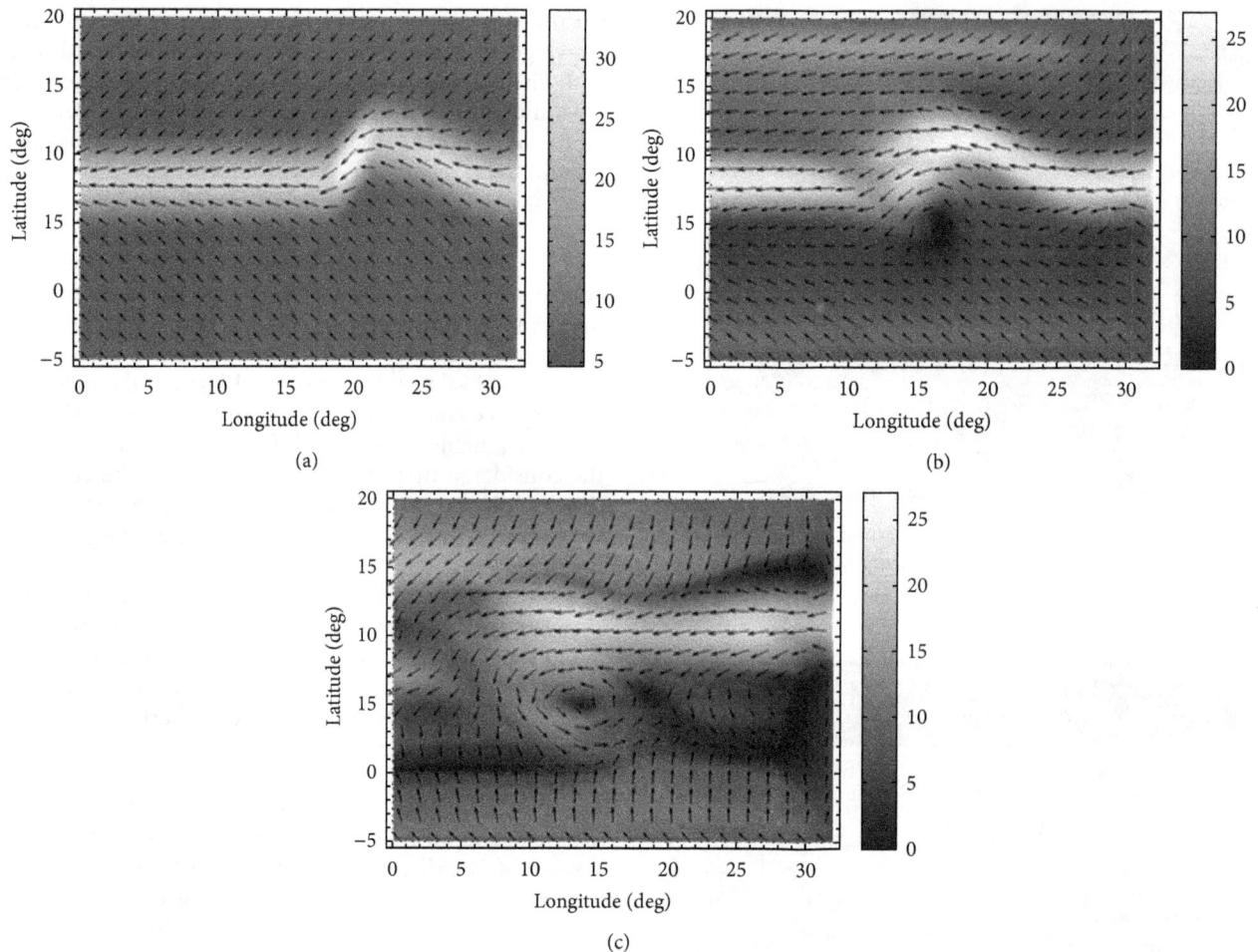

FIGURE 3: The distributions of horizontal component of the air velocity at the altitude of 600 m, assigned at the initial moment (a), computed 12 hours after the beginning of calculations (b), and computed 27 hours after the beginning of calculations (c). The results are obtained for the first initial configuration of the intertropical convergence zone. The degree of shadowing of the figures indicates the module of the velocity in m/s.

higher altitudes, where temperature is lower, energy, emitted due to phase transitions of water vapor to microdrops of water and ice particles, transforms into kinetic energy of the air flow, with the horizontal wind velocity increasing in formed cyclonic vortex. It should be emphasized that the initial degree of saturation of air by water vapor plays an important role in the formation of the calculated tropical cyclonic vortices.

3.2. The Formation of a Pair of Cyclonic-Anticyclonic Vortices. In the previous subsection, we have considered three cases of the initial configuration of the intertropical convergence zone containing the convexity in the north direction, with the east end of the convexity located at different latitudes. The common feature of these initial configurations is that the west crook of the convexity is sharp while the east crook of the convexity is gently sloping. In the present subsection, we consider cases when both west and east crooks of the

convexity are sharp and located at the same latitudes at the initial moment.

Initially, let us consider the fourth case when, at the initial moment, the intertropical convergence zone contains a convexity in the north direction. The initial form of the intertropical convergence zone may be easily seen from Figure 6(a), where it is like a light curved band. The time evolution of model parameters was numerically simulated using the mathematical model during the period of about one day. The results of time-dependent modeling are partly shown in Figure 6. As can be seen from this figure, in the course of time, the initial distribution of horizontal component of the air velocity was considerably transformed. A pair of cyclonic-anticyclonic vortices arose in the vicinity of the initial intertropical convergence zone. A cyclonic vortex arose whose center is close to the southern edge of the initial intertropical convergence zone and an anticyclonic vortex arose whose center is close to the northern edge of the initial intertropical convergence zone. The horizontal wind velocity in these vortices achieved a value of 20 m/s during the period

A Simulation Study of the Formation of Large-Scale Cyclonic and Anticyclonic Vortices in the Vicinity of the Intertropical Convergence Zone

41

(a)

(b)

(c)

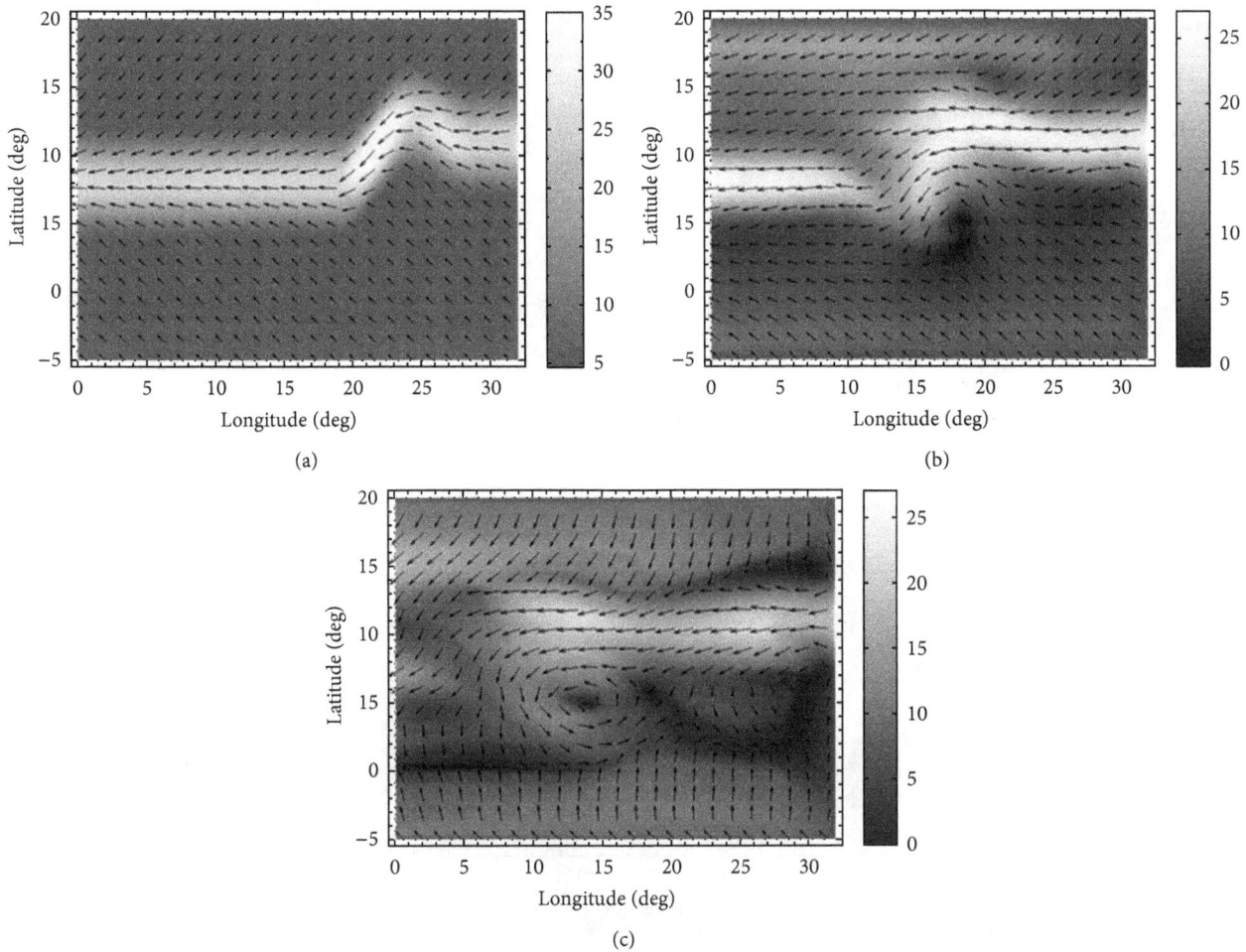

FIGURE 4: The same as in Figure 3, but obtained for the second initial configuration of the intertropical convergence zone. The degree of shadowing of the figures indicates the module of the velocity in m/s.

of twenty-seven hours. The radii of these large-scale vortices are about 400 km.

It should be emphasized that the results of satellite monitoring of the Earth's atmosphere often indicated a simultaneous origin of a cyclone-anticyclone pair. Naturally, a formation of a single cyclone or a single anticyclone was observed by satellites repeatedly. It can be noticed that, according to observations, an initially originated single cyclonic or anticyclonic vortex as well as one of vortices, belonging to a cyclone-anticyclone pair, as well as both vortices, belonging to a cyclone-anticyclone pair, sometimes, can be attenuated in the course of time and will not achieve a status of the long-live large-scale atmospheric vortices.

Let us consider the fifth case when, at the initial moment, the intertropical convergence zone contains a convexity, analogous to the convexity of the fourth case, but deviated in the south direction. The initial form of the intertropical convergence zone may be easy seen from Figure 7(a), where it is like a light curved band. The results of time-dependent modeling for the fifth case of the initial configurations of the intertropical convergence zone are partly shown in Figure 7. As can be seen from this figure, in the course of time, a

pair of cyclonic-anticyclonic vortices arose in the vicinity of the initial intertropical convergence zone. These vortices are analogous to those obtained for the fourth case.

The results of simulation indicate that the physical reason of the formation of the calculated pair of cyclonic-anticyclonic vortices is the origin of a convexity in the configuration of the intertropical convergence zone, having the specific forms. The origin of a convexity of the intertropical convergence zone leads to the beginning of an instability of stream air flow. As a consequence, a pair of cyclonic-anticyclonic vortices arises in the lower atmosphere. In the course of time, the horizontal wind velocity in the vortices increases due to a growth of kinetic energy of the air flow because of a transformation of energy, emitted owing to phase transitions of water vapor to microdrops of water and ice particles, in the flow of air moving to higher altitudes, where temperature is lower, with the initial degree of saturation of air by water vapor playing an important role in the formation of the calculated vortices.

3.3. The Formation of a Pair of Tropical Cyclonic Vortices. In the previous subsections, at the initial moment, the horizontal

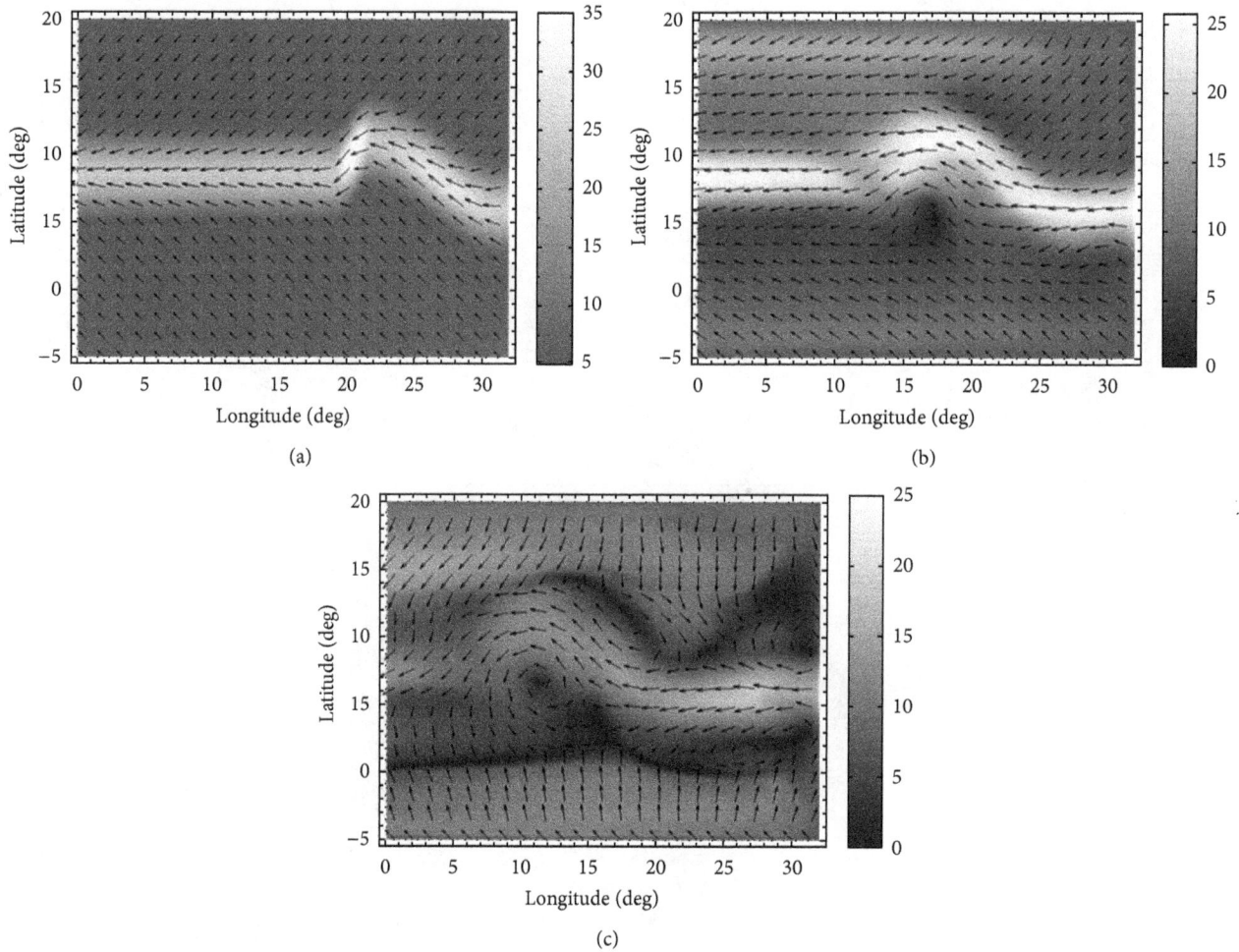

FIGURE 5: The same as in Figure 3, but obtained for the third initial configuration of the intertropical convergence zone. The degree of shadowing of the figures indicates the module of the velocity in m/s.

velocity fields were approximately symmetric relatively to the centerline of the intertropical convergence zone not only inside it but also beyond the intertropical convergence zone. In the present subsection, we consider a case when, at the initial moment, the horizontal velocity field is not symmetric relatively to the centerline of the intertropical convergence zone inside and beyond it. Moreover, in the present subsection, the dimension of the simulation domain in longitudinal direction is equal to 40°.

Let us consider the sixth case when, at the initial moment, the intertropical convergence zone contains a convexity in the north direction, with the zonal wind velocities at more northern latitudes relative to the centerline of the intertropical convergence zone being larger than at more southern latitudes relative to it. The initial form of the intertropical convergence zone may be easily seen from Figure 8(a), where it is like a light curved band. It is noted that, in the considered sixth case, the west crook of the convexity is sharp while the east crook of the convexity is gently sloping, with the east end of the convexity being 1° southern than the west crook of it. The time evolution of model parameters was numerically simulated using the mathematical model during the period

of about 3 days. The results of time-dependent modeling are partly shown in Figures 8 and 9. As can be seen from these figures, in the course of time, the initial distribution of horizontal component of the air velocity was considerably transformed. In a moment of 20 hours after the beginning of calculations, a cyclonic vortex flow arose whose center is close to the southern edge of the initial intertropical convergence zone. In a moment of 40 hours after the beginning of calculations, this cyclonic vortex has moved in the western direction for about 5 degrees. In a moment of 60 hours after the beginning of calculations, the rotational center of the formed cyclonic vortex has moved in the northwest direction for a distance of about 600 km. Simultaneously, the second cyclonic vortex arose, with its center being close to the southern edge of the initial intertropical convergence zone.

Thus, in a moment of 70 hours after the beginning of calculations, a pair of cyclonic vortices arose in the vicinity of the intertropical convergence zone. The rotational centers of these cyclonic vortices are situated near the edges of the intertropical convergence zone. The center of the first cyclonic vortex is close to the northern edge while the center of the second cyclonic vortex is close to the southern edge

A Simulation Study of the Formation of Large-Scale Cyclonic and Anticyclonic Vortices in the Vicinity of the
Intertropical Convergence Zone

43

(a)

(b)

(c)

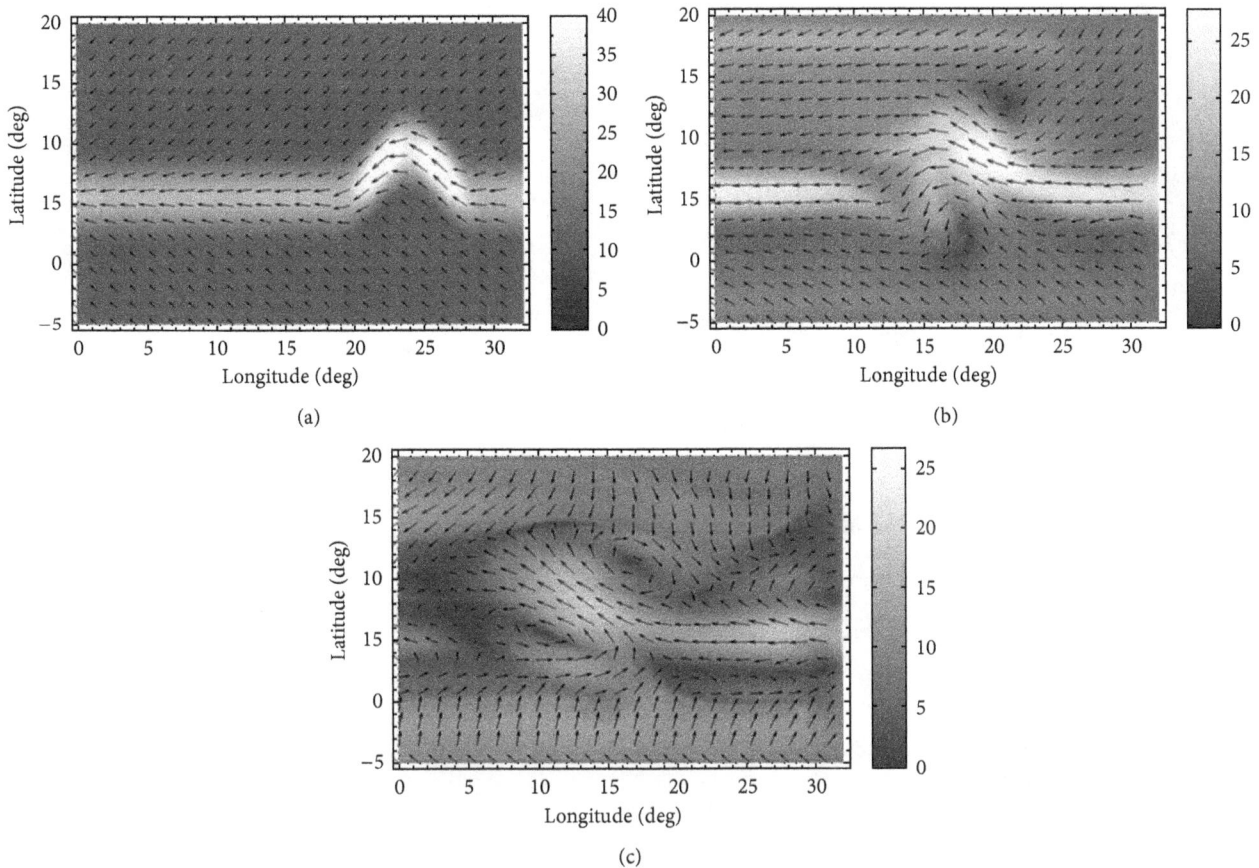

FIGURE 6: The distributions of horizontal component of the air velocity at the altitude of 600 m, assigned at the initial moment (a), computed 12 hours after the beginning of calculations (b), and computed 27 hours after the beginning of calculations (c). The results are obtained for the fourth initial configuration of the intertropical convergence zone. The degree of shadowing of the figures indicates the module of the velocity in m/s.

of the intertropical convergence zone. The horizontal wind velocity in these cyclonic vortices achieved values of 15–20 m/s during the period of seventy hours. The radii of these cyclonic vortices are about 600 km.

It can be noticed that the results of observation of the Earth's atmosphere indicated a simultaneous origin of twin tropical cyclones sometimes [25].

The results of simulation indicate that a key factor in the modeled formation of twin tropical cyclonic vortices is the origin of a convexity in the configuration of the intertropical convergence zone, having the specific form, which is accompanied by nonsymmetric horizontal velocity field in the vicinity of this zone. The pointed out factors lead to the beginning of instability of stream air flow. As a consequence, a pair of cyclonic vortices arise in the lower atmosphere in the course of time. A transformation of energy, emitted due to phase transitions of water vapor to microdrops of water and ice particles in the mixture of air and water vapor moving upward, into kinetic energy of the air flow plays an important role in the increase of the horizontal wind velocity in the course of time, with the high initial degree of saturation of air by water vapor being an important factor.

4. Summary and Concluding Remarks

A number of earlier studies have shown that the intertropical convergence zone is sometimes observed to undulate and break down into a series of tropical disturbances [3, 26, 27]. Some of these disturbances may develop into tropical cyclones, while others dissipate, and the intertropical convergence zone may reform in the original region. It has been proposed that the intertropical convergence zone may break down because of its heating-induced potential vorticity anomalies. Moreover, it has been proposed that the intertropical convergence zone breakdown results from a convectively modified form of combined barotropic and baroclinic instability of the mean flow [26]. Also, instability of the intertropical convergence zone has been studied in the works of Wang and Magnusdottir [28] and Nolan et al. [29]. The role of tropical waves in tropical cyclogenesis has been studied, too [30].

Nevertheless, in earlier studies, we have not found the idea that the transformation of the shape of the intertropical convergence zone can influence the process of the formation of tropical cyclones (with the exception of the studies by the authors of the present work). Although, the position and

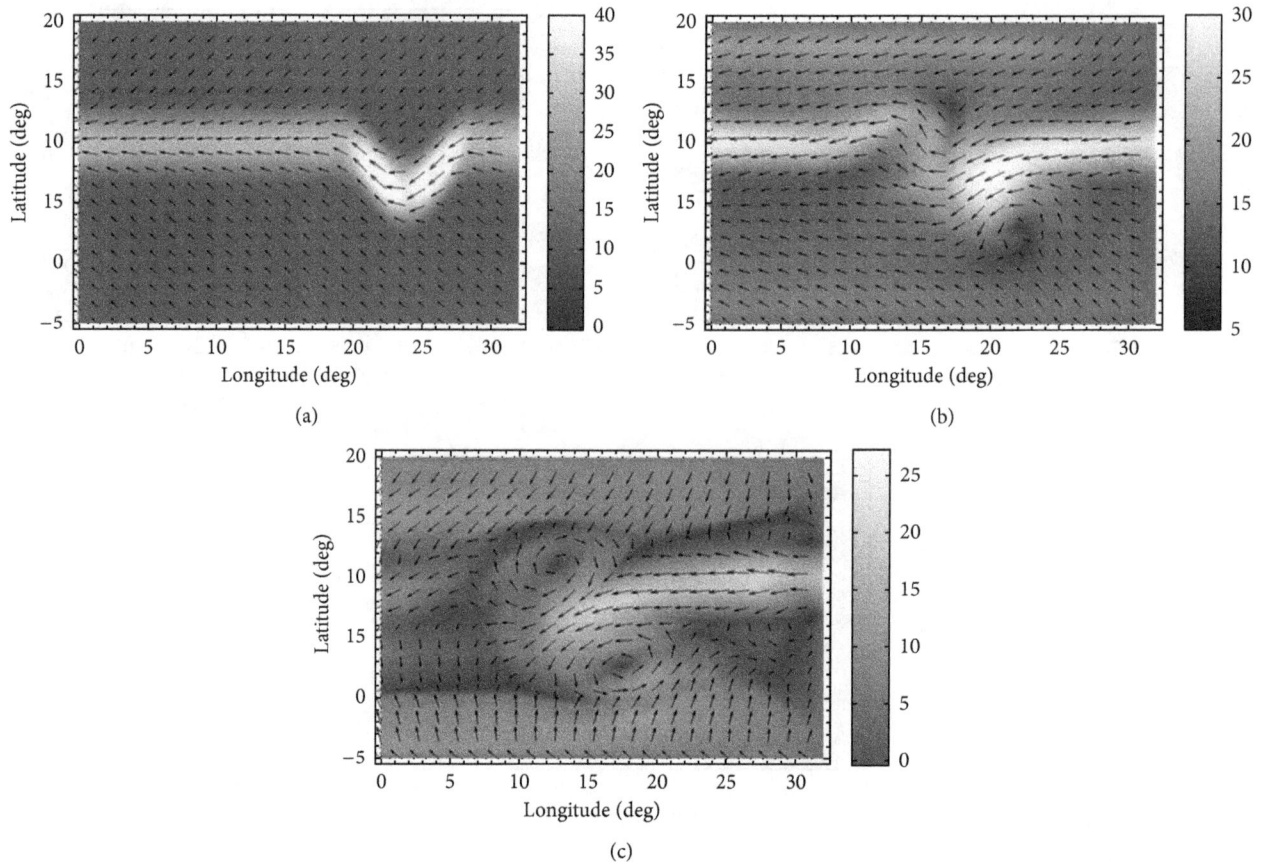

FIGURE 7: The same as in Figure 6, but obtained for the fifth initial configuration of the intertropical convergence zone. The degree of shadowing of the figures indicates the module of the velocity in m/s.

variability of the intertropical convergence zone have been studied in a number of earlier works [31–33]. It was found in these studies that the northsouth position of the intertropical convergence zone responds to changes in interhemispheric temperature contrast. An asymmetry in air-sea interactions can play an important role in forming the configuration of the intertropical convergence zone, too. In the present study, the influence of the disturbance of the configuration of the intertropical convergence zone on the process of the formation of tropical large-scale cyclonic and anticyclonic vortices was investigated numerically.

To make this investigation, the limited-area nonhydrostatic mathematical model of the wind system of the lower atmosphere, developed recently in the Polar Geophysical Institute, was utilized. The model produces three-dimensional distributions of the atmospheric parameters in the height range from land and ocean surface up to the altitude of 15 km over a limited region of the Earth's surface. The time evolution of model parameters was numerically simulated using various variants of the initial and boundary conditions which were defined as consistent with the situation when the intertropical convergence zone intersects the simulation domain in the westeast direction. Calculations were made for various cases in which the initial forms of the intertropical convergence zone were different and contained

convexities with distinct shapes, which were consistent with the results of satellite microwave monitoring of the Earth's atmosphere, included in the electron data base "GLOBAL-Field," developed in the Space Research Institute of the RAS.

The results of modeling indicated that the origin of a convexity in the configuration of the intertropical convergence zone, having the latitudinal dimension of 800–1000 km and the deviation of some hundreds of kilometers either in north or south direction, can lead to the formation of tropical large-scale vortices during the period of a few days. It was found that the initial shape of the convexity of the intertropical convergence zone plays a significant role in the process of the formation of tropical large-scale vortices.

Firstly, it was established the specific initial shapes of the intertropical convergence zone convexity, which bring forth a cyclonic vortex during the period of about one day. The common features of these initial shapes are that the convexities deviate in the north direction and that the west crook of the convexity is sharp while the east crook of the convexity is gently sloping. The east end of the convexity may be situated at more northern or at more southern latitudes than the west end of the convexity as well as at the same latitudes. The rotational center of the formed cyclonic vortex is close to the southern edge of the initial intertropical convergence zone. The cyclonic vortex has a horizontal extent

A Simulation Study of the Formation of Large-Scale Cyclonic and Anticyclonic Vortices in the Vicinity of the
Intertropical Convergence Zone

45

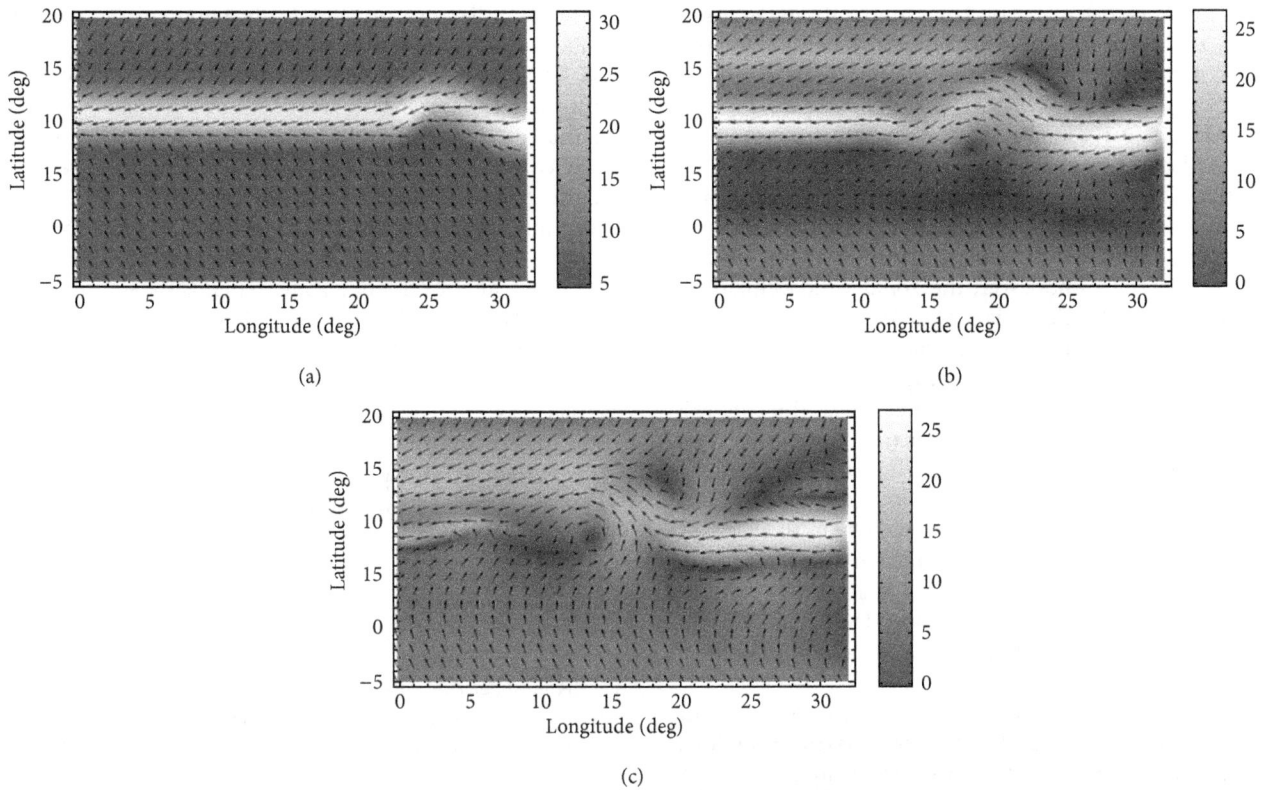

FIGURE 8: The distributions of horizontal component of the air velocity at the altitude of 600 m, assigned at the initial moment (a), computed 20 hours after the beginning of calculations (b), and computed 40 hours after the beginning of calculations (c). The results are obtained for the sixth initial configuration of the intertropical convergence zone. The degree of shadowing of the figures indicates the module of the velocity in m/s.

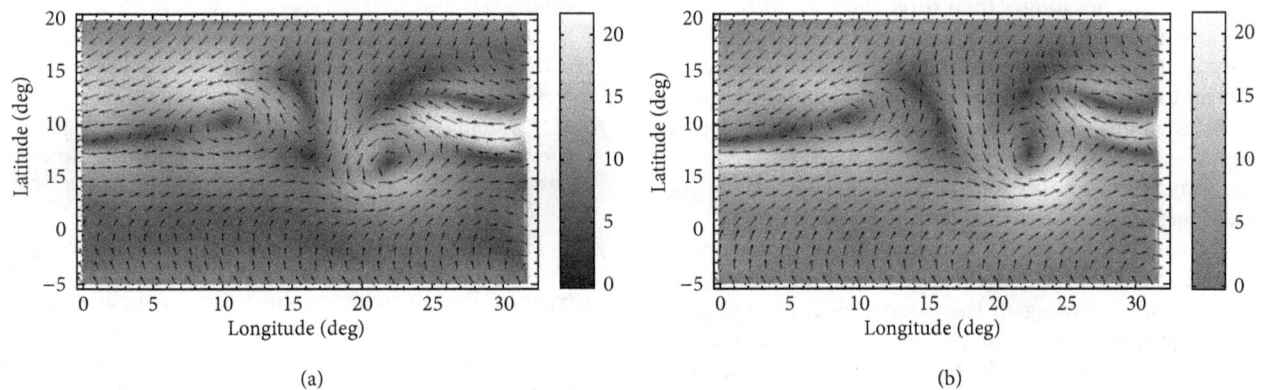

FIGURE 9: The same as in Figure 8, but computed 60 hours after the beginning of calculations (a) and 70 hours after the beginning of calculations (b). The results are obtained for the sixth initial configuration of the intertropical convergence zone. The degree of shadowing of the figures indicates the module of the velocity in m/s.

of about 600 km. The horizontal wind velocity in this cyclonic vortex can achieve a value of 20 m/s during the period of 27 hours.

Secondly, it was found the specific initial shapes of the intertropical convergence zone convexity, which bring forth a pair of cyclonic-anticyclonic vortices during the period of about one day. The common feature of these initial shapes is that both west and east crooks of the convexity are sharp

at the initial moment. The convexity may be deviated either in the north direction or in the south direction, with the west and east ends of the convexity being at the same latitudes. The rotational center of the formed cyclonic vortex is close to the southern edge of the initial intertropical convergence zone while the rotational center of the formed anticyclonic vortex is close to the northern edge of the initial intertropical convergence zone to a moment of 27 hours after the beginning of calculations. The radii of these large-scale vortices are about 400 km. The horizontal wind velocity in these vortices can achieve a value of 20 m/s during the period of twenty-seven hours.

Thirdly, it was established the specific variant of the initial and boundary conditions which, during the period of about three days, bringing forth a pair of tropical cyclonic vortices. It was supposed that, at the initial moment, the intertropical convergence zone contains the convexity in the north direction; moreover, the zonal wind velocities at more northern latitudes relative to the centerline of the intertropical convergence zone are larger than those at more southern latitudes relative to it, with the west crook of the convexity being sharp while the east crook of the convexity being gently sloping. The cyclonic vortices were formed one after another in the course of time. The rotational center of the first cyclonic vortex is close to the northern edge while the center of the second cyclonic vortex is close to the southern edge of the initial intertropical convergence zone in a moment of 70 hours after the beginning of calculations. The radii of these cyclonic vortices are about 600 km. The horizontal wind velocity in the cyclonic vortices achieved values of 15–20 m/s during the period of about three days.

It should be emphasized that, according to observations, not each cyclonic or anticyclonic vortex, arisen in the lower atmosphere, has the potential to grow up to the long-live large-scale atmospheric vortex. It is known that, sometimes, a vortex, initially arisen in the lower atmosphere, can be attenuated in the course of time and will not achieve a status of the long-live large-scale atmospheric vortex. This peculiarity may take place for the large-scale vortices arisen in the calculations of the present study, which were limited by the time intervals not longer than three days. Unfortunately, more prolonged time intervals are impossible for the utilized mathematical model because of limited sizes of its simulation domain and owing to tendency of the modeled vortices to move and to abandon the simulation domain in the course of time.

The simulation results show that a key factor in the modeled formation of tropical large-scale vortices is the origin of a convexity in the configuration of the intertropical convergence zone. As a consequence, instability of stream air flow arises, presenting in the intertropical convergence zone. This instability leads to considerable transformation of the wind field. As a result, tropical large-scale vortices may be formed in the vicinity of the initial position of the intertropical convergence zone in the course of time. In addition to that, the initial intertropical convergence zone is broken down. Energy, emitted due to phase transitions of water vapor to microdrops of water and ice particles in the mixture of air and water vapor, moving to higher altitudes, where temperature is lower, transforms into kinetic energy of the air flow, with the horizontal wind velocity increasing in the formed vortices.

It may be expected that the simulation results of the present study will be useful for tropical cyclone forecasting. The origin of a convexity in the configuration of the intertropical convergence zone, which may be observed with the help of satellite microwave monitoring of the Earth's atmosphere, is a precursor of the formation of a tropical large-scale vortex. This vortex either will grow up to the long-live large-scale atmospheric vortex, having the potential to transform to a tropical storm, or can be attenuated and vanish in the course of time. As a rule, intertropical convergence zone convexities originate during the periods of rebuilding of the global circulation of the atmosphere, in particular, from July to September over the Atlantic Ocean and Africa. Further transformation of the originated convexity of the intertropical convergence zone depends on its initial shape and on environmental conditions, in particular, on the degree of saturation of air by water vapor.

Acknowledgment

This work was partly supported by Grant no. 13-01-00063 from the Russian Foundation for Basic Research.

References

[1] K. A. Emanuel, "An air-sea interaction theory for tropical cyclones. Part I: steady-state maintenance," *Journal of the Atmospheric Sciences*, vol. 43, no. 6, pp. 585–605, 1986.

[2] M. T. Montgomery and B. F. Farrell, "Tropical cyclone formation," *Journal of the Atmospheric Sciences*, vol. 50, no. 2, pp. 285–310, 1993.

[3] C. Q. Kieu and D. L. Zhang, "Genesis of tropical storm eugene (2005) from merging vortices associated with ITCZ breakdown. Part I: observational and modeling analyses," *Journal of the Atmospheric Sciences*, vol. 65, no. 11, pp. 3419–3439, 2008.

[4] J. Mao and G. Wu, "Barotropic process contributing to the formation and growth of tropical cyclone Nargis," *Advances in Atmospheric Sciences*, vol. 28, no. 3, pp. 483–491, 2011.

[5] K. Ooyama, "Numerical simulation of the life cycle of tropical cyclones," *Journal of the Atmospheric Sciences*, vol. 26, no. 1, pp. 3–40, 1969.

[6] M. T. Montgomery and J. Enagonio, "Tropical cyclogenesis via convectively forced vortex Rossby waves in a three-dimensional quasigeostrophic model," *Journal of the Atmospheric Sciences*, vol. 55, no. 20, pp. 3176–3207, 1998.

[7] T. Li, X. Ge, B. Wang, and Y. Zhu, "Tropical cyclogenesis associated with Rossby wave energy dispersion of a preexisting typhoon. Part II: numerical simulations," *Journal of the Atmospheric Sciences*, vol. 63, no. 5, pp. 1390–1409, 2006.

[8] T. N. Venkatesh and J. Mathew, "A numerical study of the role of the vertical structure of vorticity during tropical cyclone genesis," *Fluid Dynamics Research*, vol. 42, no. 4, Article ID 045506, 2010.

[9] M. T. Montgomery, Z. Wang, and T. J. Dunkerton, "Coarse, intermediate and high resolution numerical simulations of the transition of a tropical wave critical layer to a tropical storm," *Atmospheric Chemistry and Physics*, vol. 10, no. 22, pp. 10803–10827, 2010.

[10] Y. Xu, "The genesis of tropical cyclone Bilis (2000) associated with cross-equatorial surges," *Advances in Atmospheric Sciences*, vol. 28, no. 3, pp. 665–681, 2011.

[11] S. F. Abarca and K. L. Corbosiero, "Secondary eyewall formation in WRF simulations of Hurricanes Rita and Katrina (2005)," *Geophysical Research Letters*, vol. 38, no. 7, Article ID L07802, 2011.

A Simulation Study of the Formation of Large-Scale Cyclonic and Anticyclonic Vortices in the Vicinity of the
Intertropical Convergence Zone

47

[12] K. A. Reed and C. Jablonowski, "Impact of physical parameterizations on idealized tropical cyclones in the Community Atmosphere Model," *Geophysical Research Letters*, vol. 38, no. 4, Article ID L04805, 2011.

[13] I. V. Mingalev and V. S. Mingalev, "The global circulation model of the lower and middle atmosphere of the Earth with a given temperature distribution," *Mathematical Modeling*, vol. 17, pp. 24–40, 2005 (Russian).

[14] I. V. Mingalev, V. S. Mingalev, and G. I. Mingaleva, "Numerical simulation of the global distributions of the horizontal and vertical wind in the middle atmosphere using a given neutral gas temperature field," *Journal of Atmospheric and Solar-Terrestrial Physics*, vol. 69, no. 4-5, pp. 552–568, 2007.

[15] I. V. Mingalev, O. V. Mingalev, and V. S. Mingalev, "Model simulation of the global circulation in the middle atmosphere for January conditions," *Advances in Geosciences*, vol. 15, pp. 11–16, 2008.

[16] I. V. Mingalev, V. S. Mingalev, and G. I. Mingaleva, "Numerical simulation of the global neutral wind system of the Earth's middle atmosphere for different seasons," *Atmosphere*, vol. 3, no. 1, pp. 213–228, 2012.

[17] I. V. Mingalev and V. S. Mingalev, "Numerical modeling of the influence of solar activity on the global circulation in the Earth's mesosphere and lower thermosphere," *International Journal of Geophysics*, vol. 2012, Article ID 106035, 15 pages, 2012.

[18] O. M. Belotserkovskii, I. V. Mingalev, V. S. Mingalev, O. V. Mingalev, and A. M. Oparin, "Mechanism of the appearance of a large-scale vortex in the troposphere above a nonuniformly heated surface," *Doklady Earth Sciences*, vol. 411, no. 8, pp. 1284–1288, 2006.

[19] O. M. Belotserkovskii, I. V. Mingalev, V. S. Mingalev, O. V. Mingalev, A. M. Oparin, and V. M. Chechetkin, "Formation of large-scale vortices in shear flows of the lower atmosphere of the Earth in the region of tropical latitudes," *Cosmic Research*, vol. 47, no. 6, pp. 466–479, 2009.

[20] I. V. Mingalev, N. M. Astafieva, K. G. Orlov, V. S. Mingalev, and O. V. Mingalev, "Time-dependent modeling of the initial stage of the formation of cyclones in the intertropical convergence zone of the northern hemisphere," in *Proceedings of the 33rd Annual Seminar on Physics of Auroral Phenomena*, pp. 182–185, Polar Geophysical Institute, Apatity, Russia, 2011.

[21] I. V. Mingalev, N. M. Astafieva, K. G. Orlov, V. S. Mingalev, and O. V. Mingalev, "Simulation study of the initial stage of the origin of cyclonic and anticyclonic pairs in the intertropical convergence zone," in *Proceedings of the 34th Annual Seminar on Physics of Auroral Phenomena*, pp. 189–192, Polar Geophysical Institute, Apatity, Russia, 2011.

[22] I. V. Mingalev, N. M. Astafieva, K. G. Orlov, V. S. Mingalev, O. V. Mingalev, and V. M. Chechetkin, "Possibility of a detection of tropical cyclones and hurricanes formation according to satellite remote sensing," in *Actual Problems in Remote Sensing of the Earth from Space*, vol. 8, no. 2, pp. 290–296, Space Research Institute, Moscow, Russia, 2011, in Russian.

[23] I. V. Mingalev, N. M. Astafieva, K. G. Orlov, V. M. Chechetkin, V. S. Mingalev, and O. V. Mingalev, "Numerical simulation of formation of cyclone vortex flows in the intertropical zone of convergence and their early detection," *Cosmic Research*, vol. 50, no. 3, pp. 233–248, 2012.

[24] V. S. Mingalev, I. V. Mingalev, O. V. Mingalev, A. M. Oparin, and K. G. Orlov, "Generalization of the hybrid monotone second-order finite difference scheme for gas dynamics equations to the case of unstructured 3D grid," *Computational Mathematics and Mathematical Physics*, vol. 50, no. 5, pp. 877–889, 2010.

[25] T.-C. Chen, J.-D. Tsay, M.-C. Yen, and E. O. Cayanan, "Formation of the Philippine twin tropical cyclones during the 2008 summer monsoon onset," *Weather and Forecasting*, vol. 25, no. 5, pp. 1317–1341, 2010.

[26] R. N. Ferreira and W. H. Schubert, "Barotropic aspects of ITCZ breakdown," *Journal of the Atmospheric Sciences*, vol. 54, no. 2, pp. 261–285, 1997.

[27] C.-C. Wang and G. Magnusdottir, "ITCZ breakdown in three-dimensional flows," *Journal of the Atmospheric Sciences*, vol. 62, no. 5, pp. 1497–1512, 2005.

[28] C.-C. Wang and G. Magnusdottir, "The ITCZ in the Central and Eastern Pacific on synoptic time scales," *Monthly Weather Review*, vol. 134, no. 5, pp. 1405–1421, 2006.

[29] D. S. Nolan, S. W. Powell, C. Zhang, and B. E. Mapes, "Idealized simulations of the intertropical convergence zone and its multilevel flows," *Journal of the Atmospheric Sciences*, vol. 67, no. 12, pp. 4028–4053, 2010.

[30] W. M. Frank and P. E. Roundy, "The role of tropical waves in tropical cyclogenesis," *Monthly Weather Review*, vol. 134, no. 9, pp. 2397–2417, 2006.

[31] A. J. Broccoli, K. A. Dahl, and R. J. Stouffer, "Response of the ITCZ to Northern Hemisphere cooling," *Geophysical Research Letters*, vol. 33, no. 1, Article ID L01702, 2006.

[32] A. Fedorov, M. Barreiro, G. Boccaletti, R. Pacanowski, and S. G. Philander, "The freshening of surface waters in high latitudes: effects on the thermohaline and wind-driven circulations," *Journal of Physical Oceanography*, vol. 37, no. 4, pp. 896–907, 2007.

[33] J. C. H. Chiang and A. R. Friedman, "Extratropical cooling, interhemispheric thermal gradients, and tropical climate change," *Annual Review of Earth and Planetary Sciences*, vol. 40, no. 1, pp. 383–412, 2012.

6

Effect of Fracture Aperture on P-Wave Attenuation: A Seismic Physical Modelling Study

Aniekan Martin Ekanem,[1,2,3] Xiang Yang Li,[3,4]
Mark Chapman,[2,3] Main Ian,[2] and Jianxin Wei[4]

[1] Department of Physics, Akwa Ibom State University, PMB 1167, Mkpat Enin, Nigeria
[2] School of Geosciences, University of Edinburgh, Edinburgh EH9 3JW, UK
[3] British Geological Survey, Murchison House, Edinburgh EH9 3LA, UK
[4] CNPC Geophysical Key Laboratory, China University of Petroleum, Beijing 102249, China

Correspondence should be addressed to Aniekan Martin Ekanem; anny4mart@yahoo.com

Academic Editors: E. Del Pezzo, E. Liu, F. Luzon Martinez, A. Stovas, and T. Tsapanos

We used the seismic physical modelling approach to study the effect of fracture thickness or aperture on P-wave attenuation, using a laboratory scale model of two horizontal layers. The first layer is isotropic while the second layer has six fractured blocks, each consisting of thin penny-shaped chips of 3 mm fixed diameter and same thickness to simulate a set of aligned vertical fractures. The thickness of the chips varies according to the blocks while the fracture density remains the same in each block. 2D reflection data were acquired with the physical model submerged in a water tank in a direction perpendicular to the fracture strikes using the pulse and transmission method. The induced attenuation was estimated from the preprocessed CMP gathers using the QVO method, which is an extension of the classical spectral ratio method of attenuation measurement from seismic data. The results of our analysis show a direct relationship between attenuation and the fracture thickness or aperture. The induced attenuation increases systematically with fracture thickness, implying more scattering of the wave energy in the direction of increasing aperture. This information may be useful to differentiate the effect caused by thin microcracks from that of large open fractures.

1. Introduction

Fractures open at depth tend to be aligned normal to the direction of minimum in situ stress acting on them [1], giving rise to seismic anisotropy. Over the years, seismic anisotropy has been increasingly used as a potential tool to characterize natural fractured hydrocarbon reservoirs (e.g., [2–5]). The equivalent medium theories of seismic wave propagation provide the basis of using seismic anisotropy to detect fractures from seismic data. One of such theories is the Hudson's theory [6, 7] which provides a link between fracture density and measured azimuthal anisotropy. However, the theory fails to account for the issues of fracture scale lengths. For instance, many small cracks or a few large cracks within the same volume of material can result in the same fracture density. Furthermore, equal number of cracks with the same diameter but with varying thicknesses or apertures within the same volume of material can give rise to the same fracture density.

In reservoir rocks, it is possible to have aligned fractures of the same density but at different scales and consequently, the investigation of the effects of the fracture scale lengths and thicknesses or apertures on seismic wave response may be of great interest in fractured reservoir characterization. An adequate understanding of these effects could provide useful information needed to differentiate between the effects caused by thin microcracks and large open fractures. This is particularly important because key engineering quantities such as fluid transmissivity, used in reservoir models, depend on the third power of the hydraulic aperture and even more strongly on a hydraulic aperture below a few microns, where a significant proportion of the two opposing rough fractures are in contact [8].

Wei et al. [9, 10] used the seismic physical modelling approach to examine the effects of fracture scale lengths on seismic wave velocity and amplitude for both P- and S-transmitted waves. They simulated a set of aligned fractures

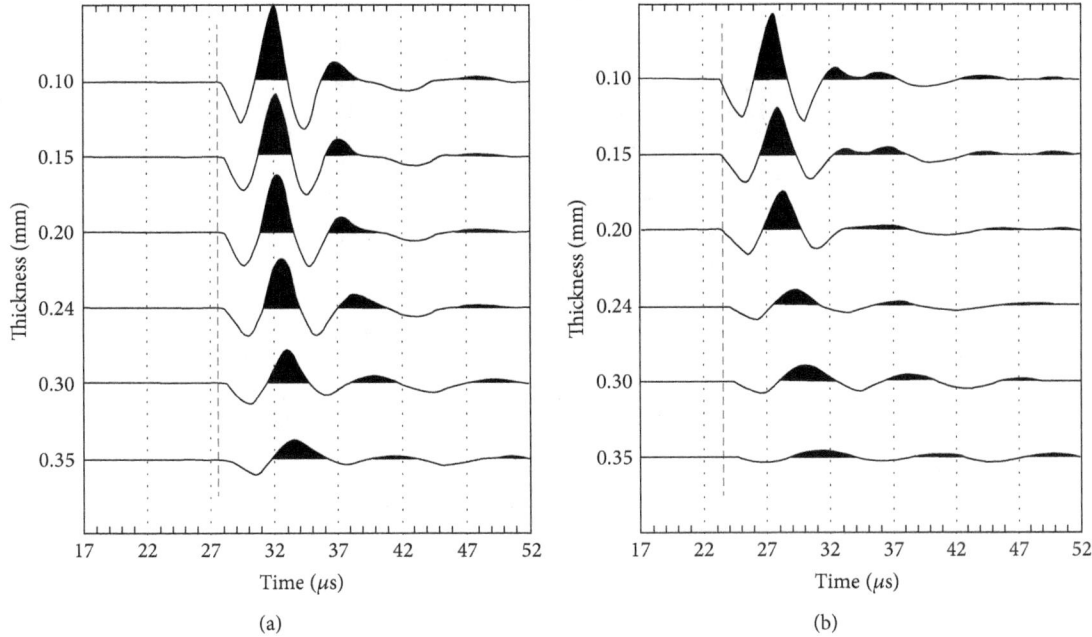

FIGURE 1: Comparison of P-waves recorded for the fractured models with different thicknesses or apertures (from [9]). (a) Parallel to the fracture strike. (b) Perpendicular to fracture strike. The wave is attenuated more in the direction perpendicular to fractures.

by embedding thin penny-shaped low-velocity chips of the same density and thickness (0.14 mm) but varying diameter (ranging from 2.5 mm to 6.0 mm) into an isotropic background material on the basis of Hudson's [6] assumption of thin penny-shaped fractures and used the pulse and transmission method to study the influence of the fracture diameter on the transmitted data. The results of their studies show that both P- and S-wave velocities increase with fracture diameter especially for wave propagation parallel to the fracture strike. They argued that as the diameter increases the number of fractures decreases to keep the density constant, resulting in a decrease in the amount of scattering and an increase in the wave velocity. Their results also reveal that the amount of shear-wave splitting decreases as the fracture diameter increases also as a result of the reduction in the number of fractures as the diameter increases. Wei et al. [9] further investigated the influence of fracture thickness or aperture on the P-waves by embedding thin round chips of the same diameter (2.1 mm) but varying thicknesses (ranging from 0.1 mm to 0.35 mm) into an isotropic background material to simulate a set of aligned fractures with different apertures. Their results show significant changes in the P-wave amplitude and waveforms with increasing thicknesses or apertures. As the thickness of the chips increases, the P-wave is significantly attenuated as illustrated in Figure 1 with more attenuation perpendicular to the fractures. However, their study fails to provide quantitative estimates of the observed attenuation from the waveforms. Chapman [11] developed a poroelastic model to account for the effects of fracture scale lengths, but there is still a lack of adequate understanding of the effects of fracture thickness or aperture on seismic P-wave attenuation. In this paper, we extend the previous studies by Wei et al. [9] to seismic reflection data

by estimating the amount of scattered attenuation caused by fractures of varying thicknesses or apertures but same density to provide more understanding of the scattering effects on P-wave amplitude.

The modelling involves first building the physical scale-model in the laboratory and then using the pulse and transmission method to record the seismic reflection response in the model. The resulting data though acquired in the laboratory have similar features as the data acquired in the field, and hence the results of the attenuation analysis could provide valuable information to differentiate between the effects caused by thin microcracks and large open fractures from field data. The setup of the seismic physical modelling experiment is inspired by Hudson's equivalent medium theory [6, 7] which considers dilute inclusions of thin, penny-shaped ellipsoidal cracks in an isotropic background medium. A known number of round thin low-velocity chips with fixed diameter but varying thicknesses or apertures are embedded into an isotropic base material to simulate a set of aligned vertical fractures with varying apertures and the fracture density is derived based on Hudson's theory [6, 7].

2. Construction of the Physical Model

The physical model is constructed from two horizontal layers (Figure 2). The first layer is made from a mixture of epoxy resin and silicon rubber and has a thickness of 38 mm, P-wave velocity of 2150 m/s, S-wave velocity of 1100 m/s, and density of 1.15 g/cm^3. The second layer is made from epoxy resin with a thickness of 75.5 mm, P-wave velocity of 2573 m/s, S-wave velocity of 1200 m/s, and density of 1.18 g/cm^3. To simulate fractures with varying apertures, thin penny-shaped chips made from a mixture of epoxy resin and silicon rubber with

(a)

(b)

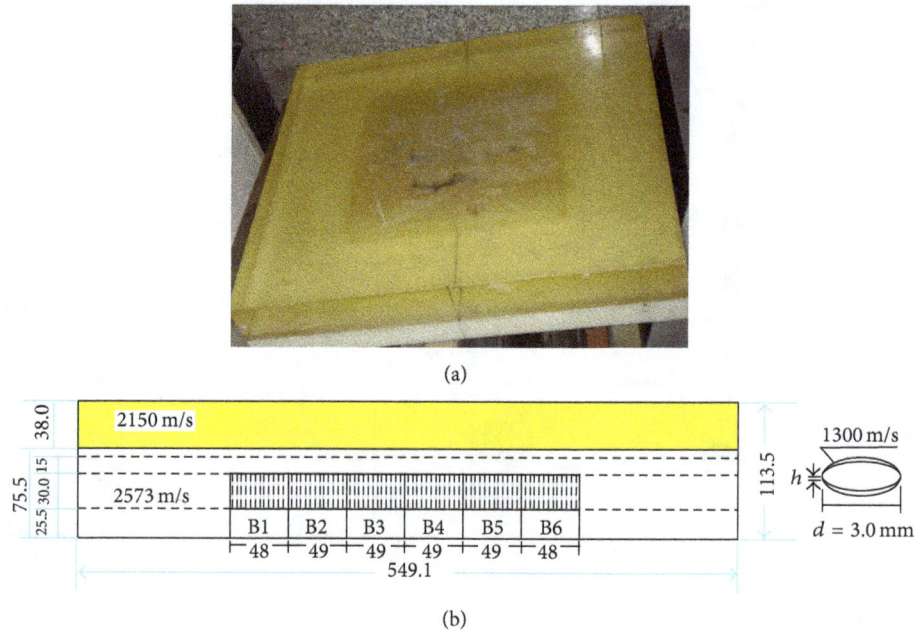

FIGURE 2: Physical model made up of two layers: (a) base model, (b) sectional view of base model. The first layer is isotropic with a P-wave velocity of 2150 m/s and density of 1.15 g/cm^3 while the second layer has six fractured blocks (each of same density), B1 to B6 in order of increasing chips' thickness. The isotropic background of this layer has a P-wave velocity of 2573 m/s and a density of 1.175 g/cm^3. The numbers shown indicate model dimensions in millimetres and the model is scaled up by 1:10,000 for spatial dimensions and time measurements.

TABLE 1: Parameters of the fracture models. The dimensions are scaled up to 1:10000. Dimensions shown are not converted to real scaling.

Model number/block	B1	B2	B3	B4	B5	B6
Number of layers	30	30	30	30	30	30
Number of chips per layer	360	360	360	360	360	360
Total number of chips in model	10800	10800	10800	10800	10800	10800
Chip thickness (mm)	0.10	0.15	0.20	0.25	0.30	0.35
Layer thickness (mm)	1.59	1.63	1.63	1.61	1.63	1.58
Radius of each chip (mm)	1.5	1.5	1.5	1.5	1.5	1.5
Volume of base material	458640	458640	458640	458640	458640	458640
Fracture density (%)	7.95	7.95	7.95	7.95	7.95	7.95

a fixed diameter of 3 mm and P-wave velocity of 1300 m/s are randomly embedded in the isotropic background of the second layer to make the layer anisotropic. The chips are very thin with very low S-wave velocity and therefore act in such a way as to provide weakness to the background material. They are arranged to form six fractured blocks with their thicknesses as 0.10, 0.15, 0.20, 0.25, 0.30, and 0.35 mm, respectively, in each block. Each block is made up of 30 layers of epoxy resin. Once a layer is laid, 360 thin chips are randomly embedded into the layer and another layer is added on the top. The whole procedure is repeated until a total of 30 layers were achieved. All the six fracture blocks have the same fracture density of 7.95%. The fracture density "FD" in this case is defined mathematically as

$$FD = \frac{na^3}{V},\qquad(1)$$

where n is the number of chips in the base material, a is the radius of each chip, and V is the volume of the base material. The fracture density for each of the fractured blocks was computed from (1). The vacuum mixing technique was utilized to create the models and the mixing was done at a very fast rate in order to control the homogeneity of the mixture. Details of the fracture model parameters are given in Table 1. The model is constructed with a scale of 1:10,000 for spatial dimensions and time measurements. This implies that the model dimensions are multiplied by a factor of 10 in order to get the corresponding field dimensions in metres. The P- and S-wave velocities of the materials in the model were measured by using the pulse and transmission method as in Wei et al. [9]. Measurements were made at an ultrasonic frequency of 250 kHz, corresponding to 25 Hz after appropriate scaling. The model is a simplified analogous representation of a fractured reservoir with varying fracture apertures. Although the simulated fractures may not be real fracture analogues in a typical fractured reservoir setting, this study is expected to provide useful information on the effects of voids in the rock on P-wave attenuation and a basis for further theoretical development since there is no theory at the moment to explain this kind of scenario.

3. Experimental Setup and Data Acquisition

The experimental setup for the data acquisition is shown in Figure 3. The physical model was submerged in a water tank for ease of data acquisition and coupling of the various parts of the model. 2D reflection data were acquired in a direction perpendicular to the fracture strikes using the pulse and transmission method. The water depth to the top of the model is 100 mm, making the net thickness of the overburden above the fractured layer to be 138 mm. The physical modelling system consists of an ultrasonic pulse source and a receiver system, an analogue/digital converter, and a motor driven positioning system with a precision of 0.1 mm (Figure 4). The pulse source (transducer) has a size of 20 mm and produces plane waves in this experiment. The dominant wavelength of the P-wave generated is 11.2 mm compared with the fracture diameter of 3 mm. The conditon of equivalence of a fractured medium to an anisotropic medium under the long wavelength approximation of the equivalent medium theory is as follows [12]:

seismic wavelength $\gg$ fracture spacing $\gg$ fracture opening

In our seismic physical modelling case, the average fracture spacing is 1.61 mm while the highest fracture thickness or aperture is 0.35 mm. This implies that 11.2 mm $\gg$ 1.61 mm $\gg$ 0.35 mm. Thus, the model satisfies the long wavelength approximation and the simulated fractures are expected to cause scattering of seismic wave energy resulting in attenuation. The source and receiver were located on the water surface. For the first shot gather located 500 mm from the edge of the tank (Figure 3), a single shot was fired into a single receiver at a minimum offset of 16 mm, the receiver was then moved a distance of 2 mm away, and another shot was fired until a total of 120 receiver positions were occupied at a spacing of 2 mm for the shot position. The shot was then moved a distance of 2 mm in the direction of the receiver and the entire procedure repeated. A total of 220 shots were made at 2 mm intervals. The ultrasonic pulse source has a centre frequency of 250 kHz and a bandwidth of 100–400 kHz. The frequency was scaled down by 10,000 : 1 to fit the scale model dimensions and velocity appropriately. The laboratory model was up-scaled using the geometry scaling. The model dimensions were multiplied by 10 to get their equivalent field scaling in metres while the frequencies were divided by 10 to get their field equivalent in Hertz. Thus, on appropriate scaling, the centre frequency of the pulse source is 25 Hz and the corresponding bandwidth is 10–40 Hz. Details of the acquisition geometry are summarized in Table 2. Aperture diffraction effect caused by the finite dimension of the transducer aperture might influence the accuracy of attenuation measurement if not corrected for (e.g., [13–16]). This extrinsic effect is mostly prominent when the aperture size is roughly similar to the wavelength. In our physical modelling experiment, the size of the transducer

TABLE 2: Summary of data acquisition geometry—all the acquisition parameters given are not converted to actual scales. The model dimensions and acquisition parameters are scaled up by 1 : 10000 while the frequency is scaled down by 10000 : 1.

Water depth to base model (mm)	100
Number of shots	220
Shot interval (mm)	2
Receiver interval (mm)	2
Number of receivers	120
Minimum offset (mm)	16
Maximum offset (mm)	254
Fold of cover	60
Number of samples	4096
Sample rate (μs)	0.1

aperture is more than the wavelength of the generated P-waves and, thus, we assume that aperture diffraction effects are negligible.

4. Data Processing

The raw data is made up of 230 shot gathers with 120 traces in each gather. On appropriate scaling to effective field dimensions, the corresponding trace spacing is 20 m, while the minimum and maximum offsets are 160 m and 2540 m, respectively. Since the data has high signal-to-noise ratio, minimal processing was done on the data in order to preserve all the amplitude information needed for attenuation analysis. Thus, the following processing sequences were applied: geometry configuration, common midpoint (CMP) sorting, trace-muting, velocity analysis, and NMO correction. Stacking was also included in the processing sequence for ease of event identification and picking of the travel times to the target layers, even though the Q values were estimated from the prestack CMP gathers. A sample CMP NMO-corrected gather is shown in Figure 5. The reflections that we analysed are those from the top of the base model (blue arrow), the top of the fractured layer (red arrow), and the bottom of the fractured layer (green arrow), with the geometry shown in Figure 2. We used an offset range of 160–940 m for all the CMP gathers analysed and all the target reflections are continuous for this offset range, respectively.

5. Attenuation Measurement

Among the various methods of measuring attenuation from seismic data, the spectral ratio method is the most common method perhaps because it is easier to use and more stable (e.g., [17–20]). In this paper, we used the QVO method (introduced by Dasgupta and Clark [20]), which is an extension of the spectral ratio method to estimate the induced attenuation (inverse of the seismic quality factor Q) in the fractured layer. Our physical modelling data have high S/N ratio and as such we assumed that interference effects caused by multiples or any other noise interference are either absent or too small to cause any significant bias on the Q estimates.

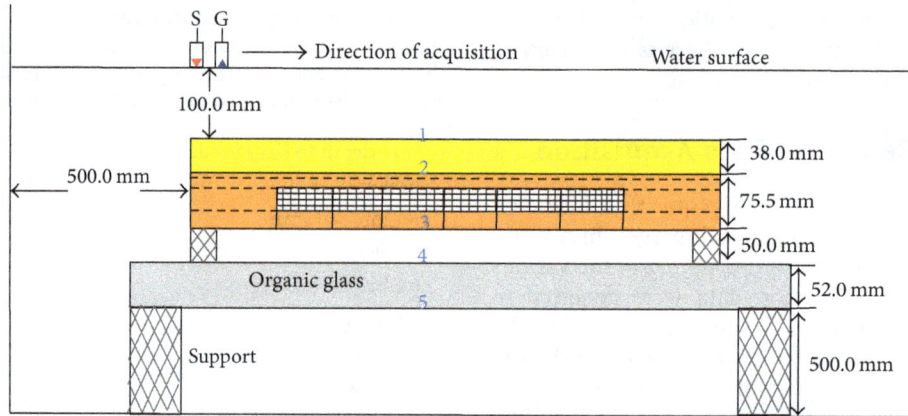

FIGURE 3: Experimental setup for data acquisition. The base model is submerged in a water tank. S and G are the shot and receiver positions, respectively. The model dimensions shown are scaled up by 1 : 10000. 2D reflection data were acquired in a direction normal to the fracture strikes. The blue numbers indicate the reflection interfaces.

FIGURE 4: The physical modelling system for 2D data acquisition in the laboratory. The base model is submerged in a water tank while the source and receiver were moved along the water surface.

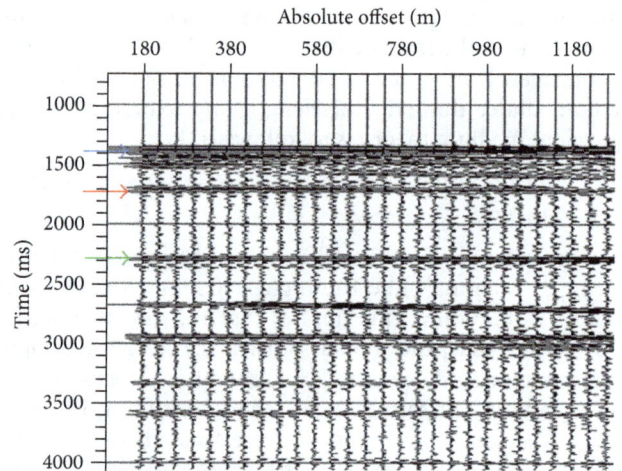

FIGURE 5: Sample NMO-corrected gather. The blue, red, and green arrows indicate the reflections from the top of the base model and the top and bottom of the fractured layer, respectively.

As a reference, we used the first trace from the top model reflection at an offset of 160 m (which is minimum offset in the data) in each preprocessed CMP gather, for comparison of the spectral ratios and a constant window length of 160 ms. We then computed the spectral amplitude ratios according to (2) and performed a least-squares regression of the logarithm of the power spectral ratio (LPSR) against frequency:

$$\ln \frac{A_2{}^2}{A_1{}^2} = \ln \frac{P_2}{P_1} = 2\ln(RG) - \frac{2\pi f}{Q}(t_2 - t_1), \quad (2)$$

where f is frequency, R is the reflectivity term, G is the geometrical spreading factor, A_2 is the spectral amplitude of the target reflection (top or bottom of fractured layer), and A_1 is the spectral amplitude of the reference trace while P_1 and P_2 are the respective spectral powers (square of amplitudes), t_1 and t_2 are the corresponding travel times, and Q is the seismic quality factor down to the reflector. The slope of the regression "p" is given by

$$p = -\frac{2\pi(t_2 - t_1)}{Q}. \quad (3)$$

Sample plots of the logarithm of the power spectral ratios versus frequency are shown in Figure 6 and are approximately linear in the frequency bandwidth of 10–40 Hz. This linearity is an approximation and the frequency bandwidth lies within the signal bandwidth of the pulse source (10–40 Hz). This bandwidth was kept constant for all the trace pairs analyzed. There is no signal below the frequency of 10 Hz and hence the scatter in the spectral plots (Figure 6). The same is applicable to frequencies beyond 40 Hz.

The best fitting slope p, defined in (3), between these limits was then obtained by least-squares regression. Following (3), this slope in the absence of a zero-offset reference trace in the data can be written as

$$p = \frac{2\pi}{Q}\left(t_{o,\mathrm{ref}} - t_o + \frac{x^2}{2}\left\{\frac{1}{t_{o,\mathrm{ref}}v^2{}_{\mathrm{ref,rms}}} - \frac{1}{t_o v^2{}_{\mathrm{rms}}}\right\}\right), \quad (4)$$

where x is offset, $t_{o,\mathrm{ref}}$ is the zero-offset travel-time of reference trace, t_o is the zero-offset travel-time of target reflection,

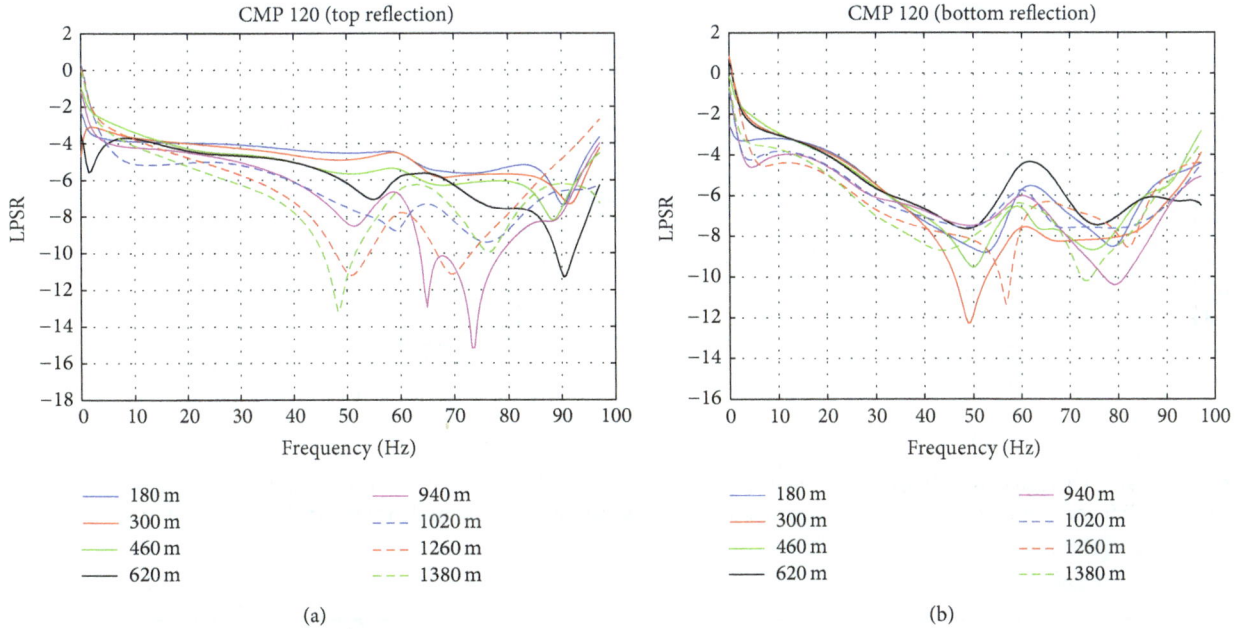

FIGURE 6: Log spectral power ratio (LSPR) against frequency plots (CMP 120). (a) Top fractured-layer reflection. (b) Bottom fractured-layer reflection. Plots are approximately linear within frequency range of 10–40 Hz.

and $V_{\text{ref,rms}}$ and V_{rms} are the root mean square velocities of the reference trace and target reflection, respectively. Equation (4) indicates a linear relationship between the spectral ratio slope and the square of offsets. Thus, we carried out another least-squares regression of the spectral ratio slopes against the square of the offsets to get the zero-offset slope (LSPR slope intercept, I) given by

$$I = \frac{2\pi \left(t_{o,\text{ref}} - t_{o,2}\right)}{Q}. \tag{5}$$

The seismic quality factor Q down to the top of the fractured layer was then computed from (5). The entire procedure was repeated for the bottom fractured-layer reflection and all the CMP gathers analyzed. With the pair of Q values computed down to the top and bottom of the fractured-layer in each CMP gather, we finally estimated the interval seismic quality factor, Q_i in the fractured-layer using the equation [20]:

$$Q_i = \frac{[t_{o,2} - t_{o,1}]}{t_{o,2}/Q_2 - t_{o,1}/Q_1}, \tag{6}$$

where Q_1 and Q_2 are the seismic quality factors down to top and bottom of the fractured-layer respectively while $t_{o,1}$ and $t_{o,2}$ are the corresponding zero-offset travel times. The zero-offset travel times were obtained by extrapolation on the time axis since the minimum offset in the data is 160 m after appropriate scaling. Sample plots of the slopes against the square of the offsets are shown in Figure 7. The red dashed lines indicate a 95% confidence interval on the best fitting (green) line in each case.

6. Results

The results of the attenuation analysis show that the magnitude of the logarithm of the spectral ratio slopes varies with the thickness of the chips, indicative of a systematic dependence of the scattering attenuation on the thickness (which is meant to model the effect of aperture in an underground crack). The absolute value of the slope increases with the chip's thickness which indicates an increase in scattering and hence attenuation. Higher interval Q values (low attenuation) are obtained for the CMP(s) at both ends of the survey line (below CMP 100 and beyond CMP 400) where there are no fractures (Figure 8). The Q values decrease systematically in the direction of increasing chips' thickness from the left edge of block B1 (CMP 120) to the right edge of block B6 (CMP 410), implying more scattering in the direction of increasing chips' thickness. This trend is shown in Figure 9 for the CMP(s) corresponding to the centres of the fractured blocks, respectively.

7. Discussion

A set of aligned fractures is known to greatly influence the propagation of seismic waves by causing scattering of the wave energy resulting in seismic coda trailing the primary reflections. Such scattering effects have been shown to be useful in providing information about the fracture properties (e.g., [21–23]). The resultant effect of the scattering is a gradual loss in the wave energy which can be observed in amplitude changes as the wave propagates through the medium. This effect can be produced by a set of fractures at different scale length and the popular Hudson theory [6, 7] fails to account for these scale length issues. The results of

(a)

(b)

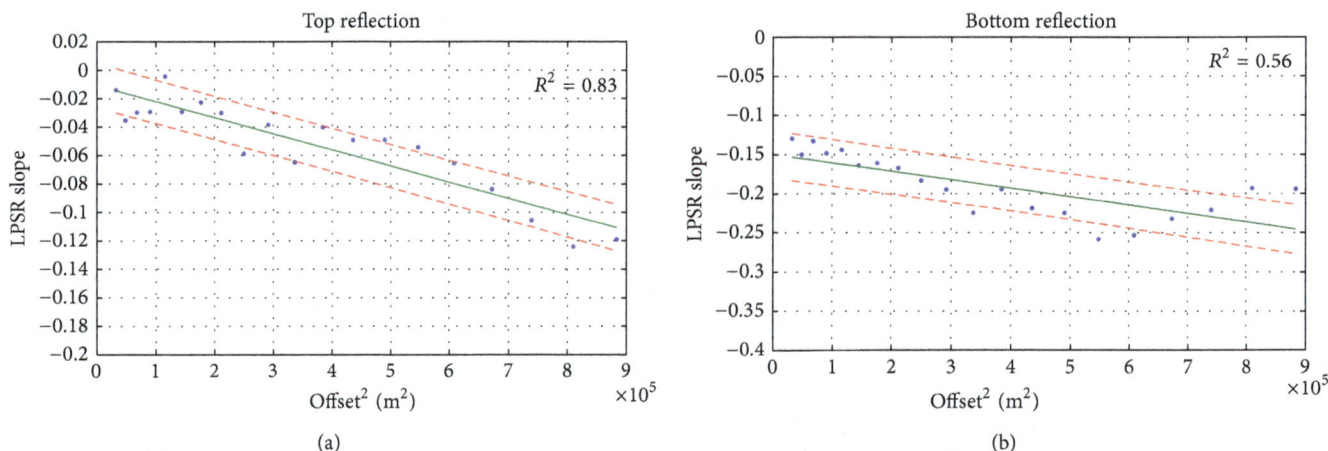

FIGURE 7: Least-squares regression of logarithm of power ratio slopes against the square of offset (CMP 260). The asterisks indicate the data point; the green line indicates the fitted line while the red dashed lines indicate a 95% confidence interval on the fitted line.

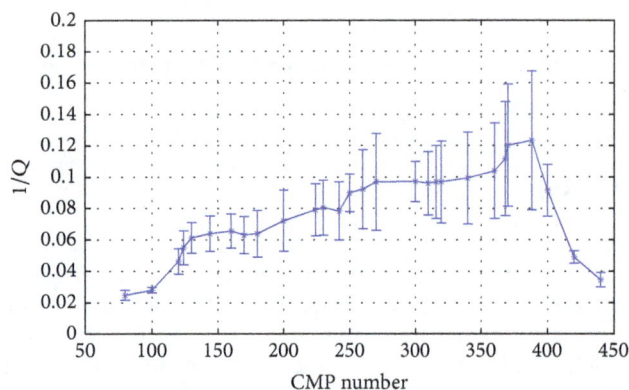

FIGURE 8: $1/Q$ results for the fractured layer against CMP numbers: fractured blocks lie between CMP(s) 120 and 410, respectively. There is a systematic increase in the induced attenuation (inverse Q) in the direction of increasing chips' thickness from CMP 120 to CMP 410. However, the attenuation is less at the ends of the line where there are no fractures.

FIGURE 9: $1/Q$ profile against chip's thickness. Q decreases with increasing chip's thickness, implying more attenuation.

the seismic physical modelling studies of Wei et al. [9, 10] demonstrate that a set of aligned fractures with different diameters but same thickness and fracture density significantly affect both P- and S-wave velocities especially for wave propagation parallel to the fracture strike. The wave velocity increases with diameter as a result of reduced scattering. Further seismic physical modelling studies by Wei et al. [9] also demonstrate that a set of aligned fractures with the same density and diameter but varying thickness or aperture has a strong influence on the P-wave amplitude and waveforms for transmitted P-wave data. The waveform is highly attenuated with increasing thickness or aperture especially for wave propagation perpendicular to the fractures.

In this paper, we have quantified the scattering effects caused by a set of fracture models with the same density and diameter but varying thicknesses or apertures for P-wave reflection data through attenuation estimates. The fracture

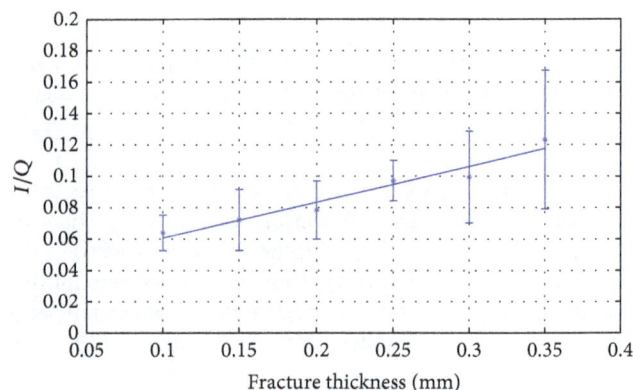

models are simulated by embedding thin penny-shaped chips into an isotropic background medium. These chips act in such a way as to cause a weakness in the medium, and thus we exploit this weakness to simulate fractures. The two layers in the base model were glued together to ensure good coupling between them. A rough estimate of the transmission coefficient at the top of the fractured layer gives a transmission coefficient >90% which is also indicative of a high-quality coupling between the two layers of the model. Data acquisition in a water tank where the model was submerged also facilitates a good coupling of the the source and receivers as well as the other components of the model. The results of our analysis show that P-wave attenuation has a direct relationship with fracture thickness or aperture. Attenuation (inverse Q) increases systematically and linearly with fracture thickness, implying proportionately more scattering of the wave energy as the wave propagates in the direction of increasing thickness or aperture. Although the simulated fractures may not be real fracture analogues in a typical fractured reservoir setting, the results provide information which might be useful in examining the effects of voids in the rock on P-wave attenuation and may provide a basis for further theoretical development to distinguish the effects caused by thin microcracks and large open fractures.

8. Conclusions

We demonstrate that a set of aligned fractures causes scattering of seismic wave energy, resulting in attenuation. A direct relationship exists between attenuation and fracture thickness or aperture, indicating the potential of using P-wave attenuation to distinguish the effect caused by thin microcracks from that caused by large open fractures. Similar studies by Wei et al. [9] on P-wave transmitted data have shown that a set of aligned fractures with the same density and diameter but different apertures or thickness cause significant attenuation in the waveform amplitude especially for wave propagation normal to the fracture strike direction. Our findings show consistency with these observations and thus provide a physical basis of using P-wave attenuation attribute to distinguish the effects caused by thin microcracks and large open fractures from seismic data.

Conflict of Interests

The authors declare that there is no conflict of interests regarding the publication of this paper.

Acknowledgments

The authors are grateful to the China National Petroleum Corporation (CNPC) for providing the funds to conduct the physical modelling experiments under a collaboration agreement between the CNPC Geophysical Key Laboratory at the China University of Petroleum and the Edinburgh Anisotropy Project (EAP) at the British Geological Survey (BGS) and the permission to publish the results. Thanks are due to Professor Wang and Professor Di both of the China University of Petroleum for being good hosts during the experimental period. This work was also partially supported by the National Natural Science Foundation of China. Special thanks are due to the Akwa Ibom State University (AKSU), Nigeria, for sponsoring Ekanem's Ph.D. studies at the University of Edinburgh.

References

[1] M. Schoenberg and C. M. Sayers, "Seismic anisotropy of fractured rock," *Geophysics*, vol. 60, no. 1, pp. 204–211, 1995.

[2] S. A. Hall, J. M. Kendall, O. I. Barkved, and M. C. Mueller, "Fracture characterization using P-wave AVOA in 3D OBS data," SEG expanded abstract, pp. 1409–1412, 2000.

[3] D. Gray and K. Head, "Fracture detection in Manderson Field: a 3-D AVAZ case history," *Leading Edge*, vol. 19, no. 11, pp. 1214–1221, 2000.

[4] X.-Y. Li, Y.-J. Liu, E. Liu, F. Shen, L. Qi, and Q. Shouli, "Fracture detection using land 3D seismic data from the Yellow River Delta, China," *Leading Edge*, vol. 22, no. 7, pp. 680–684, 2003.

[5] M. Luo and B. J. Evans, "An amplitude-based multiazimuthal approach to mapping fractures using P-wave 3D seismic data," *Geophysics*, vol. 69, no. 3, pp. 690–698, 2004.

[6] J. A. Hudson, "Overall properties of a cracked solid," *Mathematical Proceedings of the Cambridge Philosophical Society*, vol. 88, no. 2, pp. 371–384, 1980.

[7] J. A. Hudson, "Wave speeds and attenuation of elastic waves in material containing cracks," *Geophysical Journal*, vol. 64, no. 1, pp. 133–150, 1981.

[8] R. Zimmermann and I. Main, "Hydromechanical behaviour of fractured rocks," in *Mechanics of Fluid-Saturated Rocks*, Y. Gueguen and M. Bouteca, Eds., pp. 361–419, Academic Press, London, UK, 2003.

[9] J. Wei, B. Di, and X.-Y. Li, "Effect of fracture scale length and aperture on seismic wave propagation: an experimental study," *Journal of Seismic Exploration*, vol. 16, no. 2–4, pp. 265–280, 2007.

[10] J. Wei, B. Di, and Q. Wang, "Experimental study on the effect of fracture scale on seismic wave characteristics," *Petroleum Science*, vol. 5, no. 2, pp. 119–125, 2008.

[11] M. Chapman, "Frequency-dependent anisotropy due to meso-scale fractures in the presence of equant porosity," *Geophysical Prospecting*, vol. 51, no. 5, pp. 369–379, 2003.

[12] A. Bakulin, V. Grechka, and I. Tsvankin, "Estimation of fracture parameters from reflection seismic data-Part I: HTI model due to a single fracture set," *Geophysics*, vol. 65, no. 6, pp. 1788–1802, 2000.

[13] H. Seki, A. Granato, and R. Truell, "Diffraction effects in the ultrasonic field of a piston source and their importance in the accurate measurement of attenuation," *Journal of the Acoustical Society of America*, vol. 28, pp. 230–238, 1956.

[14] E. P. Papadakis, "Ultrasonic diffraction loss and phase change in anisotropic materials," *Journal of the Acoustical Society of America*, vol. 40, pp. 863–876, 1966.

[15] M. B. Gitis and A. S. Khimunin, "Diffraction corrections for measurements of the absorption coefficient and velocity of sound," *Soviet Physics*, vol. 14, pp. 305–310, 1969.

[16] E. P. Papadakis, K. A. Fowler, and L. C. Lynnworth, "Ultrasonic attenuation by spectrum analysis of pulses in buffer rods: method and diffraction corrections," *Journal of the Acoustical Society of America*, vol. 53, no. 5, pp. 1336–1343, 1973.

[17] P. S. Hauge, "Measurements of attenuation for vertical seismic profiles," *Geophysics*, vol. 46, no. 11, pp. 1548–1558, 1981.

[18] J. Pujol and S. Smithson, "Seismic wave attenuation in volcanic rocks from VSP experiments," *Geophysics*, vol. 56, no. 9, pp. 1441–1455, 1991.

[19] R. Tonn, "The determination of the seismic quality factor Q from VSP data: a comparison of different computational methods," *Geophysical Prospecting*, vol. 39, no. 1, pp. 1–27, 1991.

[20] R. Dasgupta and R. A. Clark, "Estimation of Q from surface seismic reflection data," *Geophysics*, vol. 63, no. 6, pp. 2120–2128, 1998.

[21] C. A. Schultz and M. N. Toksoz, "Reflections from a randomly grooved interface: ultrasonic modelling and finite-difference calculation," *Geophysical Prospecting*, vol. 43, no. 5, pp. 581–594, 1995.

[22] M. E. Willis, D. R. Burns, R. Rao, B. Minsley, M. N. Toksöz, and L. Vetri, "Spatial orientation and distribution of reservoir fractures from scattered seismic energy," *Geophysics*, vol. 71, no. 5, pp. O43–O51, 2006.

[23] D. R. Burns, M. E. Willis, M. N. Toksöz, and L. Vetri, "Fracture properties from seismic scattering," *Leading Edge*, vol. 26, no. 9, pp. 1186–1196, 2007.

7

Propagation and Attenuation of Elastic Waves in a Double Porosity Medium

J. S. Nandal[1] and T. N. Saini[2]

[1] *Department of Mathematics, M.D. University, Rohtak 124001, India*
[2] *Department of Mathematics, Government College Kalka, Kalka 133302, India*

Correspondence should be addressed to T. N. Saini; tns.kalka@gmail.com

Academic Editors: G. Casula, H. Perroud, and A. Streltsov

This study solves the mathematical model for the propagation of harmonic plane waves in a dissipative double porosity solid saturated by a viscous fluid. The existence of three dilatational waves is explained through three scalar potentials satisfying wave equations. Velocities of these waves are obtained from the roots of a cubic equation. Lone shear wave is identified through a vector potential satisfying a wave equation. The displacements of solid particles are expressed through these four potentials. The displacements of fluid particles in pores and fractures can also be expressed in terms of these potentials. A numerical example is solved to calculate the complex velocities of four waves in a dissipative double porosity solid. Each of the complex velocities is resolved to define the phase velocity and quality factor of attenuation for the corresponding wave. Effects of medium properties and wave frequency are analyzed numerically on the propagation characteristics of four attenuated waves. It seems that P_1 and S waves are not very sensitive to the pore/fluids characteristics, except the fracture porosity. Hence, the recovery and analysis of slower (P_2, P_3) waves become more desired to understand the fluid-rock dynamism in crustal rocks.

1. Introduction

Pores are pervasive in most of the igneous, metamorphic, and sedimentary rocks in the earth's crust. Traditional approaches to seismic exploration often make use of Biot's theory of poroelasticity. This theory has always been limited by an explicit assumption that the porosity itself is homogeneous. For acoustic analysis of many rock samples in a laboratory setting, this assumption is known to be adequate. But, in the modeling of real heterogeneous reservoirs, it may not be a realistic assumption. In fact, porosity found in the earth may have many shapes and sizes, but two types of porosity are more important. One is matrix (or storage) porosity that occupies a finite and substantial fraction of the volume of a porous rock. Other is fracture or crack porosity that may occupy very little volume, but fluid flow occurs primarily through the fracture network. However, fluid storage occurs mostly in the porous matrix.

This model identified as double porosity model [1, 2] considers a fracture network that divides the porous matrix into different blocks and the fluid in fractures surrounds the disaggregated matrix blocks supported entirely by fluid pressure. In fact, most of the near-surface rock masses are fractured to some degree. It demands to examine the coupled fluid-rock deformation through the double porosity network by extending the Biot's theory [3–5] to the composite containing fracture network in porous matrix. There are some attempts, recorded in the literature, in which fractures are incorporated into the rock models. These attempts account for the partial saturation effects and the fluid-flow during the passage of seismic waves [6–11]. In these attempts, the approach has been limited mainly to modify the elastic parameters of Biot's theory for the introduction of cracks.

The previous double porosity models are proposed by Wilson and Aifantis [12, 13], Cho et al. [14], and Bai et al. [15]. These were based on the mixture of solid and fluid phases. That means, coupling between fluid flow and solid deformations is ignored. Hence, to consider this coupling, it becomes necessary to incorporate Biot's concepts of poroelasticity into the double porosity model. Berryman and

Wang [16, 17] made efforts for a rigorous extension of Biot's poroelasticity to include fractures/cracks by making a generalization to double-porosity/dual-permeability modeling. They derived the phenomenological equations and presented the method to determine the relevant coefficients. Their discussions showed that three compressional waves in double porosity media are diffusive. Based on the volume averaging technology, Pride and Berryman [18, 19] derived the governing equations of fluid-saturated double porosity media. In addition, the fluid transport mechanism was also investigated and a symmetric dual-permeability Darcy law was established.

Viscoelasticity is a widely accepted property of many rocks in the crust and is a major cause of seismic attenuation. Moreover, in the presence of double porosity, a viscoelastic solid permeated by pores and fractures and saturated with viscous fluid represents a much realistic model for sedimentary or reservoir rocks. The present work considers the propagation of attenuated waves in this dissipative poroviscoelastic composite medium. Three scalar potentials identify three dilatational waves and a vector potential identifies the lone shear wave in the considered porous medium. Complex velocities associated to these four attenuated waves are resolved to define their propagation velocities as well as attenuation coefficients. Effects of frequency, frame anelasticity, pore-fluid viscosity, porosity, and fracture permeability are observed on the phase velocities and attenuation coefficients of the four waves.

2. Basic Equations

The double porosity medium consists of three constituents, that is, solid matrix, pore fluid, and fracture fluid, which are identified with indices $'s', 'p', 'c'$, respectively. In this porous aggregate, volume fractions of the fluid in pores (ϵ_p) and in fractures (ϵ_c) define the total porosity $f(= \epsilon_p + \epsilon_c)$ of solid skeletal and then $\epsilon_s(= 1 - f)$ is the volume fraction for solid grains. Following Berryman and Wang [16, 17], the equations of motion for low-frequency vibrations of constituent particles in double porosity media, in the absence of body forces, are given by

$$\tau_{ij,j} = \epsilon_s \rho_s \ddot{u}_i + \rho_{12} \ddot{v}_i + \rho_{13} \ddot{w}_i - b_{12} \dot{v}_i - b_{13} \dot{w}_i,$$

$$-\epsilon_p (P_p)_{,i} = \epsilon_p \rho_f \ddot{u}_i + \rho_{22} \ddot{v}_i + \rho_{23} \ddot{w}_i + b_{12} \dot{v}_i + b_{23} (\dot{v}_i - \dot{w}_i),$$

$$-\epsilon_c (P_c)_{,i} = \epsilon_c \rho_f \ddot{u}_i + \rho_{23} \ddot{v}_i + \rho_{33} \ddot{w}_i + b_{13} \dot{w}_i - b_{23} (\dot{v}_i - \dot{w}_i),$$

$$(1)$$

where τ_{ij} is the stress tensor for saturated (undrained) porous solid and P_p, P_c are fluid pressures in pore space and fracture space, respectively. ρ_s and ρ_f are intrinsic densities of solid grains and pore fluid, respectively. u_i are the components of displacement ($\mathbf{u}$) of solid particles. The components v_i and w_i denote displacements ($\mathbf{v}$ and $\mathbf{w}$) in pore fluid and fracture fluid relative to solid frame, respectively. English indices (other than s, p, c) take values 1, 2, and 3. Repetition of any of these indices implies summation. Dot over a variable implies partial derivative with time and comma before an index

implies partial space differentiation. Dynamical constants and dissipation parameters in (1) are defined as follows:

$$\rho_{12} = \frac{\left[\epsilon_c (\tau_c - 1) - \epsilon_p (\tau_p - 1) - f (\tau - 1)\right] \rho_f}{2},$$

$$\rho_{13} = \frac{\left[\epsilon_p (\tau_p - 1) - \epsilon_c (\tau_c - 1) - f (\tau - 1)\right] \rho_f}{2},$$

$$\rho_{22} = \epsilon_p \tau_p \rho_f,$$

$$\rho_{23} = \frac{\left[f (\tau - 1) - \epsilon_p (\tau_p - 1) - \epsilon_c (\tau_c - 1)\right] \rho_f}{2},$$

$$\rho_{33} = \epsilon_c \tau_c \rho_f;$$

$$b_{12} = \frac{\rho_f \epsilon_p \left(\epsilon_p \chi_{22} - \epsilon_c \chi_{12}\right) \eta}{\chi_0},$$

$$\chi_0 = \chi_{11} \chi_{22} - \chi_{12}^2,$$

$$b_{13} = \frac{\rho_f \epsilon_c \left(\epsilon_c \chi_{11} - \epsilon_p \chi_{12}\right) \eta}{\chi_0},$$

$$b_{23} = \frac{\rho_f \epsilon_p \epsilon_c \chi_{12} \eta}{\chi_0},$$

$$(2)$$

where η is shear (kinematic) viscosity of interstitial fluid and the tensor components χ_{ij} define permeability of solid frame. Tortuosity parameters relate to shape factor $r(= 0.5$ for spherical grains) and volume fractions as $\tau = 1 + r \epsilon_s / f$, $\tau_p = 1 + r \epsilon_s / \epsilon_p$, and $\tau_c = 1$.

Following Berryman and Wang [16, 17], constitutive relations for stresses in porous frame and hydrostatic pressures in pore fluid and fracture fluid are given by

$$\tau_{ij} = K_u \left(\nabla \cdot \mathbf{u} - B^p \epsilon_p \nabla \cdot \mathbf{v} - B^c \epsilon_c \nabla \cdot \mathbf{w}\right) \delta_{ij}$$
$$+ G \left(u_{i,j} + u_{j,i} - \frac{2}{3} u_{k,k} \delta_{ij}\right),$$

$$(3)$$

$$-P_p = c_{21} \nabla \cdot \mathbf{u} + \epsilon_p c_{22} \nabla \cdot \mathbf{v} + \epsilon_c c_{23} \nabla \cdot \mathbf{w},$$

$$-P_c = c_{31} \nabla \cdot \mathbf{u} + \epsilon_p c_{23} \nabla \cdot \mathbf{v} + \epsilon_c c_{33} \nabla \cdot \mathbf{w},$$

where δ_{ij} is Kronecker symbol. B^p, B^c are Skempton's coefficients [20] for fluid pressures build up in pores and fractures, respectively. K_u is bulk modulus of undrained porous solid, and G is the rigidity of porous frame. The elastic tensor c_{ij} is the inverse of a symmetric matrix $\{a_{ij}\}$.

In terms of the measurable quantities, the elements a_{ij} are given by

$$a_{11} = \frac{1}{K}, \qquad a_{12} = -\frac{\alpha^{(1)} K_s^{(1)}}{K_s K^{(1)}},$$

$$a_{13} = -\frac{\alpha}{K} - a_{12}, \qquad \alpha = 1 - \frac{K}{K_s},$$

$$\alpha^{(1)} = 1 - \frac{K^{(1)}}{K_s^{(1)}}, \qquad a_{22} = \frac{(1-\epsilon_c)\,\alpha^{(1)}}{B^p K^{(1)}},$$

$$a_{23} = -\frac{(1-\epsilon_c)\,\alpha^{(1)}}{K^{(1)}} - a_{12},$$

$$a_{33} = \frac{\epsilon_c}{K_f} + \frac{\epsilon_s + \epsilon_p}{K^{(1)}} - \frac{1-2\alpha}{K} + 2a_{12},$$

$$(4)$$

where $K, K^{(1)}(K_s, K_s^{(1)})$ are the jacketed (unjacketed) bulk moduli of porous aggregate and solid matrix, respectively. α and $\alpha^{(1)}$ are the corresponding Biot-Willis parameters. K_f is bulk modulus of saturating fluid. In terms of these coefficients, we have $K_u^{-1} = a_{11} - (a_{12} + a_{13})^2/(a_{22} + a_{33})$. The relations (3) are used to express the equations of motion, in terms of displacements and dilatations ($e_s = \nabla \cdot \mathbf{u}$, $e_p = \nabla \cdot \mathbf{v}$, and $e_c = \nabla \cdot \mathbf{w}$), as

$$\nabla\left[\left(K_u + \frac{1}{3}G\right)e_s - K_u\left(\epsilon_p B^p e_p + \epsilon_c B^c e_c\right)\right] + G\nabla^2 \mathbf{u}$$

$$= \epsilon_s \rho_s \ddot{\mathbf{u}} + \rho_{12}\ddot{\mathbf{v}} + \rho_{13}\ddot{\mathbf{w}} - b_{12}\dot{\mathbf{v}} - b_{13}\dot{\mathbf{w}},$$

$$\epsilon_p \nabla\left[c_{21}e_s + \epsilon_p c_{22}e_p + \epsilon_c c_{23}e_c\right]$$

$$= \epsilon_p \rho_f \ddot{\mathbf{u}} + \rho_{22}\ddot{\mathbf{v}} + \rho_{23}\ddot{\mathbf{w}} + (b_{12} + b_{23})\dot{\mathbf{v}} - b_{23}\dot{\mathbf{w}},$$

$$\epsilon_c \nabla\left[c_{31}e_s + \epsilon_p c_{32}e_p + \epsilon_c c_{33}e_c\right]$$

$$= \epsilon_c \rho_f \ddot{\mathbf{u}} + \rho_{23}\ddot{\mathbf{v}} + \rho_{33}\ddot{\mathbf{w}} - b_{23}\dot{\mathbf{v}} + (b_{13} + b_{23})\dot{\mathbf{w}}.$$

$$(5)$$

2.1. Viscoelastic Porous Frame. A poroviscoelastic solid saturated with viscous fluid represents a realistic homogeneous model for sedimentary or reservoir rocks. Biot-Stoll model [21] is an important mathematical model that takes into account both intergranular losses in solid frame and viscous losses in interstitial fluid. This is a very useful model to study the propagation of attenuated waves in marine sediments [22]. Sharma and Gogna [23] considered this model to study the reflection of attenuated body waves at its plane boundary. In the present problem, following Stoll [22], viscoelastic response of skeletal frame is defined by the complex transforms $K_s(1 - \iota Q_b^{-1})$, $K_s^{(1)}(1 - \iota Q_b^{-1})$ and $G(1 - \iota Q_r^{-1})$ of its elastic moduli (i.e., K_s, $K_s^{(1)}$, and G). Values of quality factors are further related to log decrement parameters (δ_G, δ_E) for rigidity (G) and Young's modulus (E) of drained porous frame. The relations, given by

$$Q_r = \frac{\pi}{\delta_G}, \qquad \frac{Q_r}{Q_b} \approx 5\frac{\delta_E}{\delta_G} - 4, \qquad (6)$$

define the attenuation from skeletal frame with the values of δ_G and δ_E/δ_G.

3. General Solution: Wave Potentials

Through the usual Helmholtz decomposition of a vector, the displacement vectors in three homogeneous isotropic constituents of porous aggregate are written as

$$\mathbf{u} = \nabla\phi + \nabla \times \mathbf{S}, \quad \nabla \cdot \mathbf{S} = 0;$$

$$\mathbf{v} = \nabla\psi + \nabla \times \mathbf{F_p}, \quad \nabla \cdot \mathbf{F_p} = 0; \qquad (7)$$

$$\mathbf{w} = \nabla\xi + \nabla \times \mathbf{F_c}, \quad \nabla \cdot \mathbf{F_c} = 0.$$

Using the above potentials, the system (5) contains third-order differential equations in the form $\nabla a + \nabla \times \mathbf{A} = \mathbf{B} = 0, \nabla \cdot \mathbf{A} = 0$, involving scalar potentials a and vector potentials $\mathbf{A}$. In particular, we can write ([24], pp. 52) $a = \nabla \cdot \mathbf{P}$ and $\mathbf{A} = \nabla \times \mathbf{P}$, for some vector function $\mathbf{P}$ derived from the integration of $\mathbf{B}$. In the present case, $\mathbf{B} = 0$ yields $\mathbf{P} = 0$, and hence we get $a = 0$ and $\mathbf{B} = 0$. This provides us two systems of (second-order) differential equations, one in scalar potentials and the other corresponding to vector potentials. Finally, in terms of displacement potentials ($\phi, \psi, \xi; \mathbf{S}, \mathbf{F_p}, \mathbf{F_c}$), the system of (5) is resolved in two subsystems as follows:

$$\left(K_u + \frac{4}{3}G\right)\nabla^2\phi - \epsilon_p K_u B^p \nabla^2\psi - \epsilon_c K_u B^c \nabla^2\xi$$

$$= \epsilon_s \rho_s \ddot{\phi} + (\rho_{12}\ddot{\psi} - b_{12}\dot{\psi}) + (\rho_{13}\ddot{\xi} - b_{13}\dot{\xi}),$$

$$\epsilon_p\left(c_{21}\nabla^2\phi + \epsilon_p c_{22}\nabla^2\psi + \epsilon_c c_{23}\nabla^2\xi\right)$$

$$= \epsilon_p \rho_f \ddot{\phi} + [\rho_{22}\ddot{\psi} - (b_{12} + b_{23})\dot{\psi}] + (\rho_{23}\ddot{\xi} - b_{23}\dot{\xi}),$$

$$\epsilon_c\left(c_{31}\nabla^2\phi + \epsilon_p c_{32}\nabla^2\psi + \epsilon_c c_{33}\nabla^2\xi\right)$$

$$= \epsilon_c \rho_f \ddot{\phi} + (\rho_{23}\ddot{\psi} - b_{23}\dot{\psi}) + [\rho_{33}\ddot{\xi} + (b_{13} + b_{23})\dot{\xi}],$$

$$(8)$$

$$G\nabla^2\mathbf{S} = \epsilon_s \rho_s \ddot{\mathbf{S}} + (\rho_{12}\ddot{\mathbf{F_p}} - b_{12}\dot{\mathbf{F_p}}) + (\rho_{13}\ddot{\mathbf{F_c}} - b_{13}\dot{\mathbf{F_c}}),$$

$$0 = \epsilon_p \rho_f \ddot{\mathbf{S}} + \rho_{22}\ddot{\mathbf{F_p}} + (b_{12} + b_{23})\dot{\mathbf{F_p}} + \rho_{23}\ddot{\mathbf{F_c}} - b_{23}\dot{\mathbf{F_c}}, \qquad (9)$$

$$0 = \epsilon_c \rho_f \ddot{\mathbf{S}} + \rho_{23}\ddot{\mathbf{F_p}} - b_{23}\dot{\mathbf{F_p}} + \rho_{33}\ddot{\mathbf{F_c}} + (b_{13} + b_{23})\dot{\mathbf{F_c}}.$$

For time harmonic ($\sim e^{-\iota\omega t}$) potentials (ϕ, ψ, ξ) to represent harmonic waves of angular frequency ω, the system (8) transforms to

$$\left[\left(K_u + \frac{4}{3}G\right)\nabla^2 + \epsilon_s \rho_s \omega^2\right]\phi - \left(\epsilon_p K_u B^p \nabla^2 + r_{12}\omega^2\right)\psi$$

$$\qquad\qquad (10)$$

$$- \left(\epsilon_c K_u B^c \nabla^2 + r_{13}\omega^2\right)\xi = 0,$$

$$\left(\epsilon_p c_{21}\nabla^2 + \epsilon_p \rho_f \omega^2\right)\phi + \left(\epsilon_p^2 c_{22}\nabla^2 + r_{22}\omega^2\right)\psi$$

$$+ \left(\epsilon_p \epsilon_c c_{23}\nabla^2 + r_{23}\omega^2\right)\xi = 0,$$

$$\qquad\qquad (11)$$

$$\left(\epsilon_c c_{31}\nabla^2 + \epsilon_c \rho_f \omega^2\right)\phi + \left(\epsilon_p \epsilon_c c_{23}\nabla^2 + r_{23}\omega^2\right)\psi$$

$$+ \left(\epsilon_c^2 c_{33}\nabla^2 + r_{33}\omega^2\right)\xi = 0,$$

where $r_{22} = \rho_{22} + (\iota/\omega)(b_{12} + b_{23})$, $r_{33} = \rho_{33} + (\iota/\omega)(b_{13} + b_{23})$, and $r_{ij} = \rho_{ij} - (\iota/\omega)b_{ij}$ $(i < j = 1, 2, 3)$.

The equations (11) are solved into two relations, given by

$$\left(A_1\nabla^4 + \omega^2 B_1\nabla^2 + \omega^4 C_1\right)\psi = \left(A_2\nabla^4 + \omega^2 B_2\nabla^2 + \omega^4 C_2\right)\phi,$$

$$\left(A_1\nabla^4 + \omega^2 B_1\nabla^2 + \omega^4 C_1\right)\xi = \left(A_3\nabla^4 + \omega^2 B_3\nabla^2 + \omega^4 C_3\right)\phi. \tag{12}$$

Using these relations in (10), we obtain

$$\left[A\nabla^6 + \omega^2 B\nabla^4 + \omega^4 C\nabla^2 + \omega^6 D\right]\phi = 0, \tag{13}$$

where

$$\begin{pmatrix} A \\ B \\ C \\ D \end{pmatrix} = \begin{bmatrix} A_1 & A_2 & A_3 \\ B_1 & B_2 & B_3 \\ C_1 & C_2 & C_3 \\ 0 & 0 & 0 \end{bmatrix} \begin{pmatrix} K_u + \dfrac{4G}{3} \\ -\epsilon_p K_u B^p \\ -\epsilon_c K_u B^c \end{pmatrix}$$

$$+ \begin{bmatrix} 0 & 0 & 0 \\ A_1 & A_2 & A_3 \\ B_1 & B_2 & B_3 \\ C_1 & C_2 & C_3 \end{bmatrix} \begin{pmatrix} \epsilon_s \rho_s \\ r_{12} \\ r_{13} \end{pmatrix};$$

$$A_1 = \delta_p^2\delta_c^2\left(c_{22}c_{33} - c_{23}^2\right),$$

$$B_1 = \delta_p^2 c_{22}r_{33} + \delta_c^2 c_{33}r_{22} - 2\epsilon_p\epsilon_c c_{23}r_{23},$$

$$C_1 = r_{22}r_{33} - r_{23}^2, \qquad A_2 = \epsilon_p\delta_c^2\left(c_{23}c_{13} - c_{12}c_{33}\right),$$

$$B_2 = \epsilon_p\delta_c^2\rho_f\left(c_{23} - c_{33}\right) + \epsilon_c c_{13}r_{23} - \epsilon_p c_{12}r_{33},$$

$$C_2 = \rho_f\left(\epsilon_c r_{23} - \epsilon_p r_{33}\right), \qquad A_3 = \delta_p^2\epsilon_c\left(c_{12}c_{23} - c_{13}c_{22}\right),$$

$$B_3 = \delta_p^2\epsilon_c\rho_f\left(c_{23} - c_{22}\right) + \epsilon_p c_{12}r_{23} - \epsilon_c c_{13}r_{22},$$

$$C_3 = \rho_f\left(\epsilon_p r_{23} - \epsilon_c r_{22}\right). \tag{14}$$

The differential equation (13) is decomposed to satisfy three Helmholtz equations, given by

$$\left(\nabla^2 + \dfrac{\omega^2}{V_i^2}\right)\phi_i = 0 \quad (i = 1, 2, 3). \tag{15}$$

The velocities (V_i) are derived from the roots of a cubic equation in V^2, given by

$$DV^6 - CV^4 + BV^2 - A = 0, \tag{16}$$

and are sorted in the descending order of their real parts. The system (15) thus implies the existence of three dilatational waves (named P_1, P_2, P_3) propagating with phase velocities V_i $(i = 1, 2, 3)$ and identified with the corresponding scalar potentials ϕ_i.

In the considered linear porous medium, the general representation of potential function for aggregate dilatation is expressed as

$$\phi = \phi_1 + \phi_2 + \phi_3, \tag{17}$$

which, on using in relations (12), yields

$$\psi = \mu_1\phi_1 + \mu_2\phi_2 + \mu_3\phi_3, \qquad \xi = \nu_1\phi_1 + \nu_2\phi_2 + \nu_3\phi_3, \tag{18}$$

where

$$\mu_j = \frac{A_2 - B_2 V_j^2 + C_2 V_j^4}{A_1 - B_1 V_j^2 + C_1 V_j^4},$$

$$\nu_j = \frac{A_3 - B_3 V_j^2 + C_3 V_j^4}{A_1 - B_1 V_j^2 + C_1 V_j^4} \quad (j = 1, 2, 3). \tag{19}$$

Similar to scalar potentials considered above, vector potentials $(\mathbf{S}, \mathbf{F_p}, \mathbf{F_c})$ are also considered as time harmonic with frequency ω. Then, for the time harmonic dependence, given by $(\sim e^{-\iota\omega t})$, the system of three equations (9) is resolved into another set of three equations, given by

$$\left(G\nabla^2 + \epsilon_s\rho_s\omega^2\right)\mathbf{S} = r_{12}\omega^2\mathbf{F_p} + r_{13}\omega^2\mathbf{F_c},$$

$$\epsilon_p\rho_f\mathbf{S} + r_{12}\mathbf{F_p} + r_{23}\mathbf{F_c} = 0, \qquad \epsilon_c\rho_f\mathbf{S} + r_{23}\mathbf{F_p} + r_{33}\mathbf{F_c} = 0. \tag{20}$$

Solving the above relations, we get a Helmholtz equation, given by

$$\left(\nabla^2 + \dfrac{\omega^2}{V_4^2}\right)\mathbf{S} = 0; \qquad V_4^2 = \frac{G}{\left(\epsilon_s\rho_s - \mu_4 r_{12} - \nu_4 r_{13}\right)},$$

$$\mu_4 = \frac{C_2}{C_1}, \qquad \nu_4 = \frac{C_3}{C_1}; \tag{21}$$

which defines the existence of a shear (or S) wave propagating with velocity V_4. The parameters μ_4 and ν_4 also relate the vector potentials as $(\mathbf{F_p}, \mathbf{F_c}) = (\mu_4, \nu_4)\mathbf{S}$.

4. Velocity and Attenuation

In the previous section, the existence of four (three dilatational and one shear) waves has been explained. The complex velocities V_1, V_2, V_3 of three dilatational waves, named P_1, P_2, P_3, respectively, are obtained from the roots of the cubic equation (16). The propagation of lone shear (or S) wave is represented through the complex velocity V_4, given by (21). The complex velocity, say V_k, of any of these four waves is resolved to define the corresponding phase (or propagation) velocity (v_k) and attenuation quality factor (Q_k) as follows:

$$v_k = \frac{|V_k|^2}{\Re(V_k)}, \qquad Q_k^{-1} = -\frac{\Im\left(V_k^2\right)}{\Re\left(V_k^2\right)}. \tag{22}$$

5. Numerical Example

Berea sandstone is considered as a physical model of the porous medium. Following Stoll [22], the values of various parameters are chosen for sandstone with water in its pores and fractures. The skeletal frame of sandstone consists of solid grains with bulk modulus $K_s = 38\,\text{GPa}$, rigidity modulus

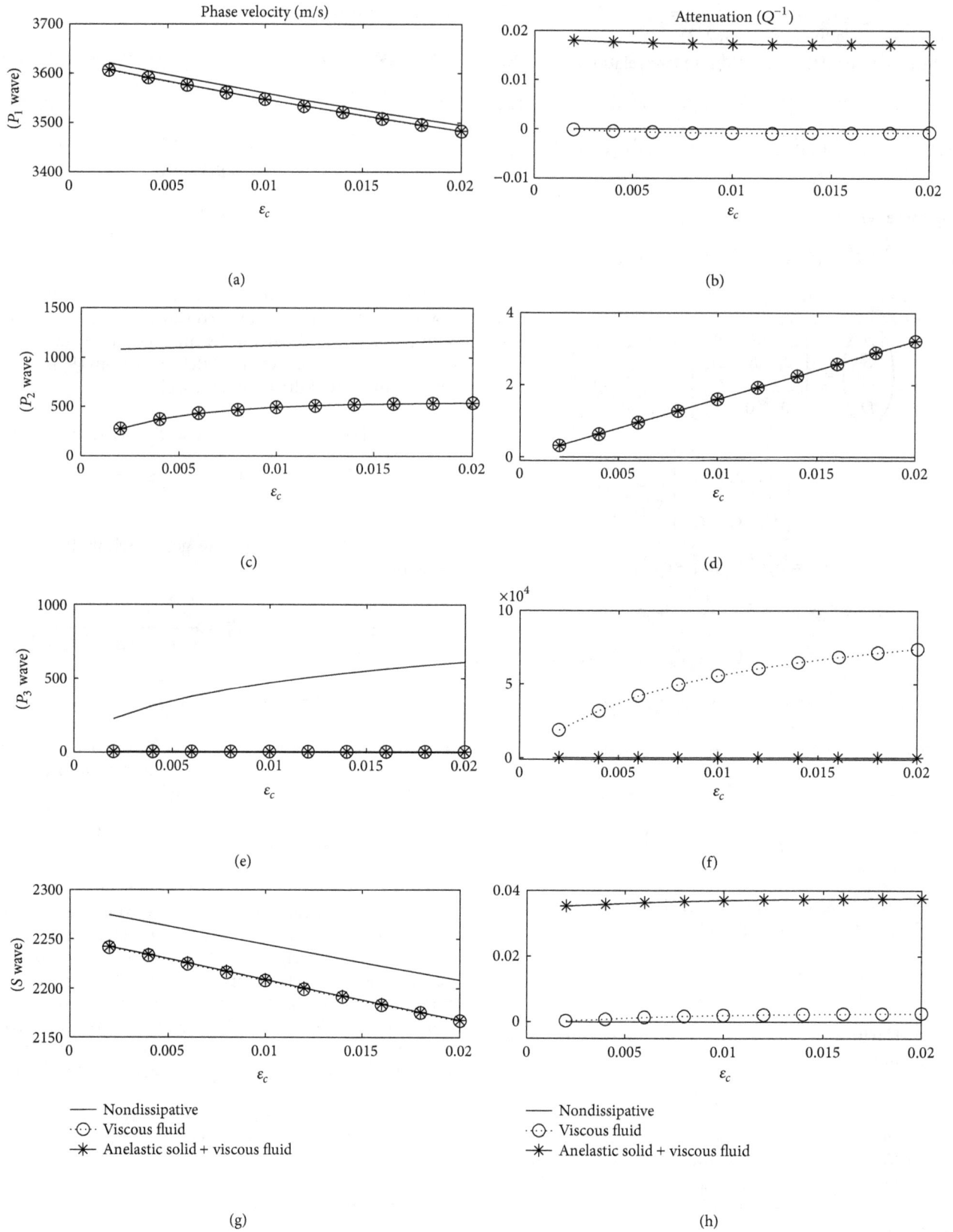

FIGURE 1: Variations of wave velocities and attenuation with fracture porosity in Berea sandstone; effect of dissipation from fluid viscosity or frame anelasticity (viscous fluid: $\eta = 0.001$ poise; pore permeability: $\chi_{11} = 10^{-16}$ m^2; fracture permeability: $\chi_{22} = 10^{-12}$ m^2; anelastic frame: $\delta_G = 0.11$, $\delta_E/\delta_G = 0.9$; wave frequency: 1000 Hz).

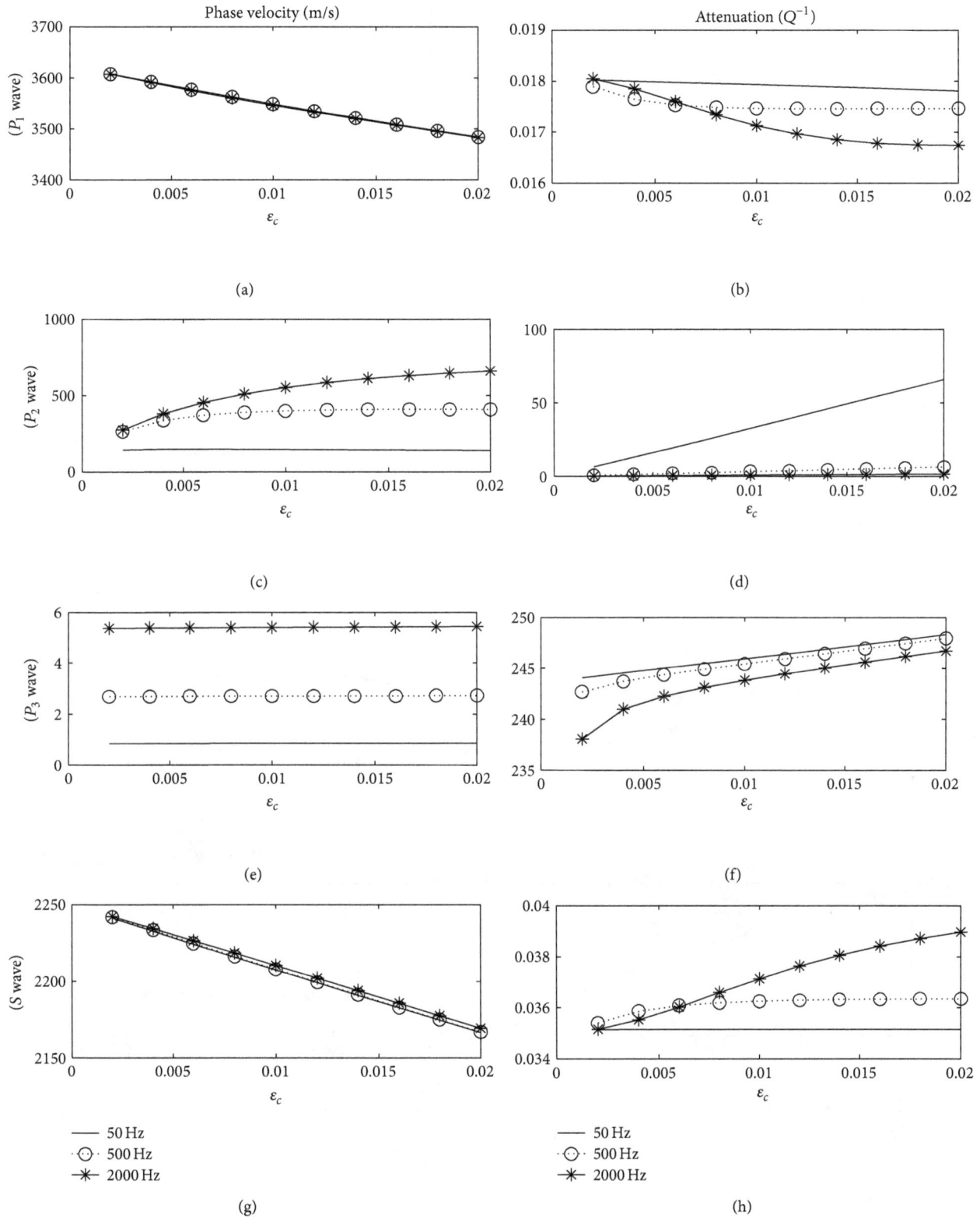

FIGURE 2: Variations of wave velocities and attenuation with fracture porosity in Berea sandstone; effect of wave frequency (viscous fluid: $\eta = 0.001$ poise; pore permeability: $\chi_{11} = 10^{-16}$ m^2; fracture permeability: $\chi_{22} = 10^{-12}$ m^2; anelastic frame: $\delta_G = 0.11$, $\delta_E/\delta_G = 0.9$).

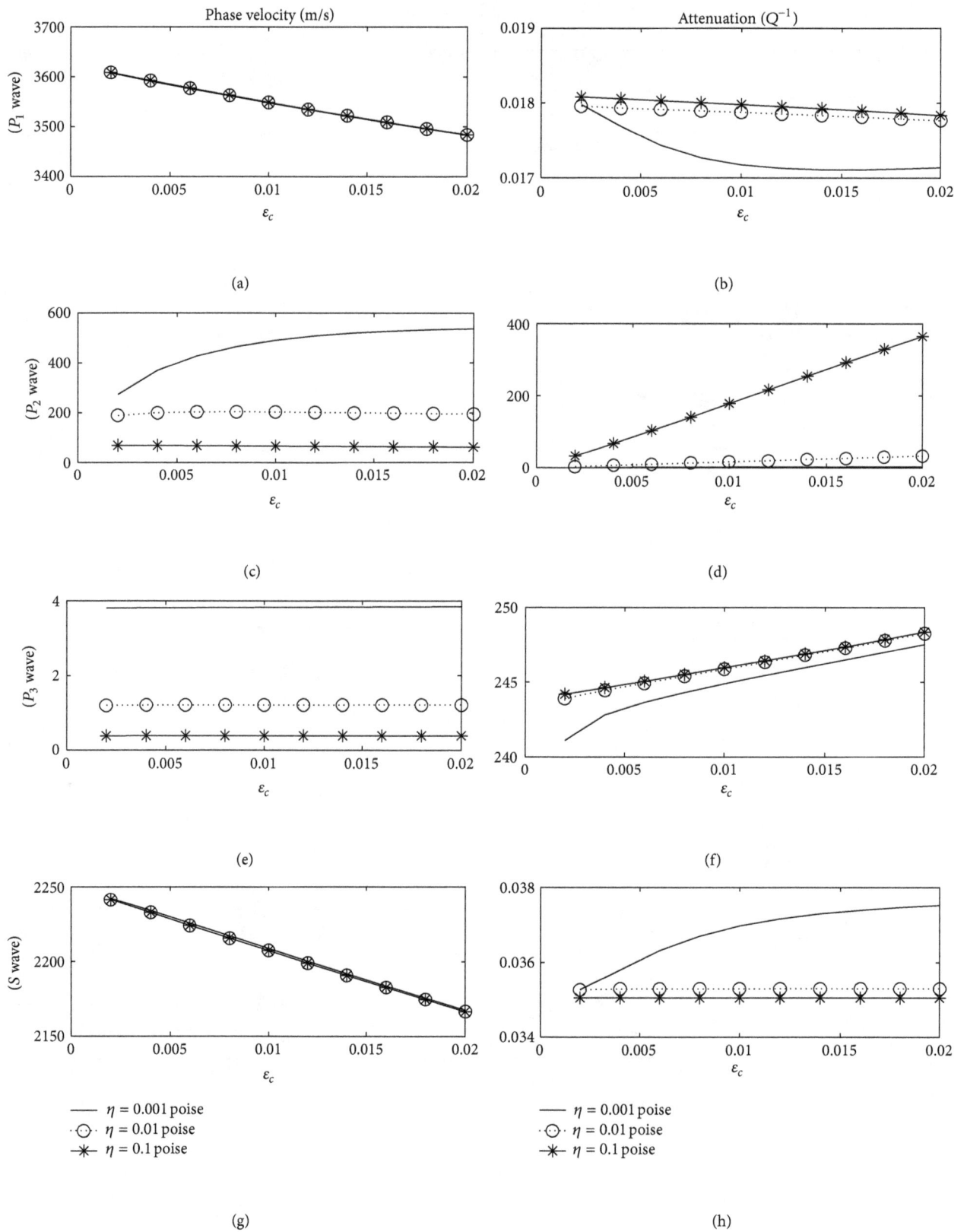

FIGURE 3: Variations of wave velocities and attenuation with fracture porosity in Berea sandstone; effect of viscosity of interstitial fluid (pore permeability: $\chi_{11} = 10^{-16}$ m^2; fracture permeability: $\chi_{22} = 10^{-12}$ m^2; anelastic frame: $\delta_G = 0.11$, $\delta_E/\delta_G = 0.9$; wave frequency: 1000 Hz).

G_s = 44 GPa, and density ρ_s = 2650 kg/m³. It supports the porosity $f = \epsilon_p + \epsilon_c$, where volume fraction of pores is fixed as $\epsilon_p = 0.2$ and volume fraction of fractures is varied up to 0.02. Both pores and fractures are filled with water of bulk modulus K_f = 2.3 GPa and density ρ_f = 1000 kg/m³. The rigidity modulus of the consolidated porous aggregate is calculated from G_s as $G = (1 - f)G_s/(1 + c_s f)$ with consolidation parameter c_s = 10 [18]. The bulk modulus of the consolidated porous aggregate is K = 6 GPa. Unjacketed bulk moduli are $K^{(1)}$ = 10 GPa and $K_s^{(1)}$ = K_s. The dynamic viscosity of water η = 0.001 poise makes the double porosity solid a dissipative one, which supports the attenuated propagation of waves. The permeability of the solid matrix to conduct the flow of fluid in pores and fractures is represented as χ_{11} = 10^{-16} m², χ_{12} = 0, and χ_{22} = 10^{-12} m². Moreover, the skeletal frame is considered viscoelastic solid with complex values for elastic constants for bulk moduli and rigidity modulus. Stoll [22] has estimated the log decrement values δ_G = 0.11 and δ_E/δ_G = 0.9, for near elastic materials. These values are used in relations (22) to calculate the quality factors for elastic moduli of porous frame. Wave frequency restricted up to 2 kHz ensures a low-frequency propagation. Skempton's coefficients for buildup of fluid pressures in pores and fractures are given by B^p = 0.6 and B^c = 0.8, respectively. For these numerical values of various parameters, the phase velocities (v_j, j = 1, 2, 3, 4) and attenuation coefficients (Q_j^{-1}, j = 1, 2, 3, 4) are calculated for the propagation of four waves in a double porosity medium. Variations of these velocities and attenuation coefficients with fracture porosity ϵ_c are plotted in Figures 1 to 4.

6. Discussion of Numerical Results

Figure 1 exhibits the variations of phase velocities and attenuations of four waves in double porosity medium (Berea sandstone) with the fracture porosity (ϵ_c) due to the presence of viscosity in saturating fluid and the anelasticity of skeletal frame. It is clear from the solid line curves that attenuation is absent when medium in nondissipative, that is, an elastic solid frame with inviscid fluid in pores and fractures. Velocities of all the waves decrease with the presence of dissipation, be it from viscous interstitial fluid or anelastic frame. The presence of attenuation comes from both, the viscous saturating fluid as well as anelastic solid frame. However, the presence of viscosity in fluid shows its effect mainly on slower P waves. On the other hand, the attenuation in faster waves (i.e., P_1 and S waves) comes mainly from anelasticity of solid frame. Velocities of these faster waves decrease with the increase of fracture porosity (ϵ_c). But the velocities of slower P waves increase with the increase of ϵ_c. Attenuation of slower P waves and S wave increase with the increase of ϵ_c but reverse may be case for fastest (P_1) wave. However, the increase is much more in case of P_2 wave. In general, faster a wave is, the lesser its attenuation is. The strange behavior is noted for the attenuation of P_3 wave, which appears very strong with viscous fluid only. The reason is that the P_3 wave is the result of the presence of fracture

porosity; therefore its extreme sensitivity to the changes in fracture properties may be expected. Note that, in case of anelastic or viscous dissipation, the velocity of this wave is negligible.

From the theoretical derivations, it is clear that complex velocities (V_j, j = 1, 2, 3, 4) depend on the angular frequency. In Figure 2, variations of phase velocities and attenuations in Berea sandstone with the fracture porosity (ϵ_c) are exhibited for three values of frequency, that is, 50 Hz, 500 Hz, and 2000 Hz. From the first column plots in this figure, it is observed that frequency has no effect on the velocity of P_1 wave. The S wave may propagate a little faster at high frequency. The slower P waves are very sensitive to the wave frequency and propagate significantly faster with an increase in frequency. However, the velocity increase with frequency increases with ϵ_c in case of P_2 wave only. The attenuation of each of the three P waves decreases slightly with the increase of frequency. It is only the S wave, which attenuates more at high frequency. Notable point is that the effect of frequency is coupled with the presence of fractures. For example, the effect of frequency on each wave (except P_3 wave) is almost absent at smaller ϵ_c and it becomes more significant with increasing ϵ_c.

With the presence of viscous fluid in pores and fracture, it becomes important to observe how the amount of this viscosity affects the velocities and attenuation of the four waves in a double porosity medium. Figure 3 illustrates this effect on the propagation and attenuation of four waves in Berea's sandstone. In general, the phase velocity of each wave decreases with an increase of viscosity. However, this decrease is negligible for P_1 wave, very slight for S wave, and very significant in case of two slower P waves. The velocity of P_2 wave appears, in general, to be sensitive to the extent of fractures in the porous medium, but in case of large fluid viscosity it loses this sensitivity. The reason may be the stronger glubeing effect of highly viscous fluid occupying a comparatively larger volume in fractures. The attenuation of each of the three P waves increases with the increase of fluid viscosity. But opposite is the case with S wave, where the attenuation decreases with increase in η. Sensitivity of attenuation to viscosity change, in general, increases with the increase in ϵ_c.

Permeability of fracture network is important in conducting the flow of fluid in a saturated porous solid. In Figure 4, the effect of fracture permeability (χ_{22}) is exhibited on the variations of velocities and attenuations of four waves in Berea's sandstone. The effect of this permeability on velocities is nearly similar to that of fluid viscosity. For example, its effect is negligible of P_1 wave, very slight on S wave, and very significant of two slower P waves. However, the velocities of these two waves increase with permeability, which is opposite to the effect of fluid viscosity. On the other hand, the effect of permeability on attenuation is different on different waves. In general, the increase in fracture permeability decreases the attenuation of all the three P waves but the attenuation of S wave increases with the increase in χ_{22}. Sensitivity of attenuation to

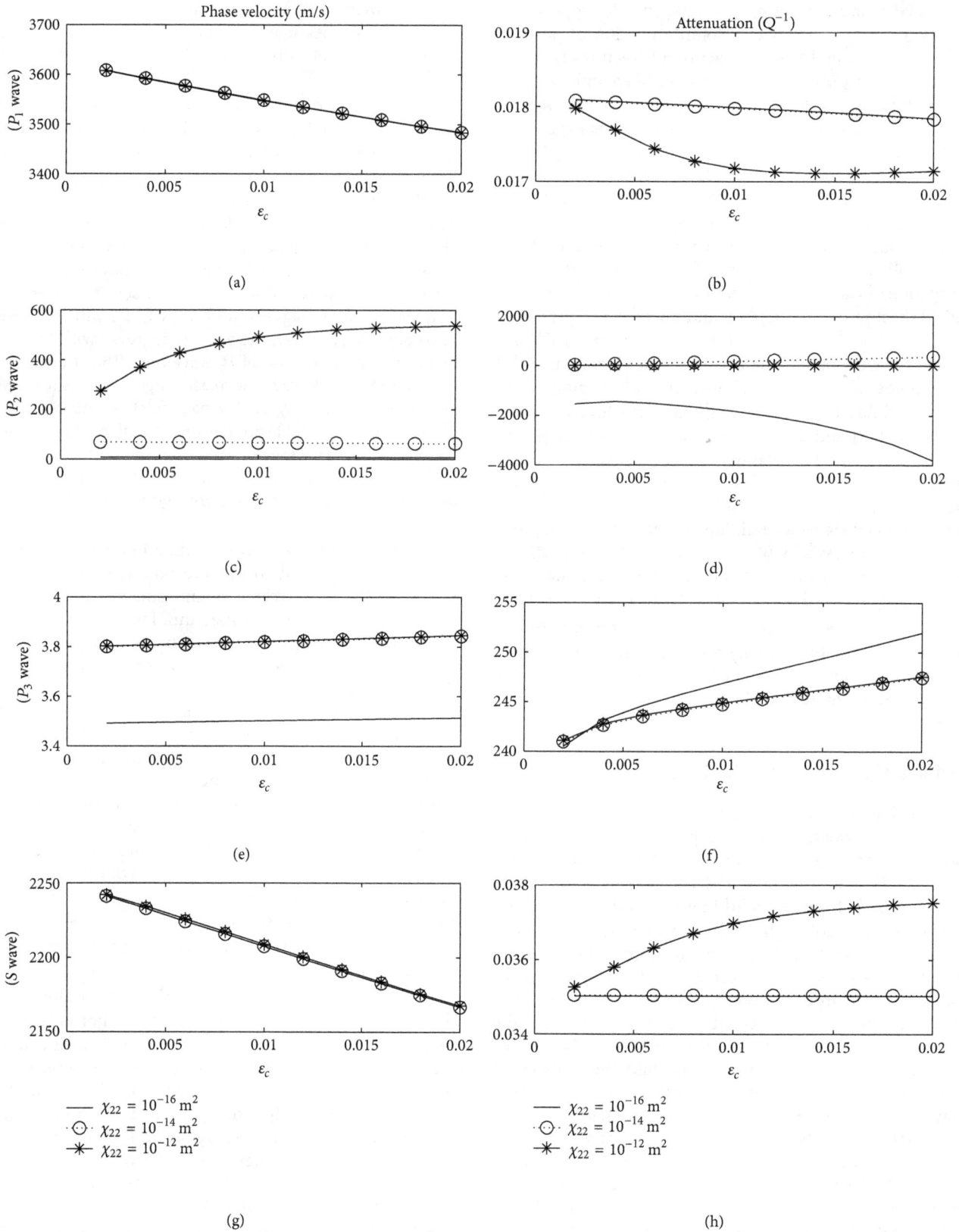

FIGURE 4: Variations of wave velocities and attenuation with fracture porosity in Berea sandstone; effect of fracture permeability (viscous fluid: $\eta = 0.001$ poise; pore permeability: $\chi_{11} = 10^{-16}$ m^2; anelastic frame: $\delta_G = 0.11$, $\delta_E/\delta_G = 0.9$; wave frequency: 1000 Hz).

fracture permeability increases with the increase in fracture porosity.

7. Concluding Remarks

The modeling and inversion procedures are used to interpret the measurable seismic quantities, like velocities and quality factors of attenuation, into the in situ properties of crustal rocks. The in-depth understanding of the relations between seismic properties and rock characteristics always helps in designing more effective mathematical models for seismic dynamism in the crust. It is well known in the phenomenology of earth materials that rocks are generally heterogeneous, porous, and often fractured or cracked. In situ, rock pores and cracks/fractures can contain oil, gas, or water. These fluid reservoirs are of great practical interest for science and economy in any society. Fracture or crack porosity may occupy very little volume but nevertheless has two very important effects on the reservoir properties. The first effect is that fractures/cracks drastically weaken the rock elastically, so that even a very small change in stress can lead to large changes in the fracture/crack apertures (and at the same time change the fracture strength for future changes). The second effect is that the fractures/cracks often introduce a high permeability pathway for the fluid to escape from the reservoir. This effect obviously is the key to reservoir analysis and the economics of fluid withdrawal.

References

[1] G. I. Barenblatt, I. P. Zheltov, and I. N. Kochina, "Basic concepts in the theory of seepage of homogeneous liquids in fissured rocks," *Journal of Applied Mathematics and Mechanics*, vol. 24, no. 5, pp. 1286–1303, 1960.

[2] J. E. Warren and P. J. Root, "The behavior of naturally fractured reservoirs," *Society of Petroleum Engineers Journal*, vol. 3, pp. 245–255, 1963.

[3] M. A. Biot, "Theory of propagation of elastic waves in a fluid-saturated porous solid. I. Low frequency range. II. Higher frequency range," *Journal of the Acoustical Society of America*, vol. 28, pp. 168–191, 1956.

[4] M. A. Biot, "Mechanics of deformation and acoustic propagation in porous media," *Journal of Applied Physics*, vol. 33, no. 4, pp. 1482–1498, 1962.

[5] M. A. Biot, "Generalized theory of acoustic propagation in porous dissipative media," *Journal of the Acoustical Society of America*, vol. 34, pp. 1254–1264, 1962.

[6] R. I. O'Connell and B. Budiansky, "Viscoelastic properties of fluid saturated cracked solids," *Journal of Geophysical Research*, vol. 82, pp. 5719–5735, 1977.

[7] G. M. Mavko and A. Nur, "Wave attenuation in partially saturated rocks," *Geophysics*, vol. 44, no. 2, pp. 161–178, 1979.

[8] G. Mavko and D. Jizba, "Estimating grain-scale fluid effects on velocity dispersion in rocks," *Geophysics*, vol. 56, no. 12, pp. 1940–1949, 1991.

[9] J. Dvorkin and A. Nur, "Dynamic poroelasticity: a unified model with the squirt and the Biot mechanisms," *Geophysics*, vol. 58, no. 4, pp. 524–533, 1993.

[10] L. Thomsen, "Elastic anisotropy due to aligned cracks in porous rock," *Geophysical Prospecting*, vol. 43, no. 6, pp. 805–829, 1995.

[11] M. D. Sharma, "Surface-wave propagation in a cracked poroelastic half-space lying under a uniform layer of fluid," *Geophysical Journal International*, vol. 127, no. 1, pp. 31–39, 1996.

[12] R. K. Wilson and E. C. Aifantis, "On the theory of consolidation with double porosity," *International Journal of Engineering Science*, vol. 20, pp. 1009–1035, 1982.

[13] R. K. Wilson and E. C. Aifantis, "A double porosity model for acoustic wave propagation in fractured-porous rock," *International Journal of Engineering Science*, vol. 22, no. 8–10, pp. 1209–1217, 1984.

[14] T. F. Cho, M. E. Plesha, and B. C. Haimson, "Continuum modelling of jointed porous rock," *International Journal for Numerical & Analytical Methods in Geomechanics*, vol. 15, no. 5, pp. 333–353, 1991.

[15] M. Bai, D. Elsworth, and J. C. Roegiers, "Modeling of naturally fractured reservoirs using deformation dependent flow mechanism," *International Journal of Rock Mechanics and Mining Sciences and*, vol. 30, no. 7, pp. 1185–1191, 1993.

[16] J. G. Berryman and H. F. Wang, "The elastic coefficients of double-porosity models for fluid transport in jointed rock," *Journal of Geophysical Research*, vol. 100, no. 12, pp. 611–627, 1995.

[17] J. G. Berryman and H. F. Wang, "Elastic wave propagation and attenuation in a double-porosity dual-permeability medium," *International Journal of Rock Mechanics and Mining Sciences*, vol. 37, no. 1-2, pp. 63–78, 2000.

[18] S. R. Pride and J. G. Berryman, "Linear dynamics of double-porosity dual-permeability materials. I. Governing equations and acoustic attenuation," *Physical Review E*, vol. 68, no. 3, Article ID 036603, 10 pages, 2003.

[19] S. R. Pride and J. G. Berryman, "Linear dynamics of double-porosity dual-permeability materials. II. Fluid transport equations," *Physical Review E*, vol. 68, no. 3, Article ID 036604, 10 pages, 2003.

[20] A. W. Skempton, "The pore-pressure coefficients A and B," *Geotechnique*, vol. 4, pp. 143–147, 1954.

[21] R. D. Stoll and G. M. Bryan, "Wave attenuation in saturated sediments," *Journal of the Acoustical Society of America*, vol. 47, pp. 1440–1147, 1970.

[22] R. D. Stoll, "Marine sediment acoustics," *Journal of the Acoustical Society of America*, vol. 77, no. 5, pp. 1789–1799, 1985.

[23] M. D. Sharma and M. L. Gogna, "Seismic wave propagation in a viscoelastic porous solid saturated by viscous liquid," *Pure and Applied Geophysics*, vol. 135, no. 3, pp. 383–400, 1991.

[24] P. M. Morse and H. Feshback, *Methods of Theoretical Physics*, McGraw-Hill, New York, NY, USA, 1953.

8

The Dependence of Electrical Resistivity-Saturation Relationships on Multiphase Flow Instability

Zoulin Liu and Stephen M. J. Moysey

Department of Environmental Engineering and Earth Science, Clemson University, Clemson, SC 29670, USA

Correspondence should be addressed to Stephen M. J. Moysey, smoysey@clemson.edu

Academic Editors: E. Liu and H. Perroud

We investigate the relationship between apparent electrical resistivity and water saturation during unstable multiphase flow. We conducted experiments in a thin, two-dimensional tank packed with glass beads, where Nigrosine dyed water was injected uniformly along one edge to displace mineral oil. The resulting patterns of fluid saturation in the tank were captured on video using the light transmission method, while the apparent resistivity of the tank was continuously measured. Different experiments were performed by varying the water application rate and orientation of the tank to control the generalized Bond number, which describes the balance between viscous, capillary, and gravity forces that affect flow instability. We observed the resistivity index to gradually decrease as water saturation increases in the tank, but sharp drops occurred as individual fingers bridged the tank. The magnitude of this effect decreased as the displacement became increasingly unstable until a smooth transition occurred for highly unstable flows. By analyzing the dynamic data using Archie's law, we found that the apparent saturation exponent increases linearly between approximately 1 and 2 as a function of generalized Bond number, after which it remained constant for unstable flows with a generalized Bond number less than −0.106.

1. Introduction

Multiphase fluid flow in porous media is an important problem for applications including petroleum production [1–3], migration of nonaqueous phase liquids (NAPLs) in soils and aquifers [4–6], and CO_2 sequestration [7]. Viscous, capillary and gravity forces interact in immiscible two phase flow systems to produce stable or unstable flow regimes [8–11]. In a stable flow regime the displacement of one fluid for another will occur along a stable front. In unstable flow regimes, fingering can occur along the displacement front. As a result, the invading fluid phase can bypass significant amounts of the original fluid phase, leaving it in place in the medium.

Electrical resistivity measurements are commonly used to investigate fluid saturations in multiphase flow systems [12–16]. The resistivity index provides an expression of resistivity for multiphase flow systems that is directly related to the degree of water saturation of the medium, S_w, through Archie's law [17]. When mineral surface conductivity is insignificant, the resistivity index I_R is equal to the ratio of the resistivity of the sample (ρ_w) measured at saturation S_w to the resistivity of the sample measured at 100% water saturation (ρ_s) (Equation (1)). The saturation exponent, n, is an empirical constant that is conceptually related to the connectivity of the electrically conductive phase, that is, water:

$$I_R = \frac{\rho_w}{\rho_s} = S_w^{-n}. \tag{1}$$

The saturation exponent is usually determined experimentally from measurements of I_R and S_w using (1). For example, Sweeney and Jennings [18] experimentally found the saturation exponent to be 1.61 for water wet carbonates, though this increased up to 12.27 for samples treated to become oil wet. Zhou et al. [16] used percolation models to show that the saturation exponent in strongly water wet materials is on the order of 1.9, whereas it can increase to over 3.5 for intermediate and oil wet systems. The saturation exponent is dependent on the presence of nonconductive fluid in the pore space and wettability of the rock [12, 19]. Although Archie's Law is widely used to determine fluid saturation from resistivity measurements, it is not always valid as the saturation-resistivity relationship depends on the wettability, saturation history, content of clay minerals, salinity of the brine phase and distribution of water and oil in the rock [3, 12, 15, 16]. Most experimental studies of the dependence between saturation and apparent resistivity were based on the assumption that fluid distribution within the sample was homogeneous, which can be rarely obtained during a transient displacement experiment. We experimentally investigate the influence of flow instability on apparent resistivity and the saturation exponent in Archie's law. To this end, multiphase flow experiments are conducted where water is used to displace mineral oil in a two-dimensional (2D) flow system. In this paper, we conduct a series of experiments where the stability of the flow is controlled by varying the water inflow rate and angle of the tank. Measurements of the bulk resistivity of the tank are obtained during the flow experiments. Transient estimates of average water saturation in the tank, S_w are derived from video collected using the light transmission method. Using these measurements we observe the relationship between saturation and resistivity for a range of flow conditions. We further evaluate whether there is a dependence of the saturation exponent in Archie's law on the flow conditions in a porous medium.

2. Background on Flow Instability

It is well known that variations in the magnitude and connectivity of permeability could lead to flow channeling in reservoirs and consequently to a reduction in oil production [20]. Even in a homogeneous medium, flow instability can cause viscous fingering that also increases the residual oil volume left behind in a reservoir [8, 9]. Flow instability is affected by the cumulative effects of capillary, buoyancy, and viscous forces. For example, viscous forces can destabilize the displacement front into narrow fingers if a less viscous fluid displaces a more viscous fluid, whereas gravity plays a stabilizing effect when a lighter fluid is on top of a denser phase [10]. The balance between forces in a two phase flow system can be quantified using the dimensionless Bond and capillary numbers along with the viscosity ratio. The viscosity ratio (M) is defined as the viscosity of displacing fluid μ_w divided by the viscosity of the displaced fluid μ_n. Viscous fingering can be observed when the viscosity ratio is less than 1 and the viscous force overcomes capillary and gravity effects.

The Bond number (B_o), given in (2), expresses the relative importance of gravitational to capillary forces in a multiphase flow system [9, 10]. In contrast, the capillary number (C_a) in (3) expresses the balance between viscous to capillary forces [9, 10, 21]:

$$B_o = \frac{\Delta P_{grav}}{\Delta P_{cap}} = \frac{\Delta \rho g a^2}{\gamma} \sin \varphi, \qquad (2)$$

$$C_a = \frac{\Delta P_{visc}}{\Delta P_{cap}} = \frac{\mu_w \nu a^2}{\gamma k}. \qquad (3)$$

In these expressions, μ_w is the viscosity of wetting fluid, ν is the filtration or Darcy velocity, a is the typical pore radius, γ is surface tension, $\Delta \rho$ is the density difference between the two fluids, g is the acceleration due to the gravity, φ is the angle of flow relative to horizontal, and k is the permeability of the porous medium [9, 10]. The capillary and Bond numbers can be combined to produce the generalized Bond number (B_o^*) given in the following [9, 10]:

$$B_o^* = B_o - C_a = \frac{a^2}{\gamma k}(\Delta \rho g k \sin \varphi - \mu_w \nu). \qquad (4)$$

The value of the generalized Bond number plays a critical role for determining the occurrence of viscous instabilities. For $B_o^* > 0$ the flow is stable and a compact and flat displacement front occurs as illustrated in Figure 1(a). However, when $B_o^* < 0$ the flow is unstable and fingering produces an uneven and often rapid movement of the infiltrating phase through the displaced phase in a porous medium (Figure 1(b); the white region in the images corresponds to oil, whereas the black areas correspond to Nigrosine dyed water). Despite the obvious contrast in macroscopic behavior, Méheust et al. note that a radical change in the local dynamics of the interface does not occur during the transition between stable and unstable displacement [10].

3. Methods and Experimental Setup

The main goal of this work is to determine the relationship between the saturation exponent in Archie's law and the degree of flow instability in a porous medium as quantified by the Bond and Capillary numbers. To achieve this objective, resistivity index curves were measured during the displacement of mineral oil by water in a 2D tank packed with glass beads. Four-wire resistance measurements were collected throughout the experiment while the light transmission method was used to simultaneously monitor changes in saturation. The effect of gravity on flow instability is controlled by changing the orientation of the tank to achieve different Bond numbers. Experiments at different flow rates were conducted to control the relative importance of viscous forces by varying the capillary number.

3.1. Experimental Setup. The fluids used in these experiments are water and mineral oil (EMD Chemicals, NJ, USA). The properties of each fluid are given in Table 1. We focus on a situation where a denser fluid with low viscosity (water) displaces a less dense, more viscous fluid (mineral oil) from below leading to a low viscosity ratio (0.015). Viscous fingering is therefore possible in this system. Negrosine (Acros Organics) dye was added to the water phase to provide contrast with the clear mineral oil to allow visual tracking of the displacement front and the development of fingers. This particular dye was selected because it did not partition from the water to oil phase in initial static tests conducted in beakers. The electrical conductivity of the water used in each experiment varied from 71.8–91.5 μs/cm.

The experiments were conducted in the specially designed 2D acrylic tank shown in Figure 2. The dimensions

(a)

(b)

FIGURE 1: Comparison of (a) stable flow conditions ($B_o^* = 0.00612$) and (b) unstable flow conditions ($B_o^* = -0.248$; experiment #12 in Table 3) for oil (white) displaced by Nigrosine dyed water (black).

TABLE 1: Properties of the fluids used in the experiments.

Wetting phase, water (with 0.05 g/L nigrosine)	
Density, ρ_w	1000 kg/m^3
Dynamic viscosity, μ_w	1.002E-3 N.s/m^2
Non-wetting phase, mineral oil	
Density, ρ_n	880 kg/m^3
Dynamic viscosity, μ_n	0.068 N.s/m^2
Interfacial tension [22], γ	0.049 N/m
Viscosity Ratio, M	0.015

TABLE 2: Physical properties of the flow system.

Length	40 cm
width	45 cm
Thickness	1.25 cm
Porosity, ε	0.30
Formation factor, F_f	3.04
Permeability, k	57 Darcy
Grain Size, D	2 mm

FIGURE 2: Side view sketch of the experimental setup for the resistivity cell with light transmission imaging and resistivity measurement systems.

of the interior flow cell of the tank are 45 cm × 40 cm × 1.25 cm. For all of the experiments in the study, the flow cell was packed with 2 mm diameter glass beads (Walter Stern). The entire cell was designed to be pressure sealed, thereby allowing for the tank to be oriented at arbitrary geometries. The outlet pressure of the tank was held at a constant positive pressure by keeping the discharge reservoir above the tank (Figure 2). Details regarding the physical properties of the tank are summarized in Table 2.

The tank could tilt to arbitrary angles so as to vary the effect of gravity on flow and control the Bond number. The component of gravity acting on the flow system is determined by $g_\varphi = g \times \sin(\varphi)$, where g is acceleration due to gravity and φ is the angle of the tank relative to horizontal. For each experiment water was injected into the tank at a constant rate selected to achieve a specified capillary number using a variable rate peristaltic pump (pump head: HV-07015-20, Master Flex). The displacing water phase is

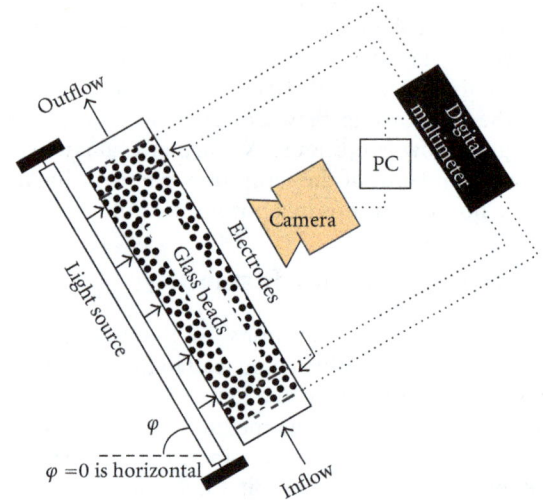

injected through a porous plastic plate covering the entire inlet surface of the beads to ensure the injection is uniform. The displaced oil phase is expelled from a similar outlet port at the opposite end of the flow cell. Both gravity effects, that is, Bond number, and flow rate, that is, capillary number, influence the stability of flow in our experiments and can be changed independently of each other. A complete listing of tank orientations and flow rates used in the experiments is given in Table 3.

The bulk DC resistance of the tank was determined using the four-electrode method [4, 24]. Two pieces of copper mesh were anchored across the inflow and outflow sides of the tank to act as current electrodes. Two additional copper strips were positioned 2 cm away from each potential electrode within the tank to act as potential electrodes (Figure 2). A National Instruments PXI system with 7.5-digit digital multimeter and multiconfiguration matrix module (NI PXI-4071, PXI-2530) were used to measure the DC

TABLE 3: Summary of the tank inclination angle (φ), pumping rate (Q), and fluid conductivity (σ_w) for each of the 34 experiments. The corresponding characteristic numbers C_a, B_o and B_o^* are also given (pore size estimated as $0.414 \times$ grain size following [23]).

Experiment index	1	2	3	4	5	6	7
Q (mL/min)	189	251	67	119	157	27	67
φ (degree)	90	90	90	90	90	90	90
σ_w (μs/cm)	98	93	90.2	93.7	81.4	126	94.7
B_o	$1.65E-02$	$1.65E-02$	$1.65E-02$	$1.65E-02$	$1.65E-02$	$1.65E-02$	$1.65E-02$
C_a	$1.39E-01$	$1.85E-01$	$4.94E-02$	$8.78E-02$	$1.16E-01$	$1.99E-02$	$4.94E-02$
B_o^*	$-1.23E-01$	$-1.69E-01$	$-3.30E-02$	$-7.13E-02$	$-9.94E-02$	$-3.47E-03$	$-3.30E-02$

Experiment index	8	9	10	11	12	13	14
Q (mL/min)	119	27	189	251	358	67	99
φ (degree)	90	90	90	90	90	90	90
σ_w (μs/cm)	82	78.7	79.5	76.7	78.4	80.1	83.1
B_o	$1.65E-02$	$1.65E-02$	$1.65E-02$	$1.65E-02$	$1.65E-02$	$1.65E-02$	$1.65E-02$
C_a	$8.78E-02$	$1.99E-02$	$1.39E-01$	$1.85E-01$	$2.64E-01$	$4.94E-02$	$7.30E-02$
B_o^*	$-7.13E-02$	$-3.47E-03$	$-1.23E-01$	$-1.69E-01$	$-2.48E-01$	$-3.30E-02$	$-5.66E-02$

Experiment index	15	16	17	18	19	20	21
Q (mL/min)	89	52	146	119	67	27	67
φ (degree)	90	90	90	90	90	90	90
σ_w (μs/cm)	74.8	77.6	74.2	79	77.6	78.3	87
B_o	$1.65E-02$	$1.65E-02$	$1.65E-02$	$1.65E-02$	$1.65E-02$	$1.65E-02$	$1.65E-02$
C_a	$6.57E-02$	$3.84E-02$	$1.08E-01$	$8.78E-02$	$4.94E-02$	$1.99E-02$	$4.94E-02$
B_o^*	$-4.92E-02$	$-2.19E-02$	$-9.13E-02$	$-7.13E-02$	$-3.30E-02$	$-3.47E-03$	$-3.30E-02$

Experiment index	22	23	24	25	26	27	28
Q (mL/min)	189	251	67	67	67	67	67
φ (degree)	90	90	30	30	30	0	45
σ_w (μs/cm)	71.8	72.8	78	91.5	90	81.1	80.5
B_o	$1.65E-02$	$1.65E-02$	$8.23E-03$	$8.23E-03$	$8.23E-03$	$0.00E+00$	$1.16E-02$
C_a	$1.39E-01$	$1.85E-01$	$4.94E-02$	$4.94E-02$	$4.94E-02$	$4.94E-02$	$4.94E-02$
B_o^*	$-1.23E-01$	$-1.69E-01$	$-4.12E-02$	$-4.12E-02$	$-4.12E-02$	$-4.94E-02$	$-3.78E-02$

Experiment index	29	30	31	32	33	34
Q (mL/min)	67	67	67	67	358	358
φ (degree)	15	60	60	60	90	90
σ_w (μs/cm)	82.8	84	79.2	84.7	89.1	88.5
B_o	$4.26E-03$	$1.42E-02$	$1.42E-02$	$1.42E-02$	$1.65E-02$	$1.65E-02$
C_a	$4.94E-02$	$4.94E-02$	$4.94E-02$	$4.94E-02$	$2.64E-01$	$2.64E-01$
B_o^*	$-4.52E-02$	$-3.52E-02$	$-3.52E-02$	$-3.52E-02$	$-2.48E-01$	$-2.48E-01$

resistivity of the tank while switching the polarity of the current electrodes to avoid electrode polarization effects. Prior to running the flow experiments, the flow cell was filled with a saline solution and measurements were taken to calibrate the geometric factor relating tank bulk resistance to resistivity. After packing the tank with the glass beads the formation factor in Archie's law was determined to be 3.04 for our experiments by measuring resistance for several different solution conductivities. Surface conductivity effects for these large glass beads are assumed to be negligible given insignificant imaginary conductivity responses measured by spectral induced polarization [25].

The system developed for the light transmission measurements [6, 26, 27] contains a light source and detector (Figure 2). In this experiment a scientific digital camera (DFK 41BU02.H USB CCD, Imaging Source) with a 5 mm lens (H0514-MP, Imaging Source) is used as the detector to quantify the intensity of light transmitted through the tank. This camera has resolution of 1280 × 960 pixels for 32 bit images, which provides a spatial resolution of 0.39 mm per pixel or about 2 pixels per pore for a distance between the camera and the tank of 40 cm. The camera is controlled by a host computer using a LabView (National Instruments) program to obtain images at a specific frame rate during the fluid displacement. The pictures that the camera takes are in raw bmp format with no compression. Images are later converted into gray scale and analyzed using MATLAB. The light transmitted through the tank is generated by an array of fluorescent bulbs (13 W each, Bi-Pin, MA) mounted to the back of the tank in a manner allowing it to move with the tank when the experimental angle is changed.

The background reference image obtained before water is injected is subtracted from each subsequent image to overcome problems related to variations in light intensity

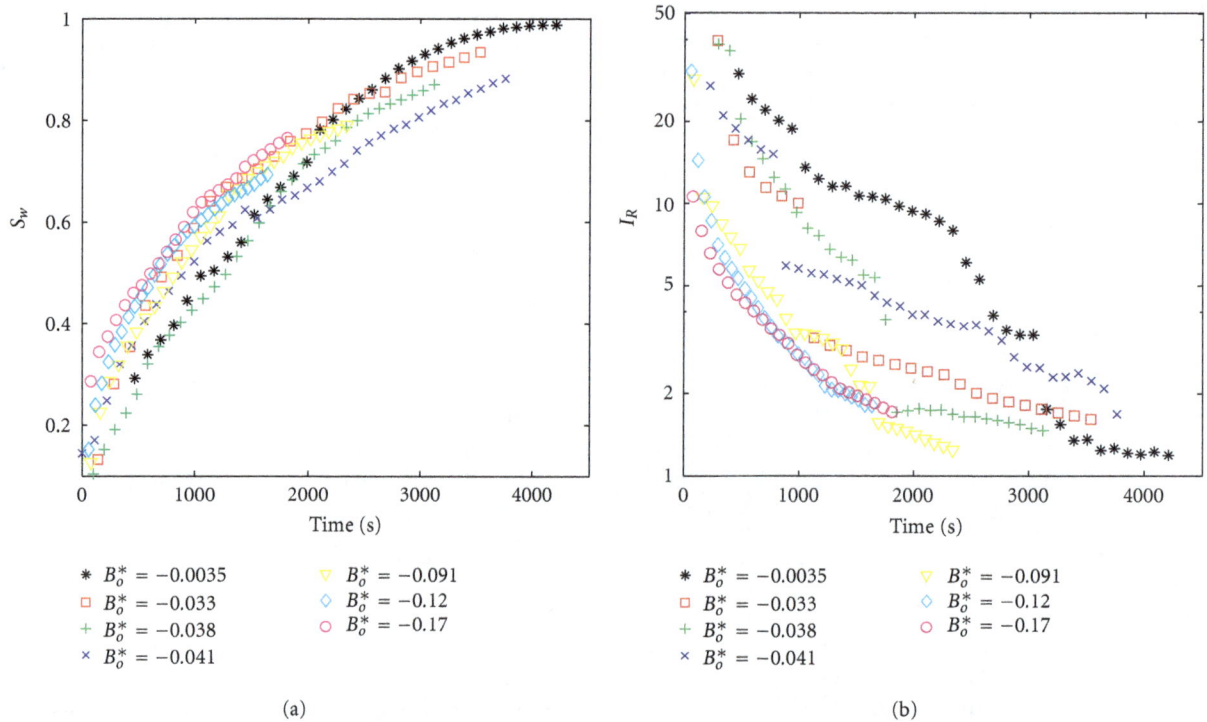

FIGURE 3: Changes in (a) average tank saturation and (b) resistivity index through time for varying values of B_o^* (data shown for experiment index = 9, 7, 28, 25, 17, 10, 11).

due to the specific arrangement of light bulbs in the array. The intensity $(I_0 - I)$ of the corrected image was found to have a linear relationship with the water saturation S_w inside the porous medium as shown in the following:

$$S_w = \frac{[1.0759 \times (I_0 - I) - 3.3557]}{100}. \qquad (5)$$

This equation is obtained from calibration experiments using a small chamber with the same material, thickness, and packing of glass beads to obtain a porosity of about 0.30, consistent with the flow cell.

With the experimental setup described above, 2 main series of 34 experiments are conducted: one series is at constant Bond number of 0.0165 and the other is at constant capillary number of 0.0494. Table 3 summarizes the experimental conditions for all experiments in terms of the orientation of the tank (φ), pumping rates (Q), water conductivity (σ), and the corresponding Capillary number (C_a), Bond number (B_o), and generalized Bond number B_o^*.

4. Results

The range of the Bond numbers that can be achieved by rotating the tank, that is, 0 to 0.0165, is smaller than the range of capillary numbers that can be achieved by changing the flow rate, that is, 0 to 0.264. Therefore, we can obtain the largest range of generalized Bond numbers by changing flow rate. The maximum generalized Bond number used in the experiments is -0.0035 because the digital multimeter was not able to read the high resistivity of the mineral oil in

completely stable situations where water uniformly displaced the oil. The lowest (i.e., most negative) generalized Bond number investigated is -0.248 as the medium tended to compact under high internal pressures if higher flow rates were applied in the closed cell.

4.1. Saturation and Resistivity Index. The average water saturation and resistivity index of the tank over time are shown in Figure 3 for different values of the generalized Bond number. Saturations change relatively smoothly in most cases as water displaces the oil. Differences between the curves are apparent, but trends for different generalized Bond numbers are not clear. In contrast, the resistivity index curves show a distinctive change in behavior with generalized Bond number. For small values of B_o^*, that is, values near zero where flow is more stable, the resistivity index curves show large, sudden drops. In contrast, for large negative values of B_o^*, in which case the flow is highly unstable with many thin fingers formed, the reduction in resistivity index over time is smooth and regular. This result is indicative of the high sensitivity of resistivity measurements to the geometry of the water phase in the medium. Note that we use resistivity index here since the fluid resistivity varied between some of the experiments (Table 3).

At small negative values of B_o^* the flow is stable and the water advances either as a uniform front or as large, individual fingers. Using the video collected during the experiment, each jump in resistivity index can be correlated with the time at which a finger of water reaches the tank's upper current electrode, thereby completing a new pathway

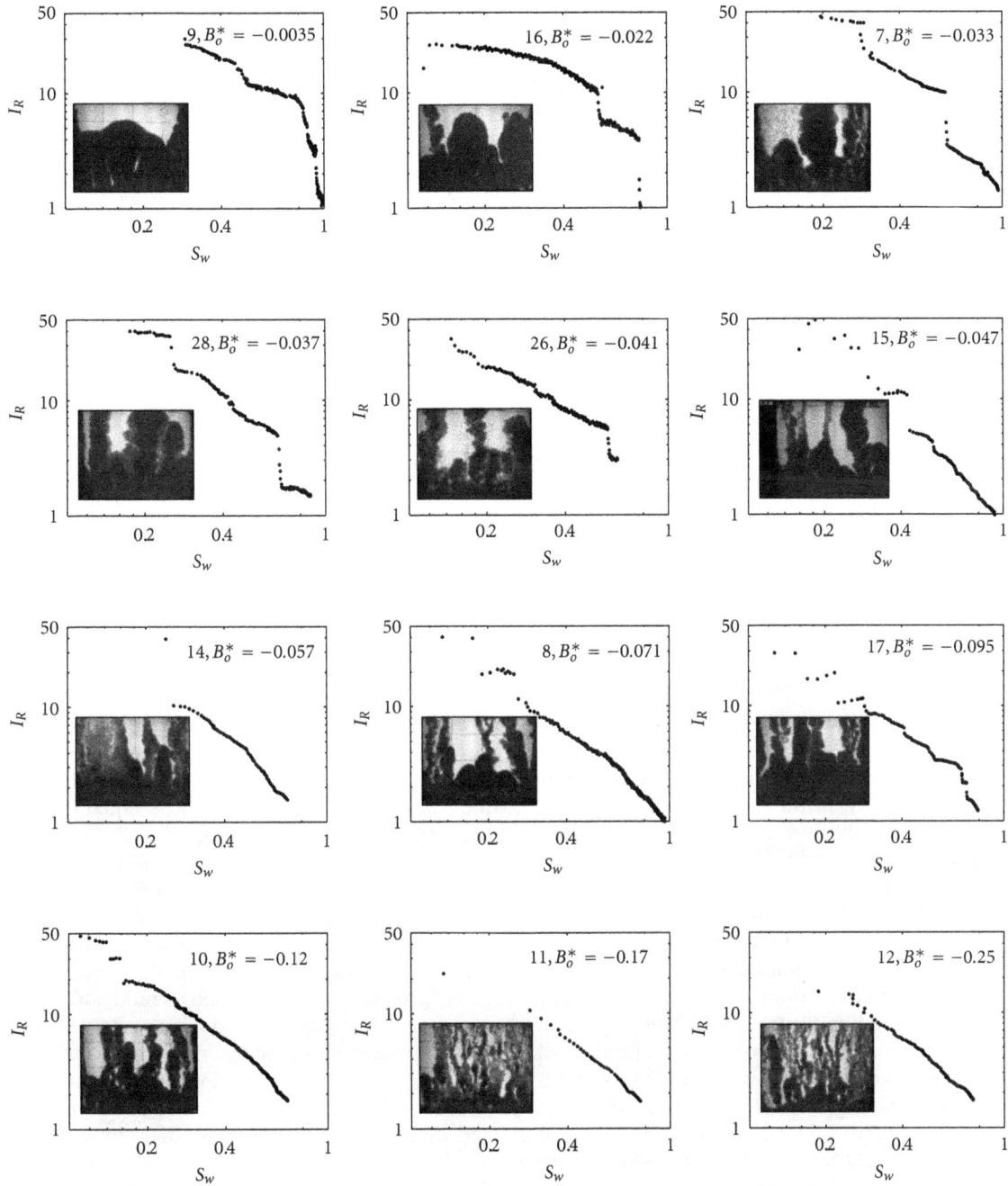

FIGURE 4: Dependence of resistivity index on saturation for different values of B_o^* (the experimental index and B_o^* are given in each figure). A representative image of the flow conditions during each experiment is shown with the inset picture.

for current to flow through the medium. At large negative values of B_o^* the flow is highly unstable, producing many thin fingers. The fingers tend to reach the tank outflow in a more uniform manner, producing the relatively smooth change in resistivity observed in Figure 3(b). The patterns of fingering observed in our experiments (Figure 4), that is, increasing number of fingers and decreasing finger thickness with increasing generalized Bond number, is consistent with observations from experiments by Løvoll et al. [9] and Méheust et al. [10] though these authors did not measure resistivity.

Based on experimental data, Méheust et al. [10] suggest a power law with an exponent of -0.55 to relate the front width of the fingering and the generalized Bond number:

$$W = B_o^{*-0.55}, \qquad (6)$$

where W is the measured front width of the finger, that is, the root mean square maximum extension perpendicular to the flow direction. Therefore bigger discrete drops in the resistivity index at small values of the generalized Bond number can be attributed to wider fingers reaching the

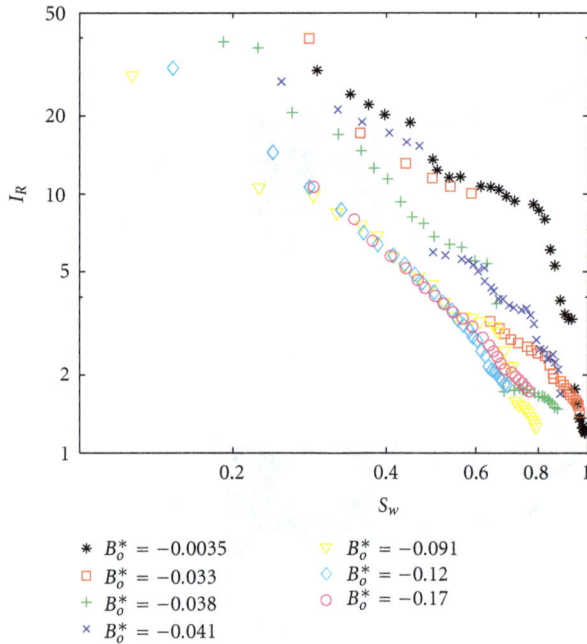

FIGURE 5: Direct comparison of resistivity index changes with saturation for different generalized Bond numbers.

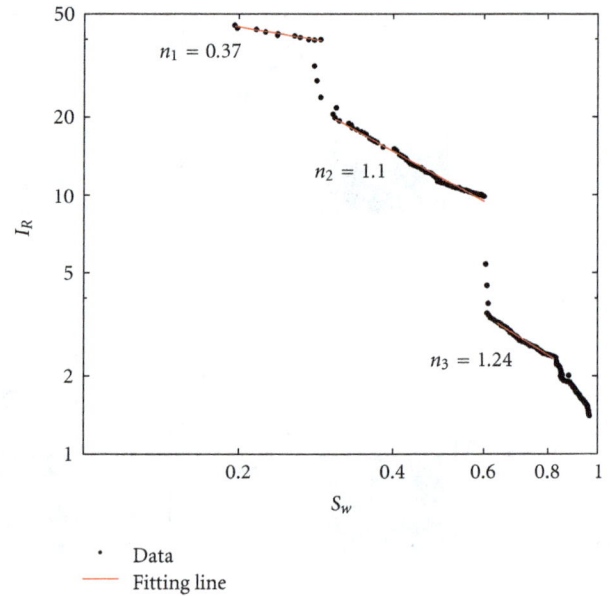

FIGURE 6: Illustration of how the saturation exponent was determined for experiments with discrete resistivity drops (data shown for experiment index 7).

electrodes and causing a larger portion of the flow cell to connect the electrodes. As the flow grows increasingly unstable, the conductive fingers have a smaller width and are more uniformly distributed in the tank, though individual fingers may form at different times. As a result, the resistivity change caused by an individual finger is small and the observed change in resistivity is smoother as a result of the progressive arrival of individual conductive fingers.

Figures 4 and 5 show the resistivity index versus water saturation for a subset of the generalized Bond numbers used in the experiments. For a given saturation value the resistivity index is generally lowest for unstable flow scenarios, that is, in Figure 5 the value of the generalized Bond number of the curves decreases (becomes more negative) from the top curve to the bottom. The slope of the resistivity index curves is flattest for small negative values of the generalized Bond number with changes in resistivity occurring as sharp drops when individual fingers reach the outflow end of the tank. As discussed previously, the magnitude of these drops in the resistivity index decrease as the generalized Bond number becomes more negative and the drops disappear when the flow is very unstable at $B_o^* = -0.123$ (Figure 5).

It is notable that the resistivity index versus saturation curves for $B_o^* = -0.123$ (Figure 5, diamond) and $B_o^* = -0.169$ (Figure 5, circle) overlap with each other, suggesting that these two unstable flow scenarios have a similar electrical behavior. In contrast, at higher generalized Bond numbers, for example, $B_o^* = -0.00347$, the flow is more stable and a very sharp drop in I_R is observed at the end of the experiment. This rapid change in resistivity index occurs because the front of the water moving to the end the tank is relatively flat (i.e., lacks fingers) compared to the other experiments. In this case, both the water saturation and

resistivity index are approximately equal to 1 at the end of the experiment, indicating that the porous medium is almost fully saturated with water (i.e., high sweep efficiency) and that the resistivity is approximately equal that observed for the 100% water saturated tank.

4.2. Influence of Flow Instability on the Saturation Exponent.
Archie's law (equation (1)) is typically assumed to apply to situations where fluids are distributed uniformly throughout a porous medium. We feel it is also valuable, however, to investigate how flow instability could affect the apparent properties of the formation during unstable flow. In Archie's law the logarithms of the resistivity index (I_R) and saturation are linearly related with a slope equivalent to the negative of the saturation exponent (n). We analyze the saturation and resistivity data obtained here in a similar way to obtain an apparent saturation exponent for each experiment. For some experiments, however, the slope of these curves varies as a function of saturation. This effect is at least partially a result of the resistivity drops discussed earlier, which are a consequence of the fact that the measurements represent a dynamic flow system where preferential flows are established within the spatially finite bounds of the tank. The saturation exponent was therefore estimated for distinct sections of the resistivity-saturation curve, ignoring the sudden drops in resistivity caused by the breakthrough of individual fingers of water (Figure 6). We acknowledge that the value of the apparent saturation exponent obtained in this way may not have the same physical meaning as in typical applications of Archie's law. Regardless, this approach still provides a way to summarize the experimental data in a way that allows for effective comparison between the experiments.

TABLE 4: Saturation exponent determined for each experiment.

B_o^*	Data				Mean	Standard deviation
	$B_o = 0.017$, Varying C_a					
−0.0035	Experiment Index	6^1	9^1	20	0.70	0.11
	n	0.59	0.80	0.71		
−0.022	Experiment Index	16			0.91	—
	n	0.91				
−0.033	Experiment Index	3	13^1	7^1 19	0.96	0.08
	n	0.96	0.90	0.90 1.08		
−0.049	Experiment Index	15			1.30	—
	n	1.30				
−0.057	Experiment Index	14			1.51	—
	n	1.51				
−0.071	Experiment Index	4	8	18^1	1.73	0.27
	n	1.73	1.99	1.46		
−0.091	Experiment Index	17			1.43	—
	n	1.43				
−0.099	Experiment Index	5			1.71	—
	n	1.71				
−0.12	Experiment Index	1^1	10	22	1.96	0.10
	n	2.06	1.97	1.86		
−0.17	Experiment Index	2	11	23	1.93	0.02
	n	1.92	1.96	1.92		
−0.25	Experiment Index	12	33	34	1.94	0.02
	n	1.93	1.96	1.94		
	$C_a = 0.049$, Varying B_o, Transition Zone					
−0.035	Experiment Index	30	31	32	1.38	0.18
	n	1.41	1.17	1.54		
−0.038	Experiment Index	28			1.22	—
	n	1.22				
−0.041	Experiment Index	24	25	26	1.28	0.19
	n	1.10	1.47	1.27		
−0.045	Experiment Index	29			1.40	—
	n	1.40				
−0.049	Experiment Index	27			1.59	—
	n	1.59				

[1]Significant resistivity drops observed in the data.

Table 4 shows the apparent saturation exponent estimated for each experiment, averaging sections of the curves for the experiments found to have large, sudden resistivity drops. We take the average of these values to estimate the saturation exponent corresponding to a given generalized Bond number. For the unstable flow experiments, that is, large negative values of B_o^*, the resistivity change is continuous so the saturation exponent is estimated by fitting the data with a single slope. We can therefore evaluate how the apparent saturation exponent in our experiments varies as a function of generalized Bond number using the measured resistivity index versus water saturation curves (i.e., Figures 4 and 5).

We find that the apparent saturation exponent increases when flow becomes increasingly unstable, that is, B_o^* becomes more negative (Figure 7). The saturation exponent reaches a constant value of 1.94 when generalized Bond number is less than −0.106; notably, this is consistent with the saturation index estimated by Zhou et al. [16] for strongly water wet materials using percolation theory. We quantify the relationship observed in Figure 7 between generalized Bond number and the saturation exponent as follows:

$$n = -9.3 \times B_o^* + 0.9 \quad \text{for} -0.106 < B_o^* < -0.00347,$$
$$n = 1.94 \quad \text{for } B_o^* < -0.106. \tag{7}$$

The observed dependence of the saturation exponent on generalized Bond number demonstrates that relationships used to estimate fluid saturation from resistivity measurements, for example, Archie's law, must be dynamic and take into account the way in which a reservoir is managed and produced. The saturation exponent is fundamentally related to the geometry of the conductive and nonconductive phases within a porous medium, specifically the connectivity of the conductive phase. Given that instability has an overwhelming influence on fluid distributions during multiphase flow, this process will also strongly influence the saturation index.

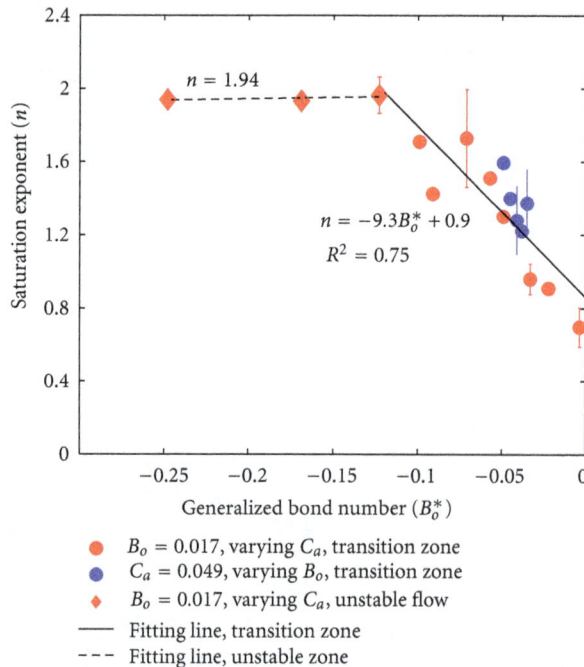

FIGURE 7: Saturation exponent as a function of generalized bond number B_o^* for all experiments with a linear fit over the range $-0.106 < B_o^* < -0.00347$. Error bars indicate the standard deviation in the saturation exponent estimate from experiments performed with the same value of B_o^*.

We identify the increasing connectivity of the water phase between the current electrodes as the primary cause for dependence of the saturation exponent on the generalized Bond number for $-0.106 < B_o^* < -0.00347$. However, when the system reaches a certain minimum degree of connectivity due to flow instability, that is, when $B_o^* < -0.106$, there no longer appears to be a dependence between the saturation exponent and generalized Bond number. These results suggest that geophysicists must collaborate with petroleum engineers to understand the dynamics of a given flow system before attempting to estimate saturation from resistivity measurements.

5. Conclusions

The influence of flow instability on electrical resistivity measurements was investigated by displacing a light, but viscous mineral oil by water in a homogeneous porous medium. The experimental setup allowed the effects of gravity and flow rate to be controlled, thereby permitting the generalized Bond number to be changed between different experiments. Video showing the distribution of the fluid phases in the tank allowed us to determine the overall tank saturation throughout the experiment, while continuous measurements of average tank resistivity were also collected throughout each experiment.

In the resulting data we observed a transition between stable and unstable displacement. By analyzing the saturation and resistivity data using Archie's law, we found that the resulting apparent saturation exponent depends linearly on

the generalized Bond number until it reaches a maximum of 1.94 for highly unstable flow systems, that is, when $B_o^* < -0.106$. These results suggest that the interpretation of fluid saturation from electrical resistivity measurements must take into account flow conditions within the subsurface, since the onset of flow instability is a primary control on the spatial distribution of the fluids in the subsurface.

Acknowledgment

Acknowledgment is made to the Donors of the American Chemical Society Petroleum Research Fund for support (or partial support) of this research.

References

[1] K. Mogensen, E. H. Stenby, and D. Zhou, "Studies of water-flooding in low-permeable chalk by use of X-ray CT scanning," *Journal of Petroleum Science and Engineering*, vol. 32, no. 1, pp. 1–10, 2001.

[2] D. Tiab and E. C. Donaldson, *Petrophysics—Theory and Practice of Measuring Reservoir Rock and Fluid Transport Properties*, Gulf Professional, 2nd edition, 2004.

[3] E. Toumelin and C. Torres-Verdín, "Object-oriented approach for the pore-scale simulation of DC electrical conductivity of two-phase saturated porous media," *Geophysics*, vol. 73, no. 2, pp. E67–E79, 2008.

[4] C. Aggelopoulos, P. Klepetsanis, M. A. Theodoropoulou, K. Pomoni, and C. D. Tsakiroglou, "Large-scale effects on resistivity index of porous media," *Journal of Contaminant Hydrology*, vol. 77, no. 4, pp. 299–323, 2005.

[5] T. W. J. Bauters, D. A. DiCarlo, T. S. Steenhuis, and J. Y. Parlange, "Preferential flow in water-repellent sands," *Soil Science Society of America Journal*, vol. 62, no. 5, pp. 1185–1190, 1998.

[6] N. Weisbrod, M. R. Niemet, and J. S. Selker, "Imbibition of saline solutions into dry and prewetted porous media," *Advances in Water Resources*, vol. 25, no. 7, pp. 841–855, 2002.

[7] N. B. Christensen, D. Sherlock, and K. Dodds, "Monitoring CO$_2$ injection with cross-hole electrical resistivity tomography," *Exploration Geophysics*, vol. 37, pp. 44–49, 2006.

[8] G. M. Homsy, "Viscous fingering in porous-media," *Annual Review of Fluid Mechanics*, vol. 19, pp. 271–311, 1987.

[9] G. Løvoll, Y. Méheust, K. J. Måløy, E. Aker, and J. Schmittbuhl, "Competition of gravity, capillary and viscous forces during drainage in a two-dimensional porous medium, a pore scale study," *Energy*, vol. 30, no. 6, pp. 861–872, 2005.

[10] Y. Méheust, G. Løovoll, K. J. Måløy, and J. Schmittbuhl, "Interface scaling in a two-dimensional porous medium under combined viscous, gravity, and capillary effects," *Physical Review E*, vol. 66, no. 5, Article ID 051603, 12 pages, 2002.

[11] J. P. Stokes, D. A. Weitz, J. P. Gollub et al., "Interfacial stability of immiscible displacement in a porous medium," *Physical Review Letters*, vol. 57, no. 14, pp. 1718–1721, 1986.

[12] W. G. Anderson, "Wettability literature survey-part 3: the effect of wettability on the electrical properties of porous media," *Journal of Petroleum Technology*, vol. 39, no. 13, pp. 1371–1378, 1986.

[13] S. Bekri, J. Howard, J. Muller, and P. M. Adler, "Electrical resistivity index in multiphase flow through porous media," *Transport in Porous Media*, vol. 51, no. 1, pp. 41–65, 2003.

[14] M. J. Blunt, M. D. Jackson, M. Piri, and P. H. Valvatne, "Detailed physics, predictive capabilities and macroscopic consequences for pore-network models of multiphase flow," *Advances in Water Resources*, vol. 25, no. 8–12, pp. 1069–1089, 2002.

[15] A. K. Moss, X. D. Jing, and J. S. Archer, "Wettability of reservoir rock and fluid systems from complex resistivity measurements," *Journal of Petroleum Science and Engineering*, vol. 33, no. 1–3, pp. 75–85, 2002.

[16] D. Zhou, S. Arbabi, and E. H. Stenby, "A percolation study of wettability effect on the electrical properties of reservoir rocks," *Transport in Porous Media*, vol. 29, no. 1, pp. 85–98, 1997.

[17] G. E. Archie, "The electrical resistivity log as an aid in determining some reservoir characteristics," *Petroleum Transactions of AIME*, vol. 146, pp. 54–62, 1942.

[18] S. A. Sweeney and H. Y. Jennings, "Effect of wettability on the electrical resistivity of carbonate rock from a petroleum reservoir," *Journal of Physical Chemistry*, vol. 64, no. 5, pp. 551–553, 1960.

[19] D. Abdassah, P. Permadi, Y. Sumantri, and R. Sumantri, "Saturation exponent at various wetting condition: fractal modeling of thin-sections," *Journal of Petroleum Science and Engineering*, vol. 20, no. 3-4, pp. 147–154, 1998.

[20] J. M. Hovadik and D. K. Larue, "Static characterizations of reservoirs: refining the concepts of connectivity and continuity," *Petroleum Geoscience*, vol. 13, no. 3, pp. 195–211, 2007.

[21] R. Lenormand, "Liquids in porous media," *Journal of Physics*, vol. 2, pp. SA79–SA88, 1990.

[22] H. Yoon, M. Oostrom, and C. J. Werth, "Estimation of interfacial tension between organic liquid mixtures and water," *Environmental Science and Technology*, vol. 43, no. 20, pp. 7754–7761, 2009.

[23] J. F. Villaume, "Investigations at sites contaminated with dense, non-aqueous phase liquids (NAPLs)," *Ground Water Monitoring Review*, vol. 5, no. 2, pp. 60–74, 1985.

[24] G. F. Tagg, "Practical investigations of the earth resistivity method of geophysical surveying," *Proceedings of the Physical Society*, vol. 43, no. 3, pp. 305–323, 1931.

[25] N. Hao, J. Waterman, T. A. Kendall, S. M. Moysey, and D. Ntarlagiannis, "Resolving IP mechanisms using micron-scale surface conductivity measurements and column SIP data," *Geochimica et Cosmochimica Acta*, vol. 74, pp. A380–A380, 2010.

[26] C. J. G. Darnault, J. A. Throop, D. A. Dicarlo, A. Rimmer, T. S. Steenhuis, and J. Y. Parlange, "Visualization by light transmission of oil and water contents in transient two-phase flow fields," *Journal of Contaminant Hydrology*, vol. 31, no. 3-4, pp. 337–348, 1998.

[27] M. R. Niemet and J. S. Selker, "A new method for quantification of liquid saturation in 2D translucent porous media systems using light transmission," *Advances in Water Resources*, vol. 24, no. 6, pp. 651–666, 2001.

9

Merge-Optimization Method of Combined Tomography of Seismic Refraction and Resistivity Data

Andy A. Bery

Geophysics Section, School of Physics, Universiti Sains Malaysia, 11800 Penang, Malaysia

Correspondence should be addressed to Andy A. Bery, andersonbery@yahoo.com.my

Academic Editors: Y.-J. Chuo and G. Mele

This paper discussed a novel application called merge-optimization method that combines resistivity and seismic refraction data to provide a detailed knowledge of the studied site. This method is interesting because it is able to show strong accuracy of two geophysical imaging methods based on many of data points collected from the conducted geophysical surveys of disparate data sets based strictly on geophysical models as an aid for model integration for two-dimensional environments. The geophysical methods used are high resolution methods. The resistivity imaging used in this survey is able to resolve the subsurface condition of the studied site with low RMS error (less than 2.0%) and 0.5 metre electrodes interval. For seismic refraction method, high resolution of seismic is used for correlation with resistivity results. Geophones spacing is 1.0 metre and the total number of shot-points is 15, which provides very dense data point. The algorithms of merge-optimization have been applied to two data sets collected at the studied site. The resulting images have been proven to be successful because they satisfy the data and are geometrically similar. The regression coefficient found for conductivity-resistivity correlation is 95.2%.

1. Introduction

The characterization of the subsurface requires a detailed knowledge of several properties of the composing rocks and fluids. Whereas some of these properties can be measured directly (seismic and borehole methods), other properties have to be estimated by indirect measurement methods such as resistivity, TEM, and magnetic. However, it is not uncommon that the geophysical data yield models of limited accuracy which may not contribute significantly to our understanding of the subsurface condition or may show incompatibilities. Thus, a new technique needs to be produced not only for better interpretation by geophysics but also for nongeophysical background people such as engineers and architects. The distribution of uncorrelated physical properties seems to be controlled by common subsurface attributes, when taken into account, able to improve and resolve the accuracy of the geophysical imaging results. An outstanding feature of the subsurface that is common to the geophysical data is the geometrical distribution of the physical properties which can be measured by the physical property changes. This condition of commonality can be incorporated in the process of estimation to obtain meaningful and more reliable subsurface imaging results.

2. Methodology for Merge-Optimization Method

In this paper, seismic refraction data using the SeisOpt@2D software were developed by Pullammanappallil and Louie [1], to obtain velocity tomograms. The tomography inversion is performed by a generalized simulated-annealing method of optimization including a controlled Monte-Carlo inversion. The first arrival times and survey geometry are used as input and no initial velocity model is needed. The apparent resistivity data were inverted using the software package RES2DINV [2]. This program solves the tomography inversion problem using a smoothness-constrained least squares method. The logarithmic average of the measured apparent resistivity values is input as the homogeneous starting model [2].

This paper adopts a merge problem formulation with resistivity-velocity cross gradients function in order to

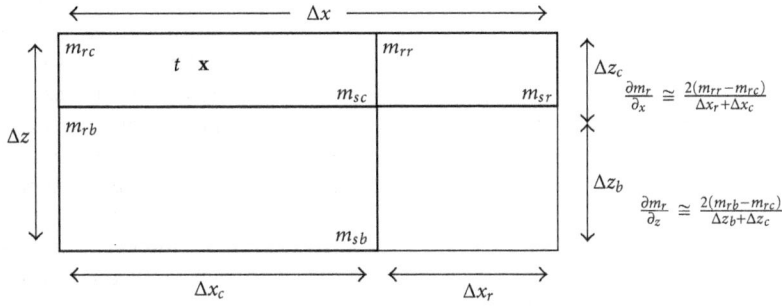

FIGURE 1: Definition of the resistivity-velocity cross gradients function and its derivatives on a rectangular grid domain. For 2D grid extending in the x and z directions, with each grid element characterised by logarithm of resistivity m_r and seismic slowness m_s, the function t is defined at the centre of a given element (marked with **x**) considering the parameters from two elements it is in contact with its right (subscripted r) and bottom (subscripted b).

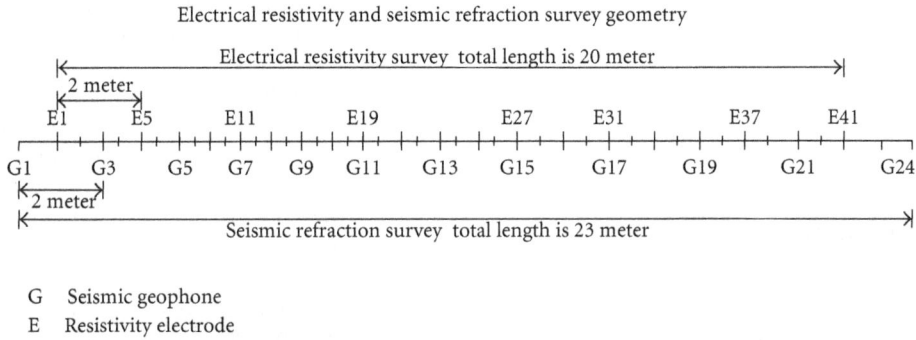

FIGURE 2: Geometry of the infield tests for both geophysical methods.

provide the required effective link between the resistivity model and the seismic velocity model. The cross product of the gradient is defined as

$$\vec{t}(x,y,z) = \nabla m_r(x,y,z) \times \nabla m_s(x,y,z), \quad (1)$$

where m_r and m_s refer to the logarithm of resistivity and P-wave slowness, respectively. In the two-dimensional case, $\vec{t}(x,y,z)$ always points in the strike direction, that is, it will be treated as a scalar t. It is incorporated as part of the objective function:

$$\Phi(\vec{m_r}, \vec{m_s}) = \left\| \begin{matrix} \vec{d_r} - \vec{f_r}(\vec{m_r}) \\ \vec{d_s} - \vec{f_s}(\vec{m_s}) \end{matrix} \right\|^2_{C_{dd}^{-1}} + \left\| \begin{matrix} \alpha_r \vec{D}\vec{m_r} \\ \alpha_s \vec{D}\vec{m_s} \end{matrix} \right\|^2$$

$$+ \left\| \begin{matrix} \vec{m_r} - \vec{m_{Rr}} \\ \vec{m_s} - \vec{m_{Rs}} \end{matrix} \right\|^2_{C_{RR}^{-1}} \quad (2)$$

Subject to $\quad \vec{t}(\vec{m}) = 0$.

Here, $\vec{d}$ represents the vector of observed data (logarithm of apparent resistivity, $\vec{d_r}$ and seismic travel-times, $\vec{d_s}$), $\vec{m} = [\vec{m_r} : \vec{m_s}]^T$ is the vector of the model parameters, $\vec{f}$ is the theoretical model respond, $\vec{D}$ is the discrete version of the

smoothing operator acting on $\vec{m}$, $\vec{m_R} = [\vec{m_{Rr}} : \vec{m_{Rs}}]^T$ is an a priori model, $\vec{C_{dd}}$ is the covariance of the field data (assumed diagonal, i.e., fully uncorrelated data), $\vec{C_{RR}}$ is the covariance of the priori model (assumed diagonal), and α_r and α_s are weighting factors that control the level of smoothing of the resistivity and seismic imaging models.

The cross gradients criterion requires the problem to satisfy the condition $\vec{t}(\vec{m}) = 0$, where any spatial changes occurring in both resistivity and velocity must point to the same point in the same or opposite direction irrespective of amplitude. In a geological sense, this implies that if a boundary exists, then it must be sensed by both geophysical methods in a common orientation regardless of the amplitude of the physical property changes. An additional flexibility of the technique is that the cross gradients constraint is also satisfied when either $\nabla \vec{m_r}$ or ∇m_s vanishes in some part of the model, thus giving the models the possibility of admitting a geological boundary which has a significant change only in the electrical resistivity or seismic velocity of the adjoining rocks.

In iteration 2D optimization approach, the subsurface model is discretised into rectangular cells of variable sizes optimized according to the natural sensitivity of each particular set of resistivity and seismic velocity measurements. We define the discrete version of (1) and the corresponding derivatives using the elements of the 3 cells scheme depicted in Figure 1. For the along strike components this simplifies to

$$t \cong \frac{4}{\Delta x \Delta z}[m_{rc}(m_{sb} - m_{sr}) + m_{rr}(m_{sc} - m_{sb}) \tag{3}$$
$$+ m_{rb}(m_{sr} - m_{sc})],$$

where the quantities Δx, Δz, m_{rc}, m_{rb}, and so forth are defined in Figure 1 above.

Here, the first order of Taylor series expression is used, and (2) is equivalent to

$$\min\left\{2\vec{n}_2^T \vec{m} + \vec{m}^T N_1 \vec{m}\right\} \tag{4}$$
$$\text{subject to} \quad t(\vec{m}_0) + \vec{B}\vec{m} - \vec{B}\vec{m}_0 = 0,$$

where

$$N_1 = \begin{bmatrix} \vec{A}_r^T \vec{C}_{ddr}^{-1} \vec{A}_r + \alpha_r^2 \vec{D}^T \vec{D} + C_{RRr}^{-1} & 0 \\ 0 & \vec{A}_s^T \vec{C}_{dds}^{-1} \vec{A}_s + \alpha_s^2 \vec{D}^T \vec{D} + C_{RRs}^{-1} \end{bmatrix},$$

$$n_2 = \begin{bmatrix} \vec{A}_r^T \vec{C}_{ddr}^{-1}\left\{\vec{d}_r - \vec{f}_r(\vec{m}_{0r}) + \vec{A}_r \vec{m}_{0r}\right\} + \vec{C}_{RRr}^{-1} \vec{m}_{Rr} \\ \vec{A}_s^T \vec{C}_{dds}^{-1}\left\{\vec{d}_s - \vec{f}_s(\vec{m}_{0s}) + \vec{A}_s \vec{m}_{0s}\right\} + \vec{C}_{RRs}^{-1} \vec{m}_{Rs} \end{bmatrix};$$

$$\tag{5}$$

$\vec{A}_r$, $\vec{A}_s$, and $\vec{B}$ are the respective partial derivatives of $\vec{f}_r$, $\vec{f}_s$, and $\vec{t}$ evaluated at the initial model, $\vec{m}_0 = [\vec{m}_{0r} : \vec{m}_{0s}]^T$. The Jacobian matrix for seismic data, $\vec{A}_s$ is computed using ray tracing as suggested by Vidale [3] and Zelt and Barton [4]. The solution to (4) used in our iterative scheme is given by

$$\vec{m} = \vec{N}_1^{-1} \vec{n}_2 - \vec{N}_1^{-1} \vec{B}^T \left(\vec{B}\vec{N}_1^{-1}\vec{B}^T\right)^{-1} \left[\vec{B}\vec{N}_1^{-1}\vec{n}_2 - \vec{B}\vec{m}_0 \right. \tag{6}$$
$$\left. + \vec{t}(\vec{m}_0)\right].$$

In the regularised solution process, the weighting factors are initially assigned large values which are then gradually reduced in subsequent iterations until the data are fitted to the required level. The merge optimization is initiated using a half-space model in the absence of reliable a priori information. The implemented cross gradients criterion and regularisation measures ensure that the resolution characteristics of the individual data sets are fully exploited the search for structurally linked models. Note that with the cross gradients criterion, there is no need to define or assume ab initio any interdependence resistivity and seismic velocity which could bias the inverse solution.

3. Seismic and Resistivity Characterization by Merge-Optimization Method

In this paper, we used data collected during recent electrical tomography and seismic refraction surveys [5]. The same data sets were used for simultaneous subsurface imaging using algorithm for improved near-surface characterization. We expect that the merge-optimization method of the data will define more accurately the main geological and features of the subsurface. Figure 2 shows the geometry used for both infield tests. The resistivity data are collected using a Wenner-Schlumberger array using with 0.5 metre electrode intervals. As for seismic refraction, 15 shot points and 1.0 metre geophone intervals are used. The purpose of this geometry is to increase the resolution of the imaging results (less noise effect and moderate sensitivity to both vertical and horizontal structures) and hence both separate data sets are more to correlate and increase the reliability of the results. The study site is located in a slope area, thus it is more challenging than the flat ground.

The seismic refraction velocity and electrical resistivity imaging derived by merge-optimization of infield tests data sets (Figures 3(a) and 4(a)) is less than 2.0% of root mean-squares (RMS) for electrical resistivity imaging results. The reconstructed distributions of the model parameters show structural similarities and hence good spatial correlation of velocity with resistivity data sets. Note that the cross gradients criterion serves for geological structural control but does not force the two models into conformity. Figures 3(b) and 4(b) show the seismic refraction results for the same area and the survey geometry. Figures 3(c) and 4(c) show the arcs formation (seismic refraction) at the subsurface in the studied site.

The resistivity images show that the subsurface of the studied sit consists of two main zones. Resistivity values lower than $900\,\Omega.m$ are indicative of residual (clayey sand soils) while values higher than $1,100\,\Omega.m$ are indicative of a weathered layer. The presence of moist zones and dry zones can be associated with loose zones and compacted soil, respectively. The seismic refraction images showed that the subsurface consists of three layers. Velocity values of 370–500 m/s are associated with loose soil mixed with boulders (high resistivity value near surface). The second layer has velocity values of 600–800 m/s associated to hard layer (unsaturated) and the third layer has velocity values greater than 1,000 m/s which are associated with a saturated layer. Comparing the results of the two geophysical methods, we can summarize that the resistivity method has the limitation of a lower depth of investigation with respect to seismic refraction. On the other hand, seismic refraction is unable to resolve well the subsurface features and has much less resolution compare to electrical resistivity. However, both data sets showed their validity and reliability when correlated together to determine the subsurface features of studied site. In this paper, 838 of data points correlated. This can be showed by relationship resistivity-velocity (Figure 5).

To ascertain the resistivity-velocity relationship for the reconstructed models, we have plotted in Figure 5 these two physical parameters for all the coincident sampling positions (838 data points) in the merge-optimization models. This plot shows a structural feature that did not come out in the results of optimization. The merge-optimization results suggest distinct subgroupings in the near-surface volume. From the correlated data points, we found an L-shaped (A-B)

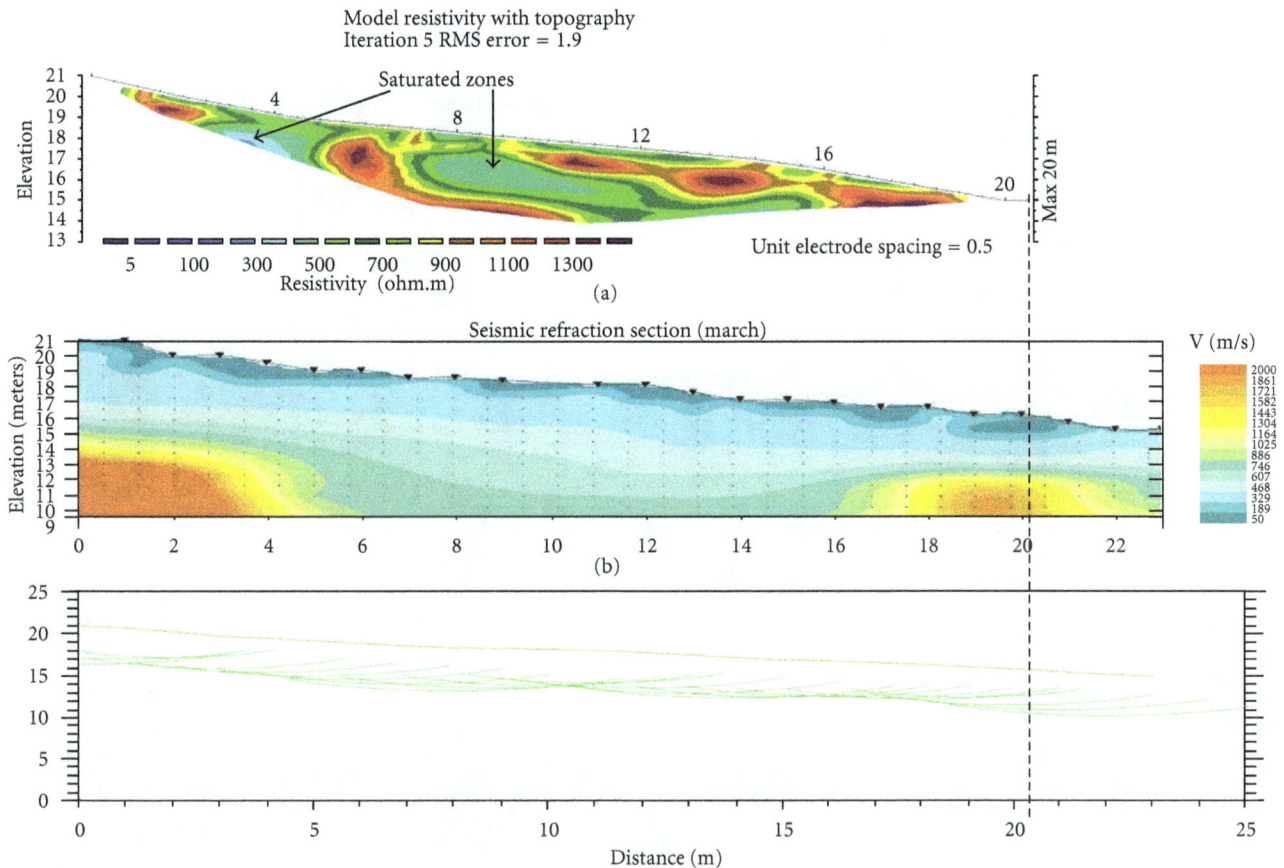

FIGURE 3: March infield's test results for optimal 2D merge resistivity and velocity models (a) resistivity model, (b) velocity model, and depth section with arcs.

trend that may be a consequence of a water table or natural divide between consolidated and unconsolidated materials.

In Figure 6, notice that the seismic rays are critically refracted at the top of the structural units mapped as layer 2, defining the possible boundary upper dry layer and moist layer where water table exist through it effect of water seepage from the top of slope. Merge-optimization thus appears to offer a tool for improved characterization of complex near-surface environments. To address the question: if the contention that electrical resistivity and seismic refraction imaging are both influenced by the same geological structures is correct, and there is some structural control on the distribution of physical properties in the subsurface, can accurately deduce any physical feature from merge structural imaging? It is possible that Figures 5 and 6 have physical implications from the external factors. Thus, before this merge-optimization method was applied, we have did a study on the physical characterizations of the soils (moisture content W, cohesion C', friction angle $\emptyset'$, porosity n, void ratio e, and saturation degree S) at various locations along the survey lines at different periods [5, 6].

4. Conclusion

The incorporation of the cross-gradients criterion in 2D optimization leads to a geologically meaningful solution by improving the near-surface structural conformity between the velocity and resistivity imaging, without forcing or assuming the form of the relationship between two geophysical methods which have their own advantage and limitation. The cross-gradients criterion also allows detecting subsurface features to which only one of the geophysical techniques is sensitive, leading to a better structural characterization. The application of these geophysical techniques combination of 2D optimization with cross-gradients to the data collected from seismic refraction and electrical resistivity field surveys has led to an improved characterization of the near-surface material and features. This study suggests that unconsolidated (possibly unsaturated and saturated) materials may be subclassified on the basis of the resistivity-velocity relationship revealed from the application of merge-optimization method. The cross-gradients approach adopted in this paper can also be used for 3D problems and for any combination of independent geophysical methods.

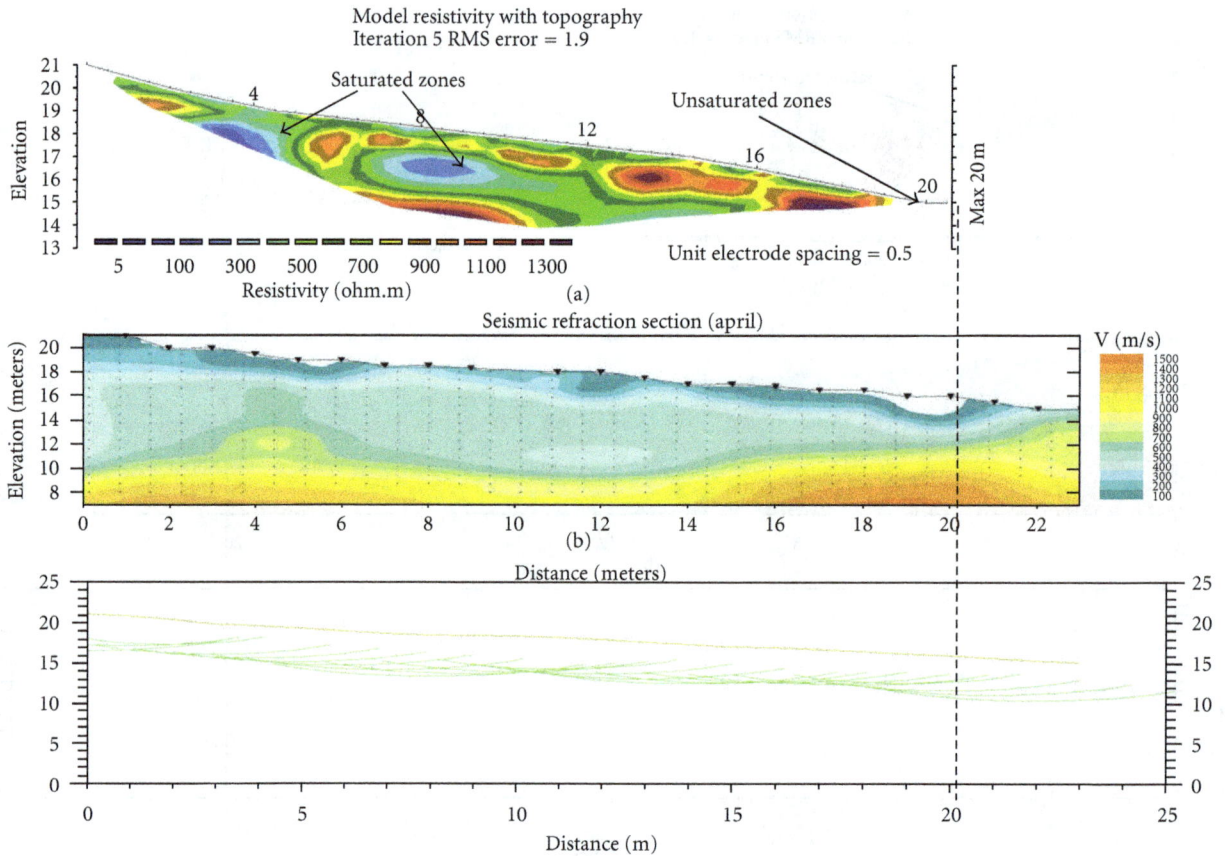

FIGURE 4: April infield's test results for optimal 2D merge resistivity and velocity models (a) resistivity model, (b) velocity model, and depth section with arcs.

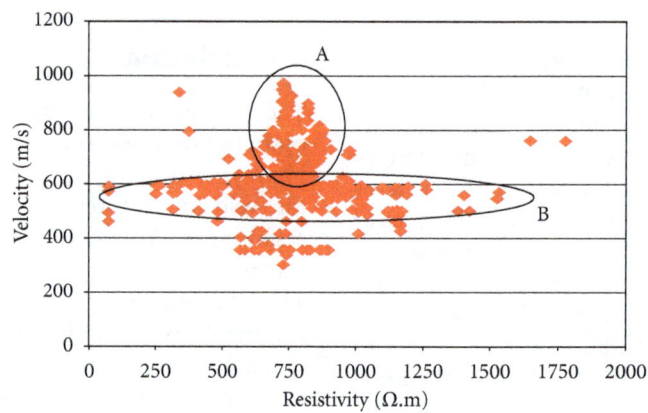

FIGURE 5: The results show an L-shaped (A-B) trend that may be a consequence of a water table or natural divide between consolidated and unconsolidated materials.

Acknowledgments

A. A. Bery would like to thank Rosli Saad, Mydin Jamal, and Nordiana Mohd Muztaza for their assistance in giving advice and data acquisition. The author also would like to thank and give appreciation to Mr. Jeff Steven and Mdm. Eva Diana for their support and advice. Lastly the author would like to thank and to express profound appreciation to anonymous reviewers for insightful comments that helped improved the quality of this paper.

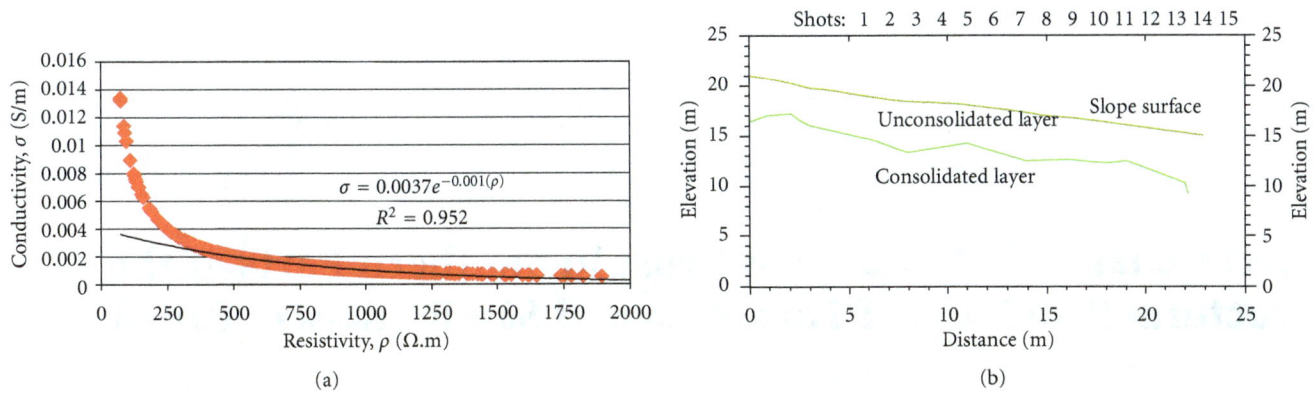

FIGURE 6: (a) Interpretative model showing empirical correlation between conductivity (S/m) and resistivity (Ω.m) is found as $\sigma = 0.0037e - 20.001(\rho)$ and strong relationship of data sets regression coefficient which is 95.2%. (b) Near-surface structure of the studied site can be recognized as a two-layer case where the upper layer is unconsolidated (weathered surface materials) and the lower layer is made of consolidated materials.

References

[1] S. K. Pullammanappallil and J. N. Louie, "A generalized simulated-annealing optimization for inversion of first-arrival times," *Bulletin of the Seismological Society of America*, vol. 84, no. 5, pp. 1397–1409, 1994.

[2] M. H. Loke and R. D. Barker, "Least-squares deconvolution of apparent resistivity pseudosections," *Geophysics*, vol. 60, no. 6, pp. 1682–1690, 1995.

[3] J. E. Vidale, "Finite-difference calculation of traveltimes in three-dimensions," *Geophysics*, vol. 55, pp. 521–526, 1990.

[4] C. A. Zelt and P. J. Barton, "Three-dimensional seismic refraction tomography: a comparison of two methods applied to data from the Faeroe Basin," *Journal of Geophysical Research B*, vol. 103, no. 4, pp. 7187–7210, 1998.

[5] A. A. Bery and R. Saad, "A clayey sand soil's behaviour analysis and imaging subsurface structure via engineering characterizations and integratedg geophysicals tomography methods," *International Journal of Geosciences*, vol. 3, no. 1, pp. 93–104, 2012.

[6] A. A. Bery and R. Saad, "Tropical clayey sand soil's behaviour analysis and its empirical correlations via geophysics electrical resistivity method and engineering soil characterizations," *International Journal of Geosciences*, vol. 3, no. 1, pp. 111–116, 2012.

10

Thermomagnetic Features of Crust in Southern Parts of the Structural Provinces of Tocantins and São Francisco, Brazil

Suze Nei P. Guimaraes and Valiya M. Hamza

Observatório Nacional, Rua General José Cristino 77, Rio de Janeiro, CEP 20921-400, RJ, Brazil

Correspondence should be addressed to Valiya M. Hamza; hamza@on.br

Academic Editors: G. Casula, Y.-J. Chuo, K. Maamaatuaiahutapu, and F. Monteiro Santos

In the present work we report results of a regional scale investigation of the thermal and magnetic characteristics of the crust in the southern sector of the geologic provinces of Tocantins and São Francisco, Brazil. Updated compilations of aeromagnetic and geothermal data sets were employed for this purpose. Use of such techniques as vertical derivative, analytic signal, and Euler deconvolution in analysis of aeromagnetic data has allowed precise locations of the sources of magnetic anomalies and determination of their respective depths. The anomalies in the Tocantins province are considered as arising from variations in the magnetic susceptibilities and remnant magnetizations of alkaline magmatic intrusions of the Tertiary period. The lateral dimensions of the bodies are less than 10 km, and these are found to occur at shallow depths of less than 20 km. On the other hand, the anomalies in the cratonic areas are related to contrasts in magnetic properties of bodies situated at depths greater than 20 km and have spatial dimensions of more than 50 km. Analysis of geothermal data reveals that the cratonic area is characterized by geothermal gradients and heat flow values lower when compared with those of the Tocantins province.

1. Introduction

The geotectonic provinces of São Francisco and Tocantins are often considered as the main structural units in the eastern part of the South American platform [1, 2]. Geological and geophysical studies performed to date have allowed mapping of structural and tectonic features near the surface of these structures, but little is known about the vertical distributions at depths in the crust.

One of the convenient ways of circumventing this difficulty is to evaluate aeromagnetic survey data and compare the results with those of geothermal studies. According to available information [3, 4] past attempts to integrate the results of aeromagnetic surveys were limited to incorporating standard corrections of technical operations. It became clear that a detailed review of aeromagnetic survey data is needed, focusing on the judicious use of corrections (leveling and microleveling, eliminating the effects of diurnal variation and the internal field) and use of advanced techniques of interpretation (vertical derivative, analytic signal, and Euler deconvolution). Also, "suture" techniques need to be

employed in integration of data from several surveys in this area in obtaining a coherent analysis of the subsurface geological significance of the results and its correlations with structural features. Another parallel objective of this work has been comparison of maps of the residual anomalies of magnetic and the geothermal fields in the area of study as a means of obtaining complementary information on the thermal state of the upper crust in the study area [5].

2. Geological Characteristics of the Study Area

The study area comprises the southern parts of the structural provinces of Tocantins and San Francisco, located in the east-central parts of the South American platform. The map in Figure 1 illustrates the geological characteristics of the study area. Basement rocks of Archean age outcrop mainly in the southern and southeastern parts. Isolated blocks have also been identified in the northwestern parts. Most of the remaining area is covered by metamorphic complexes of Proterozoic age. Phanerozoic sediment cover occurs as a set

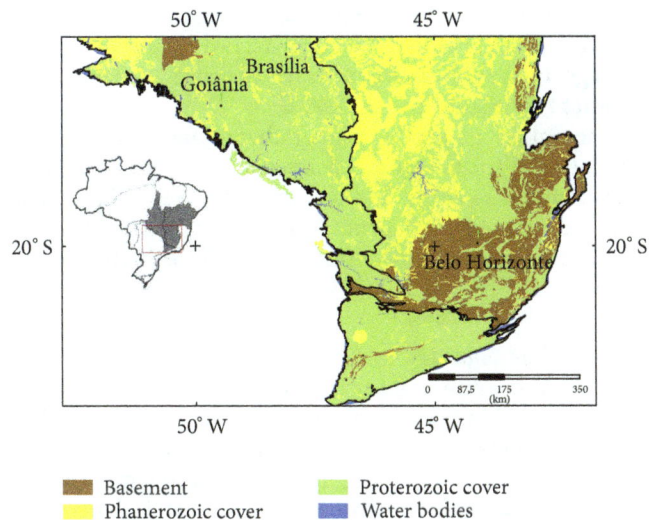

FIGURE 1: Simplified geological map of the southern São Francisco and Tocantins provinces (modified after IBGE, 2010).

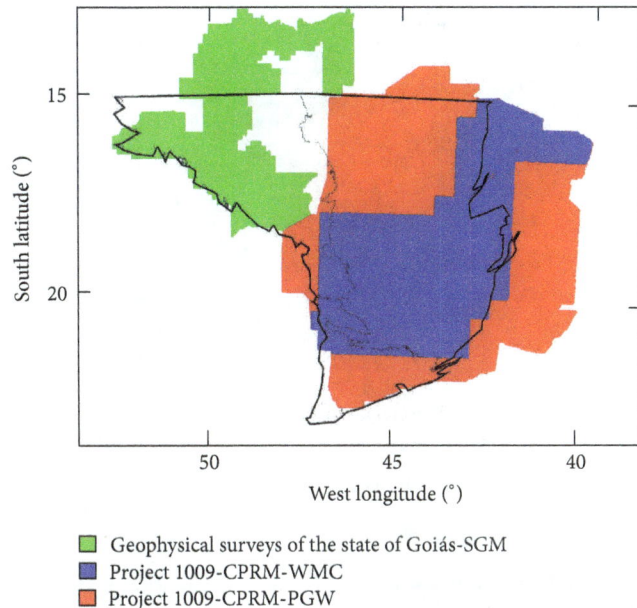

FIGURE 2: Areas covered in the three different aeromagnetic surveys of the study area.

of discontinuous blocks in the east-central parts and also in the western part of the study area.

The São Francisco structural province is a tectonically stable region, where the main geological structures have ages ranging from Mesoarchean to Paleoproterozoic [6–8]. The southern limits of this province are determined by the fold belts of the Brazilian orogeny. The Tocantins structural province is situated in the western segment of the study area. The crustal blocks in this province are composed of granitic-gneissic-granulite compartments (Goiás Median Massif) of Archean to Neoproterozoic age and metasedimentary rocks (Araguaia, Brasília, and Uruaçu belts) of Proterozoic age. A characteristic feature of this province is the presence of a significant number of alkaline magmatic intrusions of the Tertiary period. There are also indications of the occurrence of large number of nonoutcropping intrusions [9].

According to results of seismic studies [10] the thickness of the upper crust in the region of study area is 40 km in the São Francisco craton and 32 to 42 km in the Brasília belt and adjacent areas. In the Ribeira belt, located in the southern part of the study area, the crustal thickness is found to be 37 km.

3. Materials and Methods

3.1. Aeromagnetic Database. The data used in the present work is based on results of aeromagnetic surveys carried out in the study area. The earliest of these, known as Project 1009, refer to surveys carried out during 1971 and 1972 as part of Geophysical Accord between Brazil and Germany. The results of this survey, acquired initially in analog form, were later digitized by Paterson, Grant & Watson Ltd. (PGW) and Western Mining Company (WMC). These digital data sets, made available for academic research by Company of Mineral Resources Research (CPRM), were acquired for purposes of the present work. According to available information the surveys were carried out during the years of 1971 and 1972.

The nominal flight height for data acquisition was 150 m and spacing of flight lines was 2 km. The preferred direction of survey lines of this project was east-west. Integration of these data sets demanded a tedious and time-consuming work of incorporating the needed corrections and unification in the early stages of processing. The areas covered by these surveys are shown in Figure 2.

The database of the present work also includes results of geophysical surveys of the state of Goiás, carried out during the years of 2004 and 2005. This data set, made available for academic research by Superintendent of Geology and Mining (SGM) of the state of Goiás, was also acquired in the present project. High-resolution equipment for data acquisition was used in this survey, which also employed modern positioning techniques such as GPS (global positioning system). The flight height is 100 m, with line spacing of 500 m. The dominant direction of survey lines was north-south. Data was also acquired along control lines with spacing of 5 km, along directions perpendicular to the main survey lines. Data with this quality are considered as belonging to high-resolution group.

3.2. Data Processing Techniques. The bulk of aeromagnetic data processing in the present work was carried out using the computer software package Geosoft Oasis Montaj. Built-in facilities of this package allowed homogenization of the database and elaboration of maps, making use of standard procedures of interpolation, gridding,c and plotting methods [11]. The package was also used for setting geographic coordinates, which in the present work are referenced to as the datum SAD-69. The results obtained at this stage of data processing were found to be in reasonable agreement with the

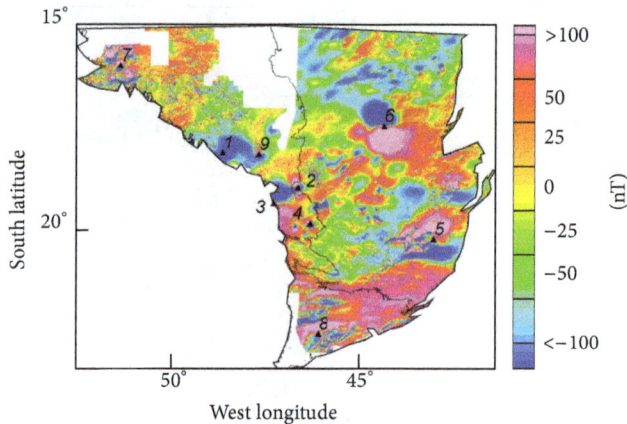

FIGURE 3: Residual magnetic field of the study area.

maps already published by CPRM in 2003 [3] and by SGM in 2009 [4].

The next step in data processing has been removal of the International Geomagnetic Reference Field (IGRF) from the records of the total field. This has been an important part of data processing in view of the significant time differences in data acquisition and the large area extent of the surveys. The main problem with the published models of IGRF is that it is meant to serve all areas of geomagnetism and, as such, is not specifically tailored to the needs of the exploration community. Hence two different approaches were adopted. In processing data from Project 1009, acquired during the decade of 1970, we used the algorithm developed by the Geomagnetism Group at the National Observatory (Rio de Janeiro, Brazil). This algorithm, which makes use of historic records of magnetic data from permanent stations and several hundred temporary stations in Brazil, provides a better representation of local magnetic field. It has often been considered as providing good correlation with low latitude geomagnetic data. As for the data of the geophysical survey of Goias, acquired during the period of 2004 to 2005, we used the built-in IGRF reference fields, available in the software package Geosoft Oasis Montaj.

The removal of the field originating in the core of the Earth and the external variations occurring mainly in the ionosphere and the sun allows derivation of the field of crustal origin. This crustal field was corrected for *heading* and incorporated into an ensemble of corrected data sets. Filters were then applied to correct directional trends. In addition, leveling and microleveling corrections were carried out to eliminate distortions of flight lines [12].

3.3. Geothermal Data Base. Current understanding of the thermal field of subsurface layers in the geologic provinces of Brazil is based mainly on results of heat flow studies, analysis of physical and chemical characteristics of thermal springs, and assessments of geothermal resources, carried out since 1970 [13–18]. Parts of this information have been gathered and organized as a modern database by the Geothermal Laboratory of the National Observatory in Rio de Janeiro.

This database refers to geothermal measurements at 1212 localities in the Brazilian territory. According to this compilation geothermal measurements have been carried out at 135 localities in the study area.

The values of geothermal gradients and heat flow in this database include those derived from results of direct measurements as well as indirect estimates. Several methods have been employed depending on the nature of primary geothermal data [15, 19]. Among these, we highlight the following.

3.3.1. Conventional: CVL [20]. This is the classical method in which temperature gradient is calculated by least square fit to borehole log data, within selected depth ranges. The use of this method is best suited for cases where the local rock formations are laterally homogeneous and have constant thermal properties. It is also important that influences of processes (such as drilling disturbances, groundwater flows, local geologic structures, and climate change effects) which induce thermal perturbations at the site are absent.

3.3.2. Bottom Hole Temperature: BHT [21]. This method makes use of bottom-hole temperatures measured in deep oil wells for determination of apparent geothermal gradients and heat flow. The data are usually corrected for disturbing effects of drilling activity.

3.3.3. Geochemical Estimates: GCL [22]. The principle of this method is based on the argument that abundances of certain chemical elements dissolved in thermal waters provide indirect information on the temperatures of geothermal reservoirs. This data along with inferences on subsurface geological structures allows estimates of geothermal gradients and heat flow at sites of thermal springs.

Hamza and Muñoz, 1996 [15], and Gomes and Hamza, 2005 [19], have provided detailed information on the use of these methods and assessment of quality and reliability of the results obtained. Standard methods are employed in interpolation and gridding of geothermal data and in deriving maps of geothermal gradients and heat flow. A major problem with the geothermal database is its relatively low data density, a limitation arising from the well-known practical difficulties in acquiring field data. Thus geothermal maps are generally useful only for identifying regional trends, but lack resolution needed for identification of small-scale structures and near surface features.

4. Results Obtained

4.1. Residual Magnetic Field. On conclusion of the correction procedures of the data sets, described in Section 3.2, techniques of *suture* were employed in unifying the corrected data sets [23]. This technique unites the data sets with a minimum of distortion along their contact boundaries, making smooth interpolations of trends present in their interior parts. The results obtained at this stage were used in deriving the map of the residual magnetic field (also called anomalous magnetic field) of the study area, illustrated in Figure 3. Referring to

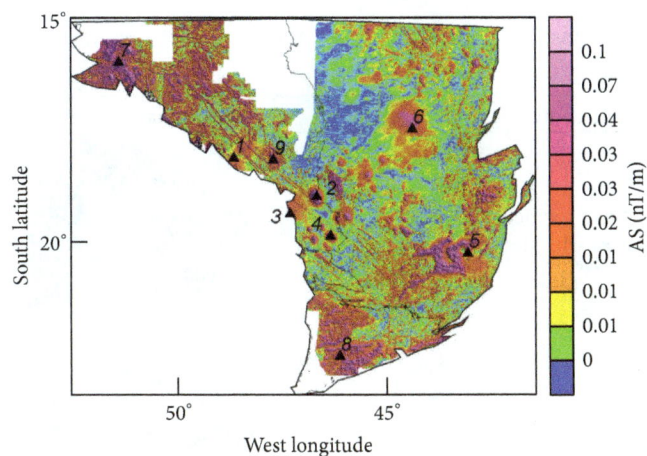

FIGURE 4: Analytic signal derived from the crustal magnetic field. Numbers refer to anomalies in the residual field, indicated in the map of Figure 3.

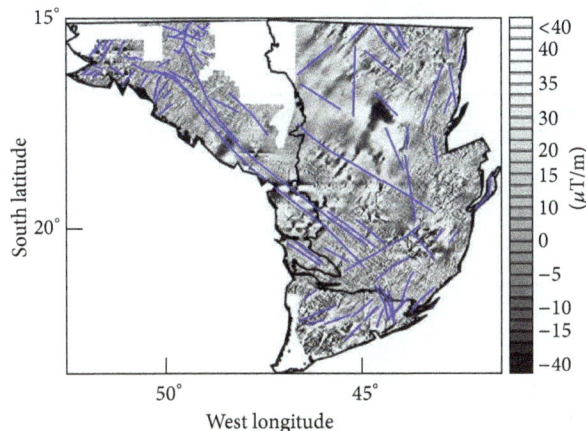

FIGURE 5: Map of the vertical derivative of crustal magnetic field. The blue lines represent the lineaments inferred from the distribution of vertical derivatives.

the map of this figure we note that there are considerable variations in the intensity of the crustal field in the study area, with both negative (lower than 100 nT) and positive values (higher than 100 nT). Most of the large-scale variations with sharp changes between positive and negative values are associated with geologic structures at relatively shallow depths. Locations of nine such structures are indicated in the map of Figure 3.

Comparison with the geology map of Figure 1 reveals that most of the large magnitude positive anomalies are situated in areas of outcrops of basement rocks. A notable exception to this trend is the SE-NW belt of anomalies cutting across the São Francisco province. This is the preferred direction of many of the structural features in the South American platform. According to geological studies, metamorphic belts along this direction are marked by expressions of intense deformation, crustal accretion, and reworking [24] which explain this strong magnetic manifestation. Hence the possibility that a belt of basement rocks occurring beneath the thin cover of Proterozoic rocks contributes to the occurrence of such anomalies cannot be ruled out. Most of the negative anomalies are found to occur along parallel belt, cutting across the southern part of Tocantins province and northern part of the São Francisco province. Some isolated negative anomalies also occur in the southeastern parts of the São Francisco province. Additional information on the tectonic context of such structures is given in the list of Table 1, which also includes maps of the individual anomalies.

4.2. *Analytic Signal*. Another technique used in interpretation of aeromagnetic data is the analytic signal [25–27], which is basically the magnitude of the second derivative in the three directions of the magnetic field. In practice, the analytical signal is regarded as the best tool for locating the edges of bodies that have magnetic contrast. When applied to the residual field magnetic anomaly the responses highlight the surface boundaries of geological bodies with contrasts in

magnetic properties relative to the surrounding rocks. Hsu, 2002 [28], suggests use of second and even higher order derivative to better highlight the bodies. However, higher orders enhance the noise leading to unrealistic solutions, especially when dealing with low quality data sets.

The geographic distribution of analytic signal for the data of the present work is illustrated in the map of Figure 4. It is clear that there are substantial variations in the magnitude of the analytic signal, the maximum value being 0.1 nT/m. As expected, most of the variations occur in localities of residual magnetic anomalies, indicated in the map of Figure 3 and listed in Table 1. These are considered as indicative of bodies with contrasts in magnetic properties and located at relatively shallow depths. A zone of low values of the analytic signal is found to occur in the north-central parts of the study area. But its interpretation is difficult in view of the lack of suitable data in the adjacent area to the west.

4.3. *Vertical Derivatives*. Following the common practice in interpretation of aeromagnetic data the technique of spatial derivative (see, e.g., Gunn, 1975 [29]) has been employed in the present work. The objective has been to highlight the features associated with high frequency variations in the residual magnetic field and mitigate features associated with low frequency variations. The map of vertical derivative of the study area is illustrated in Figure 5.

Note that the features in this map appear as linear magnetic minima, edges, and dislocations in the local field. Some of them appear as ridge-like features in the map of vertical derivatives. The geographic distributions of such ridges are found to be linear in many cases and following standard practice have been classified as magnetic lineaments. The mechanisms responsible for the connectivity of magnetic lineaments are believed to be the changes in magnetic properties and brittle behavior of the geological formations from the action of local tectonic forces [30, 31].

A total of 67 lineaments were identified. Some of the major regional lineaments are indicated in the map of Figure 5. Most of lineaments are found to be predominantly

TABLE 1: Localities and tectonic context of prominent anomalies of the residual magnetic field. Color codes of residual anomalies are given in the inset of Figure 3.

Identification	Locality	Tectonic context	Residual anomaly
1	Caldas Novas	Tocantins province. Basic intrusion in area with thermal springs.	
2	Patrocínio	Tocantins province. Alkaline intrusion with remnant magnetism.	
3	Pedrinópolis	Tocantins province. Intense anomaly in metamorphic fold belt.	
4	Araxá	Tocantins province. Alkaline intrusion with remnant magnetism.	
5	Itabirito	São Francisco province. Area with extensive iron ore deposits.	
6	Pirapora	São Francisco province. Basic intrusion of the Espinhaço sequence.	
7	Santa Fé de Goiás	Tocantins province. Located in the Magmatic Arc of Goiás.	
8	Poços de Caldas	Tocantins province. Area with extensive iron ore deposits.	

TABLE 1: Continued.

Identification	Locality	Tectonic context	Residual anomaly
9	Catalão	Tocantins province. Alkaline intrusion with remnant magnetism.	

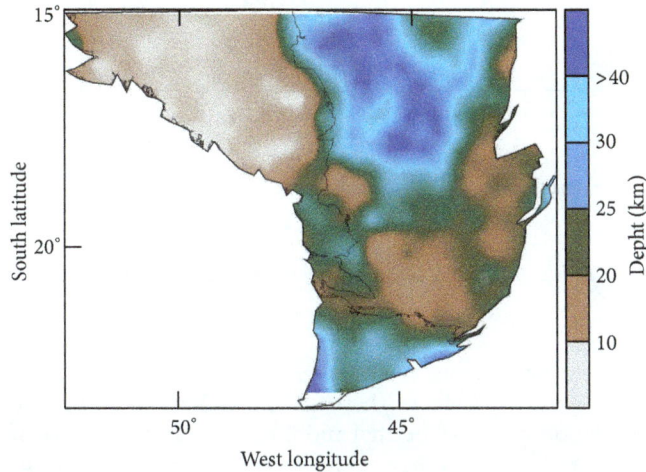

FIGURE 6: Depth of the anomalous magnetic sources from Euler deconvolution results.

in the NE-SW direction, coincident with the preferred directions identified in the geological surveys (e.g., Blum [32]). These are found to cross the two structural provinces of Tocantins and São Francisco, which has also been identified in geological studies (e.g., Heilbron and Machado [33]).

4.4. Euler Deconvolution. The Euler deconvolution technique [34] employed in this work has the objective of extracting information about the depths of magnetic sources. The result is independent of the direction and inclination of the main magnetic field and the orientation of the magnetic sources. Thus, the method is relatively insensitive to small-scale distortions of the field. A total magnetic anomaly (T) without correction of regional values produced by a set of three-dimensional sources satisfies the homogeneous field equation of Euler:

$$(x - x_0)\frac{\partial}{\partial x}T + (y - y_0)\frac{\partial}{\partial y}T + (z - z_0)\frac{\partial}{\partial z}T = -\eta T, \quad (1)$$

where x_0, y_0, and z_0 are the initial coordinate positions of the anomalous source and η is the structural index as defined in [34].

Following the practice adopted in the literature the variations in the degree of homogeneity of the field are associated with a set of structural indices, which specifies the type of magnetic source. According to Thompson [26] a structural index of zero represents contact between different types of

rocks, while indices of 0.5 and 1 represent, respectively, linear features such as faults and dykes. Also, the structural indices 2 and 3 are indicative of cylindrical and spherical structures, respectively. The success of the method is evaluated in a qualitative way, through accumulation of solutions associated with the different indices.

In an attempt to minimize biased interpretations in determining the depths, the following guidelines were adopted in the use of this technique.

(1) The established limit for uncertainty in the depth value, calculated by the Euler equation, is 10%.

(2) The magnetic anomalies modeled are dipolar features, consistent with the current knowledge of deep geological structures. Since the focus of the study is on regional scale variations, we follow the practice of Hsu [28] and use the structural index three in the solution of Euler equation.

(3) The size of the interpolation window of the program used to calculate the Euler deconvolution is 15 km. According to standard practice this value of window size corresponds to sources of 3 to 6 km extension.

The results obtained indicating the depths of sources are illustrated in the map of Figure 6. In the present case, the map derived is based on results of using the *krigrid* interpolation method [23], with a grid spacing of 0.5 degrees (~55.7 km). The grid-averaged depths of anomalous magnetic sources are in the ranges of 10 to 50 km, as indicated in the color scale of this figure.

Note that the depths of bodies in the north-west sector of the Tocantins province and in the southeastern potion of São Francisco craton are in the range 10 to 20 km. On the other hand, the depths of bodies in northern parts of the São Francisco province and in the southern parts of the Tocantins province are in the range of 25 to 50 km. There are indications that the systematic differences in the depths of magnetic sources arise from mechanisms related to the thermal regime of the crust.

4.5. Maps of Geothermal Gradients and Heat Flow. The geothermal data compiled in the present work has been employed in interpolation of values with grid spacing of 0.5 degrees (55.7 km). The maps of geothermal gradients and terrestrial heat flow derived from the gridded data sets are illustrated in Figures 7(a) and 7(b). Also indicated in these maps are localities of geothermal measurements.

The geothermal gradient map (Figure 7(a)) reveals a zone of relatively high values (greater than 18°C/km) along

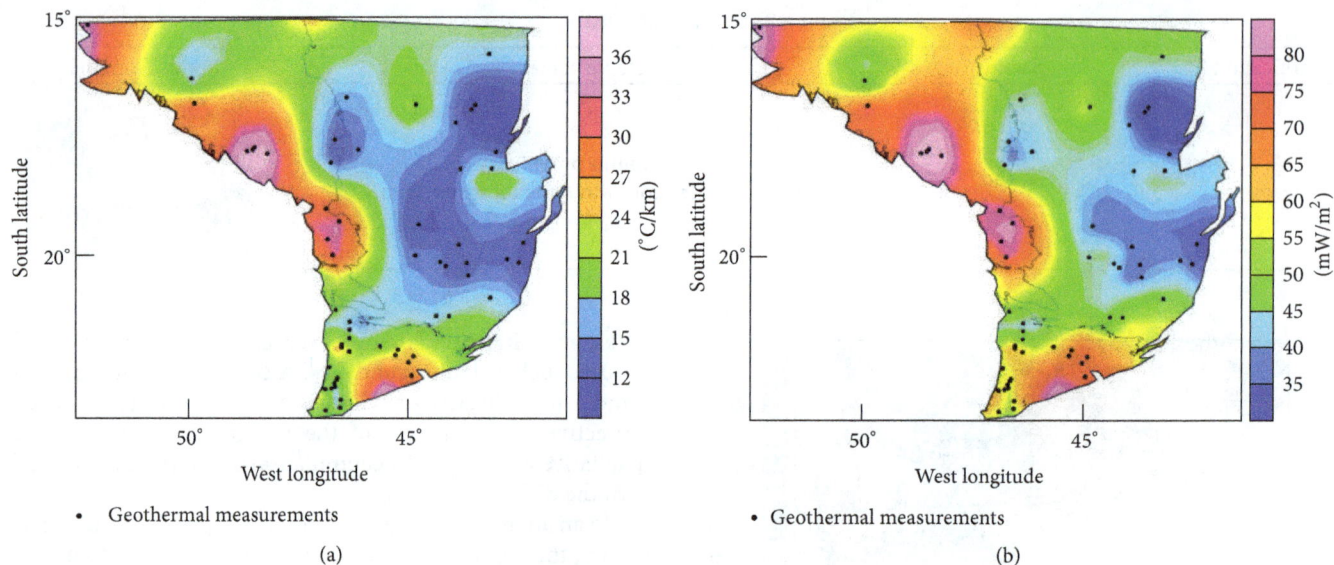

FIGURE 7: Maps of geothermal gradient (a) and heat flow (b) in the study area. The black dots indicate localities of geothermal measurements.

a narrow belt in the western and extreme southern parts of the Tocantins province. On the other hand, the São Francisco province is characterized by relatively low thermal gradients in the range 10 to 18°C/km. The heat flow map of the study area shown in Figure 7(b) reveals features similar to those of the geothermal gradient. Thus, heat flow values in excess of 50 mW/m^2 occur along narrow belts in the western parts of the Tocantins province, whereas most of the São Francisco province is characterized by heat flow values of less than 50 mW/m^2. Comparative analysis of features discernible in Figures 6 and 7 points to an inverse correlation between thermal and magnetic characteristics of the study area. Thus, areas of high geothermal gradients and heat flow are characterized by sources of magnetic anomalies at relatively shallow depths and vice versa.

5. Discussion and Conclusions

The present work constitutes a regional scale investigation of the thermal and magnetic characteristics of the crust in the southern sector of the geologic provinces of Tocantins and São Francisco, Brazil. Use of such techniques as vertical derivative, analytic signal, and Euler deconvolution in analysis of aeromagnetic data has allowed a better understanding of the sources of magnetic anomalies and estimates of their respective depths. There are indications that anomalies in the Tocantins province arise from variations in the magnetic susceptibilities and remnant magnetizations of alkaline magmatic intrusions of the Tertiary period. The lateral dimensions of the bodies are less than 10 km, and these are found to occur at shallow depths of less than 20 km. On the other hand, the anomalies in the cratonic areas are related to contrasts in magnetic properties of bodies situated at depths greater than 20 km and have spatial dimensions of more than 50 km. Analysis of geothermal data reveals that the cratonic area is characterized by geothermal gradients

and heat flow values lower when compared with those of the Tocantins province. There are indications of an inverse correlation between thermal and magnetic characteristics of the study area. Thus, areas of high geothermal gradients and heat flow are characterized by sources of magnetic anomalies at relatively shallow depths and vice versa.

Acknowledgments

The authors thank the Mineral Resources Research Company (CPRM), the National Commission of Nuclear Energy (CNEN), and the Superintendent of Geology and Mining of Goiás (SIC/SGM-GO) for the donation of the geophysical data, and the Coordenação de Aperfeiçoamento de Pessoal de Nível Superior-CAPES for financial support.

References

[1] F. F. M. de Almeida, "O Cráton do São Francisco," *Revista Brasileira De Geociências*, vol. 7, pp. 349–364, 1977.

[2] F. F. M. de Almeida, Y. Hasui, B. B. Brito Neves, and R. A. Fuck, "Brazilian structural provinces: an introduction," *Earth-Science Reviews*, vol. 17, pp. 1–19, 1981.

[3] CPRM-Serviço Geológico do Brasil, "Geologia, Tectônica e Recursos Minerais do Brasil: texto, mapas & SIG," 2003.

[4] SGM-Superintendência de Geologia e Mineração, "Geologia do Estado de Goiás e Distrito Federal," *Goiânia*, p. 143, 2008.

[5] C. H. Alexandrino and V. M. Hamza, "Estimates of heat flow and heat production and a thermal model of the São Francisco craton," *International Journal of Earth Sciences*, vol. 97, no. 2, pp. 289–306, 2008.

[6] F. F. Alkmim and S. Marshak, "Transamazonian orogeny in the Southern São Francisco Craton Region, Minas Gerais, Brazil: evidence for Paleoproterozoic collision and collapse in the Quadrilátero Ferrífero," *Precambrian Research*, vol. 90, no. 1-2, pp. 29–58, 1998.

[7] M. A. Santos Pinto, *Le Recyclage de la Crôute Continentale Archéene: Exemple du Bloc du Gavião—Bahia, Brasil [Thèse de doctorat]*, Université de Rennes I, França, 1996.

[8] W. Teixeira, P. Sabaté, J. S. F. Barbosa, C. M. Noce, and M. A. Carneiro, "Archean and Paleoproterozoic Tectonic evolution of the São Francisco Craton, Brazil," in *Proceedings of the 31st International Geologic Congress*, U. G. Cordani, E. J. Milani, A. Thomas Filho, and D. A. Campos, Eds., pp. 101–137, Rio de Janeiro, Brazil, 2000.

[9] M. M. Pimentel and R. A. Fuck, "Neoproterozoic crustal accretion in central Brazil," *Geology*, vol. 20, no. 4, pp. 375–379, 1992.

[10] G. S. L. A. de França, *Estrutura da Crosta no sudeste e centro-oeste do Brasil, usando função do receptor Local: São Paulo [Ph.D. thesis]*, Instituto de Astronomia e Geofísica, Universidade de São Paulo, 2003.

[11] I. C. Briggs, "Machine contouring using minimum curvature," *Geophysics*, vol. 39, pp. 38–48, 1974.

[12] B. S. R. Minty, "Simple micro-leveling for aeromagnetic data," *Exploration Geophysics*, vol. 22, pp. 591–592, 1991.

[13] V. M. Hamza, S. M. Eston, and R. L. C. Araujo, "Geothermal energy prospects in Brazil: a preliminary analysis," *Pure and Applied Geophysics*, vol. 117, no. 1-2, pp. 180–195, 1978.

[14] V. M. Hamza, S. M. Eston, R. L. C. Araujo, I. Vitorello, and N. Ussami, "Brazilian Geothermal Data Collection-Series I," Publication no. 1109, Instituto de Pesquisas Tecnológicas do Estado de São Paulo, São Paulo, Brazil, 1978.

[15] V. M. Hamza and M. Muñoz, "Heat flow map of South America," *Geothermics*, vol. 25, pp. 599–646, 1996.

[16] S. J. Hurter, *Aplicação dos geotermômetros químicos em águas de fontes brasileiras na determinação de fluxo geotérmico [M.Sc. thesis]*, Instituto de Astrononia e Geofísica, Universidade de São Paulo, São Paulo, Brazil, 1987.

[17] S. J. Hurter, S. M. Eston, and V. M. Hamza, "Brazilian Geothermal Data Collection Series 2-Thermal Springs," Publication no. 1233, Instituto de Pesquisas Tecnológicas do Estado de São Paulo s/a-IPT, 1983.

[18] I. Vitorello, V. M. Hamza, and H. N. Pollack, "Terrestrial heat flow in the Brazilian highlands," *Journal of Geophysical Research*, vol. 85, no. 7, pp. 3778–3788, 1980.

[19] A. J. L. Gomes and V. M. Hamza, "Geothermal gradient and heat flow in the state of Rio de Janeiro," *Brazilian Journal of Geophysics*, vol. 23, pp. 325–347, 2005.

[20] A. E. Beck, "Techniques of measuring heat flow on land," in *Terrestrial Heat Flow*, W. H. K. Lee, Ed., Geophysical Monograph Series 8, pp. 24–57, AGU, Washington, DC, USA, 1965.

[21] H. Da Silva Carvalho and V. Vacquier, "Method for determining terrestrial heat flow in oil fields," *Geophysics*, vol. 42, no. 3, pp. 584–593, 1977.

[22] C. A. Swanberg and P. Morgan, "The linear relation between temperatures based on the silica content of groundwater and regional heat flow: a new heat flow map of the United States," *Pure and Applied Geophysics*, vol. 117, no. 1-2, pp. 227–241, 1978.

[23] A. G. Journel and J. C. Huijbrogts, *Mining Geostatistics*, Academic Press, London, UK, 1978.

[24] I. Endo and R. Machado, "Reavaliação e novos dados geocronológicos (Pb/Pb e K/Ar) da região do Quadrilátero Ferrífero e adjacências," *Geologia USP*, vol. 2, no. 1, pp. 23–40, 2002.

[25] M. N. Nabighian, "The analytic signal of two dimensional magnetic bodies with polygonal cross-section: its properties and use for automated anomaly interpretation," *Geophysics*, vol. 37, no. 3, pp. 507–517, 1972.

[26] D. T. Thompson, "EULDPH: a new technique for making computer-assisted depth estimates from magnetic data," *Geophysics*, vol. 47, pp. 31–37, 1982.

[27] R. J. Blakely and R. W. Simpson, "Approximating edges of source bodies from magnetic or gravity anomalies," *Geophysics*, vol. 51, no. 7, pp. 1494–1498, 1986.

[28] S. K. Hsu, "Imaging magnetic sources using Euler's equation," *Geophysical Prospecting*, vol. 50, pp. 15–25, 2002.

[29] P. J. Gunn, "Linear transformation of gravity and magnetic fields," *Geophysical Prospecting*, vol. 23, no. 2, pp. 300–312, 1975.

[30] P. H. O. Costa, A. R. F. Andrade, G. A. Lopes, and S. L. Souza, "Projeto Lagoa Real. Mapeamento Geológico, 1:25.000," Nuclebrás/CBPM, Salvador, Bahia, 1985.

[31] L. S. Osako, *Estudo do potencial mineral do depósito uranífero de Lagoa Real, BA, com base em dados geológicos, aerogeofísicos e sensoriamento remoto [M.Sc. thesis]*, Instituto de Geociências, Universidade Estadual de Campinas, 1999.

[32] M. Blum, *Processamento e Interpretação de Dados de Geofísica Aérea no Brasil Central e sua aplicação à Geologia Regional e à Prospecção Mineral [Ph.D. thesis]*, Instituto de Geociências, Universidade de Brasília, 1999.

[33] M. Heilbron and N. Machado, "Timing of terrane accretion in the Neoproterozoic-Eopaleozoic Ribeira Orogen (se Brazil)," *Precambrian Research*, vol. 125, no. 1-2, pp. 87–112, 2003.

[34] A. B. Reid, J. M. Allsop, H. Granser, A. J. Millett, and I. W. Somerton, "Magnetic interpretation in three dimensions using Euler deconvolution," *Geophysics*, vol. 55, no. 1, pp. 80–91, 1990.

11

Heat Flow in the Campos Sedimentary Basin and Thermal History of the Continental Margin of Southeast Brazil

Roberta A. Cardoso[1,2] and Valiya M. Hamza[1]

[1] *Observatório Nacional-ON/MCT, Rio de Janeiro, Brazil*
[2] *Superintendence for Petroleum and Gas/DPG, Empresa de Pesquisa Energética (EPE), Rio de Janeiro, Brazil*

Correspondence should be addressed to Valiya M. Hamza; hamza@on.br

Academic Editors: E. Liu and A. Tzanis

Bottom-hole temperatures and physical properties derived from geophysical logs of deep oil wells have been employed in assessment of the geothermal field of the Campos basin, situated in the continental margin of southeast Brazil. The results indicate geothermal gradients in the range of 24 to 41°C/km and crustal heat flow in the range of 30 to 100 mW/m^2 within the study area. Maps of the regional distributions of these parameters point to arc-shaped northeast-southwest trending belts of relatively high gradients and heat flow in the central part of the Campos basin. This anomalous geothermal belt is coincident with the areas of occurrences of oil deposits. The present study also reports progress obtained in reconstructing the subsidence history of sedimentary strata at six localities within the Campos basin. The results point to episodes of crustal extension with magnitudes of 1.3 to 2, while extensions of subcrustal layers are in the range of 2 to 3. Thermal models indicate high heat flow during the initial stages of basin evolution. Maturation indices point to depths of oil generation greater than 3 km. The age of peak oil generation, allowing for variable time scales for cooling of the extended lithosphere, is found to be less than 40 Ma.

1. Introduction

The continental margin of southeast Brazil is composed of three major sedimentary basins: Espirito Santo, Campos, and Santos (e.g., [1–3]). The structural highs of Victoria and Cabo Frio are usually considered as representing the limits of the Campos basin in the continental platform region. The submerged parts of this basin have an area of approximately 100.000 km^2, while that of its continental part is only 500 km^2. The relative locations of these basin segments are indicated in the map of Figure 1. The Campos basin is the most prolific oil producing basin in the western South Atlantic, with more than sixty hydrocarbon accumulations, currently accounting for about 80% of Brazilian oil production. The locations of the main oil fields of the Campos basin are indicated as green colored areas in the map of Figure 1. Note that most of the known oil fields are located in the central rift zone of the Campos basin.

Most of the earlier studies on the geothermal field of the Campos basin have been based on results of bottom-hole temperature measurements in oil exploration wells (e.g., [4–6]). Ross and Pantoja [7] reported results of a work with focus on local subsurface temperature fields. Jahnert [8] presented maps of regional variations deep isotherms and thermal gradients in areas of oil fields and discussed their eventual correlations with regional scale structural features and gravity anomalies. Unfortunately, the details of data base employed by [7, 8] are not available for public domain analysis. In addition, very little information is available in these reports on the methods employed in data reduction and error analysis. In fact, Cardoso and Hamza [9] pointed out to potential systematic errors in earlier calculations of geothermal gradients, arising from the use of inappropriate values for ocean bottom temperatures. Such problems turned out to be major obstacles in carrying out independent assessments of earlier works on the thermal field of the Campos basin.

More recently, Gomes and Hamza [10] reported geothermal data for the adjacent coastal region of Rio de Janeiro. In addition, Vieira et al. [11] and Vieira and Hamza [12] reported

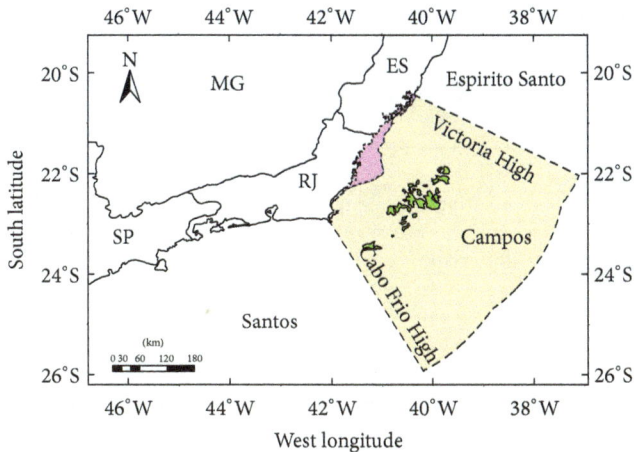

FIGURE 1: Locations of sedimentary basins (Santos, Campos, and Espirito Santo) in the continental margin of southeast Brazil. The ash and pink colored areas indicate the onshore and offshore segments of the Campos basin and the dotted lines their approximate limits. The green colored patches are locations of the main oil fields in the Campos basin.

estimates of heat flow values for the oceanic crust adjacent to the Campos basin. In this context, the focus of the present work is on the analysis of updated geothermal data for the offshore segment of the Campos basin that also take into consideration the supplementary data sets for the oceanic and continental areas. The purpose is to gain better insights into the thermal structure of the continental margin of southeast Brazil. The present work also examines the implications of the results of this new analysis in improving the assessments of thermal maturation indices of sedimentary strata in the Campos basin.

2. Geologic Context

According to interpretations of geologic data, the stratigraphic and structural evolution of the Campos basin has been strongly influenced by the breakup of the Pangaea super continent and formation of oceanic crust between South American and African lithospheric plates (e.g., [13–19]). Subsequent development of the basin seems to have been determined by stretching and thinning events of the local crust. There are indications of an initial short-period phase of fault-controlled subsidence, followed by a relatively long period of thermal subsidence (e.g., [20–23]). Several studies have been carried out on the structural framework of the continental shelf and upper slope of the Brazilian marginal basins, describing faults, structural alignments, and extension of fracture zones (e.g., [15, 17, 21, 24]). Ponte and Asmus [25] provide a summary of geologic knowledge regarding these features, emphasizing their structural-stratigraphic framework and tectonic evolution. According to these studies (see also [26, 27]), the main phases to be considered in the evolution of the Brazilian marginal basins are prerift, rift, transitional, and drift. During the prerift phase (Late Jurassic to Early Cretaceous), continental sediments were deposited

in peripheral intracratonic basins. In the rift phase (during Early Cretaceous), the breakup of the continental crust of the Gondwana continent gave rise to a central graben and rift valleys where lacustrine sediments were deposited. The transitional phase (during Aptian) developed under relative tectonic stability, when evaporitic and clastic lacustrine sequences were deposited. In the drift phase (during Albian to Holocene), a regional homoclinal structure developed, consisting of two distinct sedimentary sequences, a lower clastic-carbonate and an upper clastic.

Seismic surveys have identified the presence of structures in deep strata, which have been interpreted as intrusive magmatic activities of early Tertiary times. Several geologic studies (e.g., [27–29]) have pointed out evidences of the occurrence of sporadic magmatic activity during the Eocene and Santonian-Campanian. Volcanic building associated with intrusive and extrusive rocks has been observed that have local impacts on the mineralogical and sedimentary composition of the siliciclastic deposits [30]. Occurrences of subaerial and subaqueous volcanism are well identified and its characteristics are described in terms of seismic, log, and lithologic evidences [31]. However, no evidences have been found of volcanic or magmatic activities since Miocene times. Also, there are no reports of occurrences of hydrothermal fluid circulation processes in the ocean floor. The main rock formations of the Campos basin, episodes of tectonothermal activities, and their respective ages are listed in Table 1.

The hydrocarbon accumulations are distributed throughout the stratigraphic column of the basin, from Neocomian to Miocene. The reservoirs range from fractured basalts and porous bioclastic limestone (coquinas) in the Lagoa Feia Group to limestone and sandstone in the Macaé Group and sandstones in the Campos Group. Detailed geochemical analyses show that almost all the hydrocarbon accumulations discovered to date originate mainly from lacustrine calcareous black shale deposited in a closed Upper Neocomian lake system.

3. Temperature and Heat Flow Data

Information on geothermal characteristics of subsurface layers in the study area is derived almost exclusively on bottom-hole temperature (BHT) data acquired in boreholes and oil wells, in addition to sea floor temperature data in oceanic regions of the Campos basin. Given below are brief descriptions of the data sets compiled and procedures adopted in the determination of geothermal gradients and heat flow in the study area.

3.1. Sea Floor Temperatures. The oceanographic data sets provided by the Directorate of Hydrograph and Navigation (DHN) [32] of the Brazilian Navy have been useful in determining sea water temperatures in the coastal area of southeast Brazil. The relative accuracy of sensors used in deep sea water temperature measurements is often considered as better than 0.1°C.

Analysis of these data sets points to the presence of systematic trends of decreasing temperatures with depth of

TABLE 1: Sedimentary formations in the Campos and Santos basins (modified after [76]) and periods of main thermal activity.

Basins			Age (Ma)	Tectonothermal activity
Campos		Santos		
Group/formation	Member	Group/formation		
		Sepetiba		
Campos	Ubatuba	Marambaia	Tertiary	Depositional sequences controlled by thermal
Emboré		Iguape		subsidence and occurrence of intrusions associated
Campos	Tamoios	Itajai-Açu, Santos	Cenomanian to Maastrichtian	with hot-spot activity
Emborê		Juréia	Santonian to Maastrichtian	
	Outeiro	Itanhaém	Neo-Albian	
Macaé	Quissamã	Guarujá	Eo-Albian	
	Goytacas	Florianópolis	Albian	Depositional sequences controlled by fault-bounded
	Retiro	Ariri	Neo-Alagoas	subsidence
Lagoa Feia		Guaratiba	Aratu-EoAlagoas	
Cabiunas		Camboriú	Eo-Cretaceous	
Basement		Basement	Pre-Cretaceous	Extensive magmatic activity associated with early rifting episodes

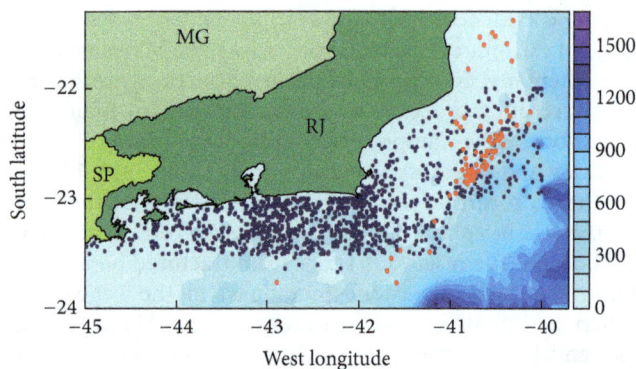

FIGURE 2: Locations of sea floor temperature measurements (indicated as blue dots) in the bathymetry map of the coastal area of southeast Brazil. The red dots refer to locations of oil wells with BHT measurements reported in the present work. The side bar indicates sea floor depth in meters.

sea floor in the study area. The rates of decrease are relatively high at depths less than 100 meters but drop sharply at deeper levels. At depths of more than 200 meters, the rate of decrease in water temperatures is found to be less than 0.02°C/m. Under such conditions, temperatures measured in the lower parts of the water column close to the sea floor are nearly identical to temperatures at the sediment—water interface at the sea floor. Hence, shallow water temperature data sets can be used to obtain approximate estimates of the temperatures of the sea floor. In the present work, data sets for the continental platform area of southeast Brazil have been employed in deriving an empirical relation between the height of the water column (z) and temperatures at the sea floor (SFT):

$$SFT = 8 \times 10^{-9} z^3 + 3 \times 10^{-6} z^2$$
$$- 3.01 \times 10^{-2} z + 22.505. \tag{1}$$

Equation (1) was used in determination of sea floor temperatures at more than 1,360 localities in the Campos basin. Illustrated in the map of Figure 2 are the locations of oil wells in which BHT measurements have been carried out.

3.2. BHT Data for Oil Wells. Zembruscki [33] reported bottom-hole temperature (BHT) data for 96 oil wells in the area of the Campos basin. Most of such wells are terminated at the depths hydrocarbon reservoirs, usually in the range of 2,000 to 3,500 meters. The locations of a selected set of 76 wells are indicated in the map of Figure 2 along with the sites of sea floor temperature measurements. The sensors employed in BHT measurements have accuracies of no better than 1°C.

Analysis of the details of this data set revealed some difficulties in the determination of temperature gradients. In most cases, the records make reference to single temperature measurement at the bottom of selected wells. In such cases, it is common practice to make use of available information on sea floor temperatures in the determination of gradient values. In the present work, results of oceanographic studies in the coastal area of southeast Brazil (discussed in the previous section) were employed in determination of temperatures at the sediment—water interface.

Another difficulty with the BHT data reported in [33] is that it does not provide auxiliary information necessary for incorporating corrections for drilling disturbances. Hence, the empirical correction procedure of AAPG [34] was selected as the best option available. Procedures for such corrections have been discussed extensively in the literature (see, e.g., [35]). Use of AAPG method has been found to lead to corrections in BHT values of less than 10%, but, for wells with depths greater than 1000 meters, alterations in temperature gradient values are found to be less than 5% (see, e.g., [36]). Similar conclusions were also reached by [37] in the analysis of BHT data for the Anadarko basin (southern USA) and by [38] Blackwell and Richards (2004)

TABLE 2: Summary of measured and AAPG corrected bottom-hole temperatures for 76 oil wells in the Campos basin. Values in the 7th column refer to estimated errors in BHT. The last two columns provide depths of water column (Z_{SF}) and sea floor temperatures (T_{SF}) at the sites of wells.

Well ID	Latitude	Longitude	Depth (m)	BHT (°C) Measured	BHT (°C) Corrected	Error	Z_{SF} (m)	T_{SF} (°C)
1-RJS-5B	−22.2042	−40.3867	3897	111.1	124.4	6.2	80	16.4
1-RJS-13	−22.6970	−40.8284	2209	67.8	75.1	3.8	90	16.2
1-RJS-23	−22.6376	−40.8345	2972	85.6	96.0	4.8	93	16.1
1-RJS-26	−22.5577	−40.8612	2585	72.2	81.1	4.1	71	16.7
1-RJS-27	−22.2646	−40.8448	1687	66.7	71.9	3.6	56	17.1
1-RJS-28A	−22.7517	−40.6958	2672	81.7	90.9	4.5	145	14.8
1-RJS-32	−22.5125	−40.4232	3825	111.7	124.8	6.2	298	12.2
1-RJS-33	−23.7614	−42.8883	3135	87.8	98.8	4.9	149	14.7
1-RJS-36	−22.3514	−40.8904	1951	65.6	71.9	3.6	56	17.1
1-RJS-38	−22.7094	−40.6531	2848	88.9	98.8	4.9	180	14.1
1-RJS-41	−22.6953	−40.7176	3541	107.2	119.6	6.0	108	15.7
4-RJS-42	−22.4372	−40.4360	3359	96.1	107.9	5.4	136	15.0
1-RJS-43	−22.7396	−40.8700	3530	95.0	107.3	5.4	100	15.9
1-RJS-44	−22.6128	−40.6573	2804	97.8	107.6	5.4	118	15.5
1-RJS-45	−22.6980	−40.6007	3282	89.4	100.9	5.0	284	12.4
1-RJS-46	−22.4773	−40.5134	3344	95.6	107.4	5.4	125	15.3
1-RJS-47E	−22.5792	−40.7234	3819	116.7	129.8	6.5	105	15.8
1-RJS-48	−22.3053	−40.9468	3545	110.0	122.4	6.1	50	17.3
1-RJS-49	−22.7846	−40.8009	3192	88.9	100.1	5.0	105	15.8
1-RJS-50	−22.4540	−40.4731	3218	107.2	118.5	5.9	132	15.1
1-RJS-52	−22.2519	−40.3531	3520	107.2	119.5	6.0	130	15.2
1-RJS-53	−22.6296	−40.5474	3300	90.6	102.2	5.1	284	12.4
1-RJS-54	−22.5700	−40.5058	3494	101.7	113.9	5.7	254	12.8
4-RJS-55	−22.8131	−40.7775	3075	77.8	88.6	4.4	114	15.6
1-RJS-56	−21.5938	−40.3867	5000	154.4	168.1	8.4	32	17.8
1-RJS-57	−21.4934	−40.5444	4202	148.3	162.1	8.1	28	17.9
1-RJS-58	−21.6038	−40.6354	3796	125.0	138.1	6.9	24	18.0
3-RJS-59	−22.4731	−40.4689	3252	99.4	110.8	5.5	150	14.7
1-RJS-60	−22.4050	−40.4520	3704	109.4	122.2	6.1	118	15.5
4-RJS-62A	−22.7722	−40.7675	3201	87.8	99.1	5.0	109	15.7
1-RJS-63A	−22.7945	−40.7344	3206	80.0	91.3	4.6	122	15.4
1-RJS-64A	−22.8344	−40.7518	3205	81.1	92.4	4.6	132	15.1
1-RJS-65	−22.7305	−40.7669	3449	97.2	109.3	5.5	103	15.8
1-RJS-66	−22.3251	−40.1457	4490	128.9	143.0	7.1	200	13.7
1-RJS-68	−22.9722	−41.0108	2667	80.6	89.8	4.5	98	16.0
1-RJS-69	−22.5098	−40.9939	2297	77.2	84.9	4.2	57	17.1
1-RJS-70	−22.2301	−40.9954	2355	83.3	91.2	4.6	39	17.6
1-RJS-71	−22.3170	−40.9319	3504	106.7	119.0	5.9	54	17.2
1-RJS-73	−22.7714	−40.8014	3187	91.1	102.3	5.1	130	15.2
1-RJS-74	−22.7512	−40.7989	3189	99.4	110.6	5.5	101	15.9
1-RJS-75	−22.3034	−40.4504	3605	109.4	122.0	6.1	106	15.8
1-RJS-76	−22.5733	−40.6008	4870	134.4	148.3	7.4	122	15.4
1-RJS-78	−22.7419	−40.8122	3240	98.3	109.7	5.5	110	15.7
1-RJS-79	−21.5245	−40.5132	3761	133.9	146.9	7.3	26	18.0
3-RJS-80	−22.4446	−40.4783	3170	102.8	114.0	5.7	121	15.4
1-RJS-82	−21.8235	−40.8087	2282	86.1	93.7	4.7	24	18.0

TABLE 2: Continued.

Well ID	Latitude	Longitude	Depth (m)	BHT (°C)			Z_{SF} (m)	T_{SF} (°C)
				Measured	Corrected	Error		
1-RJS-83	−22.5043	−40.5401	3252	106.1	117.5	5.9	124	15.3
1-RJS-84	−22.4229	−40.5046	3326	105.0	116.7	5.8	115	15.5
1-RJS-85	−22.6102	−40.6079	3994	118.3	131.8	6.6	141	14.9
1-RJS-88	−22.4628	−40.4658	3257	99.4	110.9	5.5	170	14.3
1-RJS-89	−22.8458	−40.9379	2760	86.1	95.7	4.8	117	15.5
1-RJS-90	−22.3342	−40.2792	3503	103.3	115.6	5.8	126	15.3
1-RJS-91	−23.2109	−41.1755	2890	95.6	105.7	5.3	111	15.6
1-RJS-92	−22.7929	−40.8364	3127	91.1	102.1	5.1	99	15.9
1-RJS-93	−22.8412	−40.7963	2988	90.0	100.5	5.0	115	15.5
1-RJS-94	−21.5334	−40.6822	2704	111.1	120.5	6.0	23	18.1
1-RJS-95	−22.6685	−40.6278	4439	128.9	143.0	7.1	179	14.1
1-RJS-96A	−21.3834	−40.3158	4003	144.4	157.9	7.9	53	17.2
1-RJS-97C	−21.7498	−40.3318	3992	121.7	135.2	6.8	46	17.4
1-RJS-99	−23.7681	−41.5889	4245	107.2	121.1	6.1	154	14.6
1-RJS-100	−23.5431	−41.6448	3122	84.4	95.4	4.8	146	14.8
1-RJS-101	−22.4765	−40.5828	4603	167.2	181.3	9.1	114	15.6
1-RJS-102A	−22.4480	−40.5660	5049	152.2	165.8	8.3	108	15.7
1-RJS-105	−23.4693	−41.5478	3380	103.9	115.8	5.8	131	15.2
1-RJS-106	−22.7641	−40.7404	2780	99.4	109.1	5.5	115	15.5
1-RJS-107	−23.6564	−41.6900	3898	118.9	132.2	6.6	150	14.7
1-RJS-108	−22.3445	−40.6733	4958	147.2	161.0	8.0	67	16.8
1-RJS-111	−22.7978	−40.8058	3168	96.1	107.2	5.4	130	15.2
1-RJS-113	−22.9523	−40.9423	2971	85.0	95.4	4.8	100	15.9
1-RJS-114	−22.8488	−40.8432	3072	95.0	105.8	5.3	112	15.6
1-RJS-115	−22.8682	−40.8145	3096	84.4	95.3	4.8	113	15.6
1-RJS-116	−22.7294	−40.6195	3990	118.3	131.8	6.6	296	12.2
1-RJS-117	−22.2137	−40.1260	4087	114.4	128.1	6.4	119	15.4
3-RJS-120	−22.5806	−40.5271	3180	98.9	110.1	5.5	233	13.1
4-RJS-121	−22.5718	−40.5546	3217	96.1	107.4	5.4	178	14.1
1-RJS-125	−23.4852	−41.2235	2496	81.1	89.6	4.5	168	14.3

in the discussion of error estimates in the heat flow map of North America.

In the present work, corrections to BHT values were made using the relation

$$\Delta T = az + bz^2 - cz^3 - dz^4, \qquad (2)$$

where ΔT is the correction for temperature, z is the depth in meters, and a, b, c, and d are the polynomial coefficients with values of 1.878×10^{-3}, 8.476×10^{-7}, 5.091×10^{-11}, and 1.681×10^{-14}, respectively. The details of data on well depths and bottom-hole temperatures are provided in Table 2, along with AAPG corrected values and estimates of errors in the relevant correction procedure. Also, given in this table are data on thicknesses of water columns and sea floor temperatures at the sites of wells along with complementary information on locations and well identification.

3.3. Geothermal Gradients. Geothermal gradients values were calculated for sites of 76 oil wells considered in

the present work. The relation used in determination of gradient values (Γ) for sites of oil wells is

$$\Gamma = \frac{T_{BHT} - T_{SF}}{Z_{BHT} - Z_{SF}}, \qquad (3)$$

where T_{BHT} is the corrected bottom-hole temperature at depth Z_{BHT} and T_{SF} is the sea floor temperature at depth Z_{SF}. The values of gradient calculated using (3) are found to fall in the range of 10 to 30°C/km. A summary of values of geothermal gradients for the set of wells considered in the present work is provided in Table 5.

As can be noted in the map of Figure 2, most of well sites are located along the central rift zone of the Campos basin, with the data density being poor in the coastal zone and also in the eastern portion of the study area. A convenient means of improving data density in the coastal area is to consider geothermal gradients values reported in [10] for 22 sites in the state of Rio de Janeiro. The problem arising from poor data density in the eastern segment of the Campos basin has been minimized by considering estimates of geothermal gradients

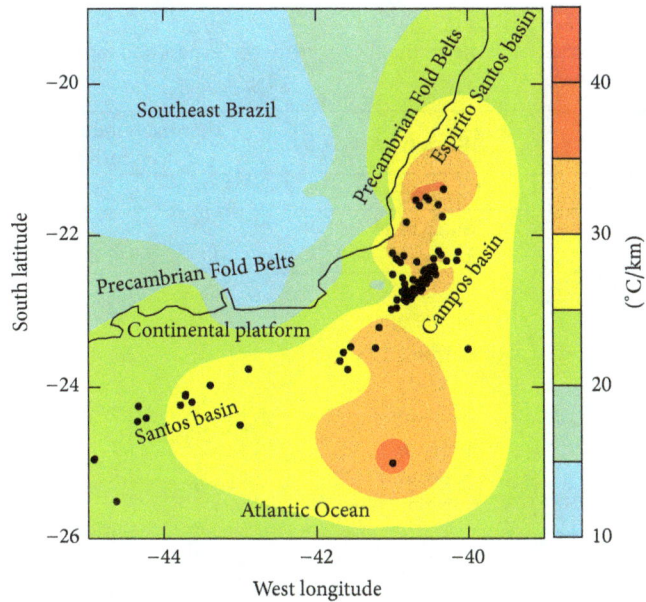

FIGURE 3: Map of geothermal gradients in the region comprising the Campos basin in the continental margin of southeast Brazil. The black dots are locations of geothermal gradient measurements. The position of the coastal zone is indicated by the dark curve. Also indicated are the locations of adjacent Santos and Espirito Santo basins.

for southwestern part of the Atlantic Ocean, reported in [11, 12, 39].

The regional distribution of gradient values is illustrated in the map of Figure 3. It reveals that the gradients are higher than 30°C/km along a northeast-southwest trending arcuate belt. This belt of relatively high gradients is roughly coincident with the rift zone in the continental margin of southeast Brazil, identified in geologic studies. It is also coincident with the region of oil and gas fields, indicated in Figure 1.

3.4. Thermal Conductivity. The difficulties in obtaining suitable samples of deep sedimentary strata turned out to be the main problem in thermal conductivity studies of the Campos basin. Since core samples and cuttings from drilling operations of oil wells are rarely available for academic research, the availability of thermal property data is limited to results of isolated efforts. Marangoni and Hamza [40] carried out thermal conductivity measurements of ocean floor sediment samples in the coastal region of southeast Brazil. Gomes and Hamza [41] have carried out thermal conductivity measurements on samples of cuttings recovered in drilling operations of water wells in the state of Rio de Janeiro. More recently, del Rey and Zembruscki [42] reported results of thermal conductivity measurements on samples from the adjacent Espirito Santo basin. A summary of thermal conductivity values derived from results of these earlier studies is given in Table 3.

Viana [43] reported estimates of thermal conductivity for 33 sites in the Campos basin based on a procedure that

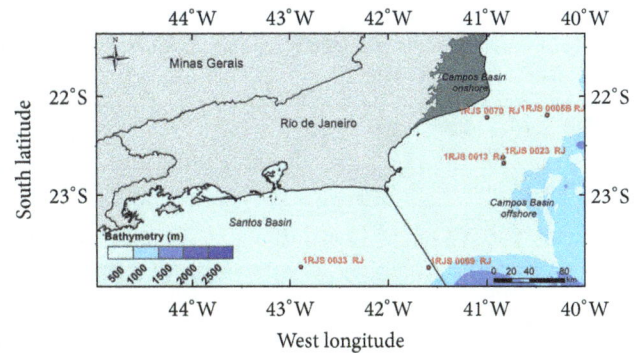

FIGURE 4: Locations of six wells with geophysical logs, employed for determination of thermal conductivity.

makes use of information on relative proportions of solid rock matrix and of the fluids occupying the pore spaces of samples collected during drilling operations. The procedure adopted makes use of literature values of thermal conductivity of the rock matrix and pore fluids (see, e.g., [44]) in deriving estimates of effective thermal conductivities.

In the present work, additional determinations of thermal conductivity were made making use of log data provided by the National Agency of Petroleum (ANP) for six wells in the Campos basin. In addition, lithologic descriptions of samples collected during drilling operations were also made available by ANP. The locations of these wells are indicated in the map of Figure 4.

The procedure for thermal conductivity determinations adopted in the present work is essentially the same as that adopted by Viana [43], with the exception that porosity values were derived from well log data. Thermal conductivity values (k) of the fluid saturated medium with porosity ϕ were calculated using the relation proposed by Woodside and Messmer [45]:

$$k(z) = k_s^{1-\phi(z)} k_w^{\phi(z)}. \qquad (4)$$

In (4), k_s is the thermal conductivity of solid matrix and k_w is that of the fluid in the pore space. As an illustrative example, we present in Figure 5 results obtained from analysis of lithologic descriptions of samples collected from well RJS-23. Note that overall distributions of thermal conductivity values are in the general range of 1 to 4 W/m/°C. As expected, the rock formations rich in silt and shale fractions are characterized by relatively low values of thermal conductivity, while sections rich in carbonates and evaporates have relatively high values. A summary of the values of the main rock types obtained by this method is provided in Table 4.

The determinations of thermal conductivity were also carried out using geophysical well log data provided by ANP. The logs provide vertical distributions of sonic velocity, gamma ray, bulk density, and electrical resistivity. Among these, logs of sonic velocity were found most useful for the present work. The procedure adopted here is based on a modified form of the empirical relation proposed by Houbolt

TABLE 3: Thermal conductivity values reported in earlier studies. The names in brackets in the second column are corresponding formations in the adjacent Campos basin.

Region	Layer/formation	Rock type	Thermal conductivity (W/m/K)
Platform area [40]	Ocean floor	Sediments	1.5–2.1
Continental part of Campos basin [41]	Drill cuttings	Shaly sand	2.2–3.5
		Sandy shale	2.5–3.2
Campos ([63, 64]) and Espirito Santo [42] basins	Rio Doce (Emborê)	Shale	1.6
		Sandstone	1.9
	Caravelas (Ubatuba)	Carbonates	2.3
	Urucutuca (Campos)	Shale	1.9
		Sandstone	2.7
	Barra Nova (Macaé)	Shale	1.7
		Sandstone	2.5
		Carbonates	3.0
	Mariricu (Lagoa Feia)	Shale	2.4
		Sandstone	2.8
		Carbonates	2.9
		Anhydrite	5.7
Basement [77]	Precambrian	Mafics	2.1
		Metamorphics	3.9

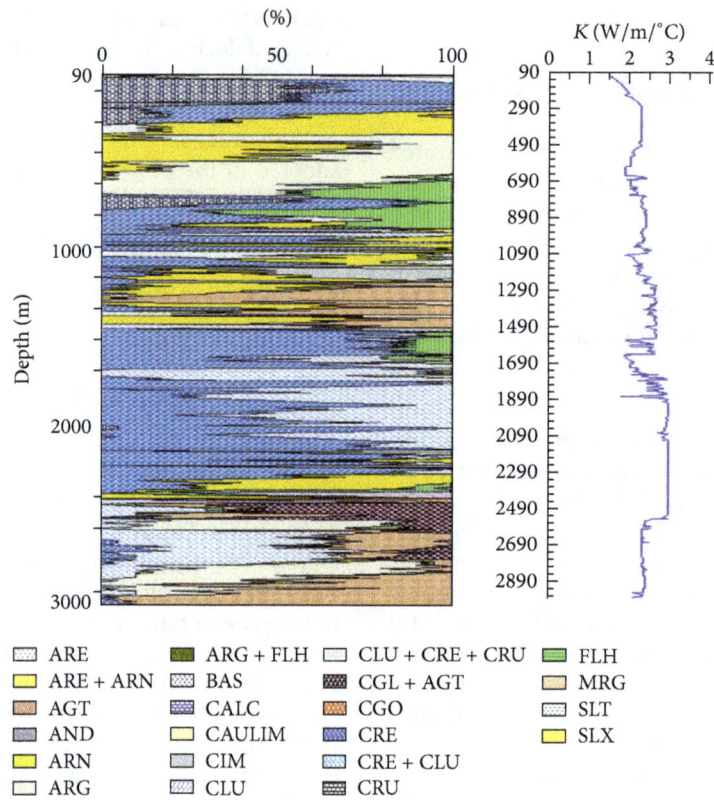

FIGURE 5: Lithologic sequences encountered in well RJS23 (left-hand part of the upper panel) and thermal conductivity values (right-hand part of the upper panel). The legend for lithologic sequences is indicated in the lower panel. The rock types identified are ARE: sand, ARN: sandstone, AGT: argillite, AND: anhydrite, ARG: clay, BAS: basalt, CALC: limestone, CAULIM: kaolin, CIM: cement, CLU: calcilutite, CGL+AGT: conglomerate with argillite, CGO: conglomerate, CRE: calcarenite, CRU: calcirudite, FLH: shale, MRG: marga, SLT: siltstone, and SLX: silex.

TABLE 4: Mean thermal conductivities of sedimentary formations in the Campos basin, derived from lithologic descriptions of drill cuttings.

Stratigraphic unit	Dominant lithology	Thermal conductivity (W/m/K)
Emborê formation	Sandstones	2.0 ± 0.4
Mb. Grussaí	Carbonates	1.9 ± 0.3
Mb. Siri	Carbonates	2.2 ± 0.4
Macaé formation	Carbonates	2.9 ± 0.5
Mb. Ubatuba	Sandstones	2.4 ± 0.4
Mb. Carapebus	Sandstones	2.5 ± 0.5
Lagoa Feia formation	Shale (Coquina)	2.8 ± 0.6
Cretaceous sediments	Sandstones and shale	2.3 ± 0.5

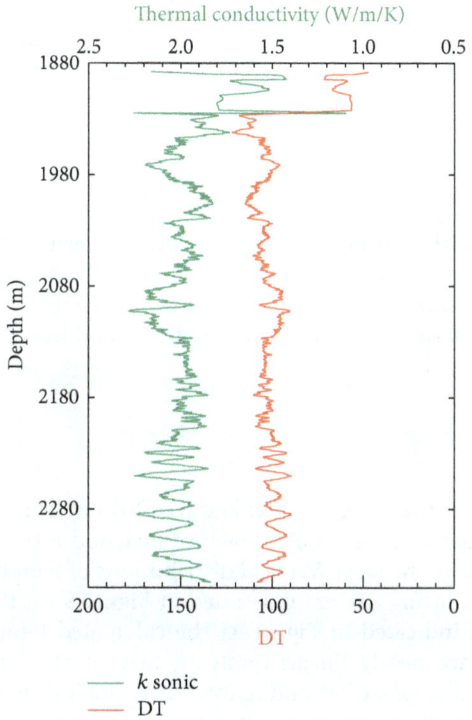

FIGURE 6: Illustration of the relation between transit time (DT) profiles (red curve) and thermal conductivity (green curve) derived from empirical relation [46], for depth interval from 1880 to 2300 m in well RJS-99.

and Wells [46] between sonic velocity (V_P) and thermal conductivity of saturated medium (k_s):

$$k_s = \frac{(d + a) V_p}{b (c + T)}, \qquad (5)$$

where a, b, and c are empirical constants, T the temperature, and d an off-set parameter specific to the particular sonic log. The values of the off-set parameter need to be adjusted in order to obtain thermal conductivity values that are physically realistic and compatible with the range of values

listed in Table 4. The numerical values of off-sets that provide best results have been found to be 33 for well RJS-33, 30.5 for well RJS-70, and 75 for well RJS-99. The values of temperature T at depth Z were derived from the relation

$$T (Z) = T_{\text{BHT}} - \Gamma (Z_{\text{BHT}} - Z), \qquad (6)$$

in which Γ is the value of the geothermal gradient. The sonic log of well RJS-99, presented in Figure 6, illustrates the inverse relation between transit time (which is the reciprocal of sonic velocity) and thermal conductivity. This procedure has been employed in determining vertical distributions of thermal conductivity for the set of six wells indicated in Figure 4. The final results are presented in Figure 7.

3.5. *Heat Flux.* The procedure employed in calculating heat flow (q) followed the practice set out in [47]. It makes use of the relation between temperatures at bottom-hole (T_{BHT}) and at sea floor (T_{SF}):

$$q = \frac{(T_{\text{BHT}} - T_{\text{SF}})}{\sum_{i=1}^{N} R_i Z_i}, \qquad (7)$$

where N is the number of layers and R_i is the thermal resistivity (inverse of thermal conductivity) of the ith layer with thickness Z_i. The summation over N layers allows determination of the cumulative thermal resistance up to the depth at which BHT measurement was carried out.

Heat flow values were calculated using (7) for 76 localities in the oceanic segment of the Campos basin. For the sites of six wells, indicated in Figure 4, thermal conductivity values derived from sonic logs were employed in calculating heat flow. For the sites of the remaining 70 wells, heat flow values were calculated using estimates of thermal conductivity reported in [43]. A summary of values of geothermal gradients, thermal conductivity, and heat flow for the set of wells considered in the present work is provided in Table 5.

As in the case of temperature gradients discussed earlier, the data density is poor in the coastal zone and also in the eastern portion of the Campos basin. Heat flow values reported in [10] for 22 sites in the state of Rio de Janeiro served as constraints in interpolation schemes used for deriving maps for areas adjacent to the coastal zone. Also, the undesirable effects of poor data density in the eastern segment of the Campos basin have been minimized by considering estimates of heat flow for the southwestern part of the Atlantic Ocean, reported in [11, 12, 39].

The regional distribution of heat flow values is illustrated in the map of Figure 8. It reveals the presence of a region with heat flow greater than 70 mW/m² in the northern part of the Campos basin. This zone of high heat flow has the shape of an arcuate belt, similar to the zone of anomalous geothermal gradients indicated in Figure 3. Its position is closer to the continent in the northern part, near the border with Espirito Santo basin. In the south, at the border with Santos basin, it is nearly 200 km away from the coast. Note that the width of the high heat flow belt is in the range of 100 to 150 km. The nature of tectonic processes responsible for the origin of the anomalous heat flux is unknown. But it certainly represents

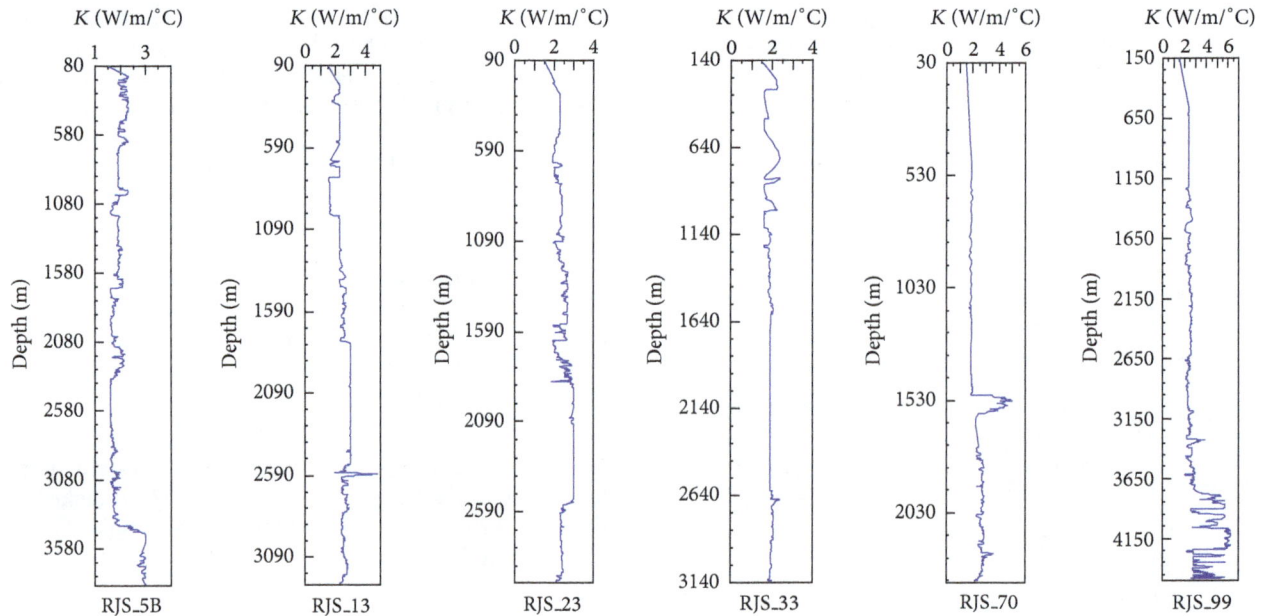

FIGURE 7: Vertical distributions of thermal conductivity, derived from lithologic descriptions of samples collected during drilling operations, at sites of six wells indicated in Figure 4.

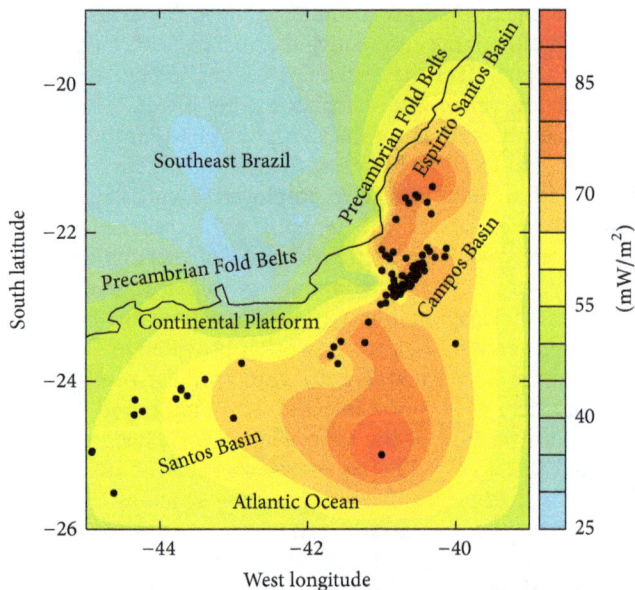

FIGURE 8: Map of terrestrial heat flow of the region comprising the Campos basin in the continental margin of southeast Brazil. The black dots are locations of geothermal measurements. The position of the coastal zone is indicated by the dark curve. Also indicated are the locations of adjacent Santos and Espirito Santo basins.

3.6. Vertical Distribution of Temperatures. Vertical distributions of temperatures at the sites of six wells, referred to in Figure 4, have been calculated making use of data on bottom-hole temperatures, thermal conductivity, and heat flow. The relation used is

$$T(z) = T_0 + \int_0^z qR(z)\,dz, \qquad (8)$$

where T_0 is the sea floor temperature, Z the thickness of the layer under consideration, q the heat flux, and R the thermal resistance of the layer. Vertical distributions of temperatures, calculated using (8), are illustrated in Figure 9 for the set of six wells indicated in Figure 4. The calculated temperature profiles are nearly linear, implying that most of the heat transfer takes place by conduction. There are indications that departures from linearity are related to thermal property variations associated with changes in lithologic sequences.

Nevertheless, vertical distribution of BHT data for the remaining wells reveals the presence of a curvature that is convex towards the depth axis. In the absence of thermal refraction effects, possible mechanisms that can produce such curvatures are either systematic increase in thermal conductivity with depth or heat transfer associated with upflow of fluids. Large-scale variations in the sea floor temperatures can also lead to curvatures in subsurface temperature profiles but this mechanism seems unlikely. Vertical variations of thermal conductivity encountered in the wells (see Figure 7) rule out the possibility of systematic increase with depth. This leaves advection heat transfer by upflow of fluids as the most likely mechanism.

At this point, it is convenient to examine the thermal effects of fluid flow in permeable media. Lu and Ge [48] derived a solution for the problem of heat transfer in a

a relatively recent thermal reactivation episode, unrelated to previous magmatic events. In fact, the relatively narrow width of the anomaly is an indication that the heat source is located at relatively shallow crustal depths and the time elapsed is no more than 10 Ma.

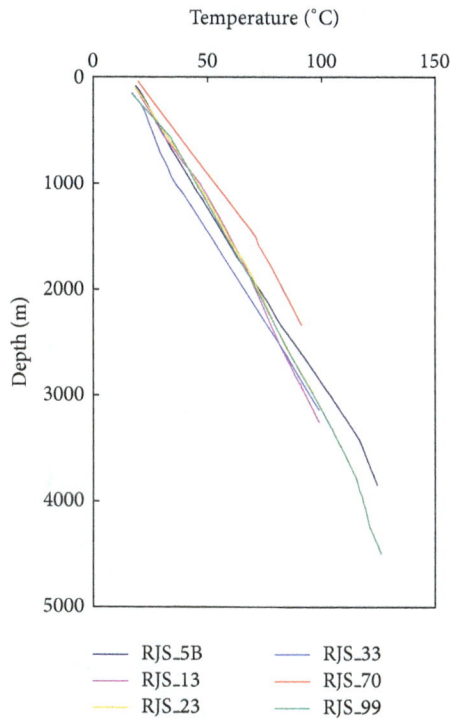

FIGURE 9: Vertical distribution of temperatures at sites of six wells, indicated in the map of Figure 4.

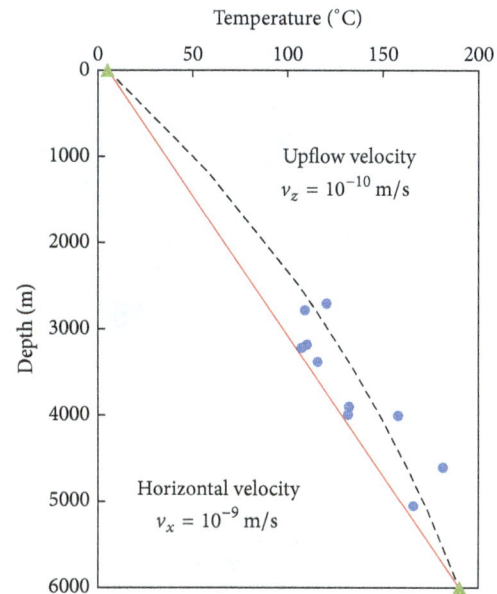

FIGURE 10: Illustration of nonlinearity in temperature variation with depth in the Campos basin. The dashed curve is the fit to the observational data based on the model of Lu and Ge [48].

permeable medium, allowing for the effects of advective fluid flows in both horizontal and vertical directions. For a layer with thickness L, in which fluid flow takes place with velocities v_x and v_z in the x and z directions, respectively, the solution for temperature T at depth z within the medium may be expressed as

$$\frac{T - T_0}{T_L - T_0}$$

$$= \left\{ \frac{\exp\left[\beta\left(z/L\right)\right] - 1}{\exp\left(\beta\right) - 1} + \frac{\alpha\gamma}{\beta\eta} \left[\frac{\exp\left[\beta\left(z/L\right)\right] - 1}{\exp\left(\beta\right) - 1} - \frac{z}{L} \right] \right\}. \tag{9}$$

In (9), T_0 and T_L refer, respectively, to temperatures at the top and bottom boundaries of the flow domain. The terms α ($\alpha = c_w \rho_w v_x L / \kappa$) and β ($\beta = c_w \rho_w v_z L / \kappa$) represent dimensionless Peclet numbers, and the terms η and γ are, respectively, the vertical and horizontal temperature gradients. Note that, in the absence of fluid movement in the horizontal direction (and/or horizontal temperature gradient), (9) is simplified to the case of vertical flow discussed in [49]. The left-hand side of this equation represents the dimensionless temperature (θ). Its variation with depth z allows the use of curve-matching methods for determination of the flow parameter β and the quantity δ ($= \alpha\gamma/\beta\eta$). These results in turn may be used for determination of the velocity components, as pointed out by [50] and more recently by [51].

An example of the curve-matching procedure is illustrated in Figure 10 for BHT data for a selected set of ten wells located at sites of turbidity deposits. In this case, the vertical

and horizontal velocities of fluid flow are found to be of the order of 10^{-10} and of 10^{-9} m/s, respectively. The red line in this figure refers to the case where no flow takes place. It is possible that hydraulic head of groundwater trapped in turbidity deposits provides favorable conditions for advective upflow of fluids in the central rift zone of this basin.

4. Subsidence History of the Campos Basin

The sequence of tectonic events that gave rise to formation of the Campos basin has been the object of a number of investigations over the last few decades. Most of them are focused on the geological characteristics of events associated with early continental rift (e.g., [52–56]) and influence of hot-spot activities in the South Atlantic (e.g., [57, 58]). Thermomechanical aspects of basin evolution have also been considered in several studies (e.g., [59–61]), but again with emphasis on geological aspects. Only recently analysis of geothermal data has been taken up as a complementary tool in model studies of subsidence of the Campos basin [62–64]. In the present work, updated data sets on temperature gradients and heat flow are employed along with available information on lithologic sequences and geophysical logs of deep oil wells in obtaining better insights into the thermotectonic evolution of the Campos basin.

It is customary in model studies of subsidence history to start off with reconstruction of depositional sequences, the results of which are employed subsequently in determining paleothermal conditions of sedimentary strata. This standard approach has also been adopted in the present work where attention is focused initially on determining the characteristics of subsidence driven by sediment loads. Results of this initial stage are employed later in model simulations

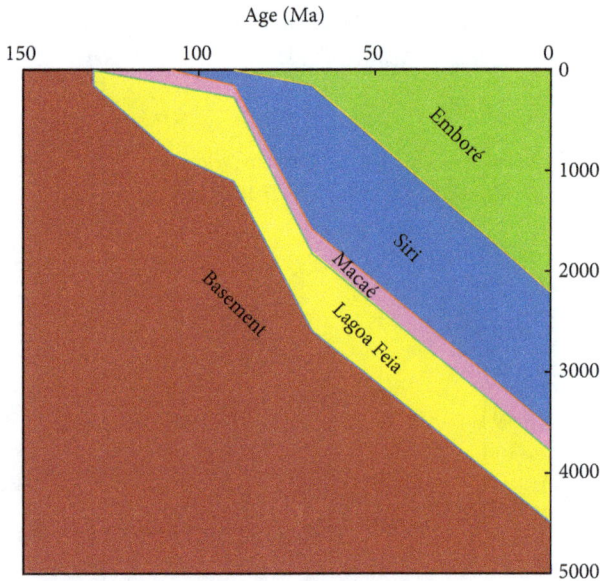

FIGURE 11: Example of backstripping procedure applied to lithologic sequences in well RJS-99.

for determining the characteristics of deep seated thermal processes.

4.1. Sediment-Loaded Subsidence. It is common practice in the relevant literature (e.g., [65]) to employ methods of backstripping in analysis of geohistory of sediment-loaded subsidence. The standard procedure involves examining the effects of sequential removal of sedimentary strata within the framework of isostatic response to unloading. The ensuing changes in thickness, density, and porosity values of sedimentary layers are then calculated using appropriate relations for sediment decompaction (e.g., [66]). In the present work, available information on lithologic sequences and geophysical well logs was employed in determining subsidence history at sites of the six wells, indicated in Figure 4. An illustrative example of the backstripping procedure is presented in Figure 11, for the site of the well RJS-99. In this figure, the sequential line segments indicate the nature of age-depth relations during the depositional history of individual formations. The point of intersection of the final line segment with the vertical depth axis indicates the present thickness of the formation, while the breaks in the lines indicate changes in the depositional sequences. The colored areas in this figure indicate the age-depth relations of the main sedimentary formations (Lagoa Feia, Siri, Macaé, and Emboré) during the evolutionary history of subsidence. Note that the thicknesses of formations decrease with depth due to compaction. The degree of compaction is relatively large for shale rich formations. For example, the original thickness of shale rich Lagoa Feia formation is 831 meters at the depositional age of 110 Ma, while the present thickness is 714 meters, a reduction of approximately 14%. Similar rates are also found for Siri formation.

A summary of the results obtained in applying the backstripping procedure to data sets derived from drill records is provided in Table 6, for the six wells indicated in Figure 4. It includes values calculated for the thicknesses of the formations and their respective porosities, at the main stages of evolution of the Campos basin. Note that, for any particular formation, the thicknesses and porosities decrease with elapsed time. However, the rates of changes vary from one well site to another. In general, shale rich formations were found to have relatively high compaction rates.

4.2. Thermal Subsidence. According to extensional models of basin evolution (e.g., [67]), stretching processes taking place in the crust and upper mantle play significant roles in the evolutionary histories of sedimentary basins. In deriving estimates of deep seated stretching, the usual practice is to start off with determinations of *thermal subsidence*, which is the subsidence discounted for the effects of sediment loading. It is calculated using the relation described in [67] for elevation (*e*):

$$e(t) = \left(\frac{a\rho_m\alpha T_m}{\rho_m - \rho_w}\right)\frac{4}{\pi^2}\left[\left(\frac{\beta}{\pi}\text{sen}\frac{\pi}{\beta}\right)e^{-t/\tau}\right], \quad (10)$$

where *a* is the thickness of the lithosphere, ρ_m the mantle density, ρ_w the density of sea water, α the linear thermal expansion coefficient, T_m the temperature of the mantle, β the stretching factor, *t* the time elapsed after the stretching event, and τ the thermal time constant of the lithosphere. This last parameter is defined in [67] as

$$\tau = \frac{a^2}{\pi^2\kappa} \quad (11)$$

in which κ is the thermal diffusivity of the lithosphere. The thermal subsidence is the difference between elevation (*e*) and that above the datum for thermal equilibrium. The values of the parameters given in (10) and (11) are listed in Table 7.

In the model proposed by [67], the stretching factor (β), which determines the thermal subsidence, is assumed to be constant at all depths in the lithosphere. Royden and Keen [68] proposed a model in which the rate of crustal stretching (δ) is different from that of subcrustal lithosphere (β). In this latter case, the relation for thermal subsidence is derived from the relation

$$e(t) = \left(\frac{2a\rho_m\alpha T_m}{(\rho_m - \rho_w)\pi}\right)\frac{2}{\pi^2}$$
$$\times\left[(\beta-\delta)\sin\left(\frac{\pi T_m}{a\beta}\right) + \delta\sin\left(\frac{\pi}{\gamma}\right)e^{-t/\tau}\right]. \quad (12)$$

These two models (designated hereafter as constant stretching (CS) and variable stretching (VS), resp.) were employed in studies of thermal subsidence in the present work. Note that, for $\beta > \delta$, the value of thermal subsidence calculated for the VS model is always smaller than that for the CS model.

An illustrative example is presented in Figure 12 for thermal subsidence curves at the site of well RJS-23, calculated on

TABLE 5: Summary of geothermal gradient (Γ), thermal conductivity (λ), and heat flow (q) values for 76 wells in the Campos basin. Error vales (σ) refer to uncertainties in the methods used for determination of the relevant parameters.

Well ID	Γ (°C/km)		λ (w/m/K)		q (mW/m^2)	
	Mean	σ_Γ	Mean	σ_λ	Mean	σ_q
1-RJS-5B	31.8	2.2	2.1	0.2	66.7	6.7
1-RJS-13	31.3	2.2	2.5	0.2	79.1	7.9
1-RJS-23	31.4	2.2	2.5	0.2	76.9	7.7
1-RJS-26	29.2	2.0	2.6	0.2	75.8	7.6
1-RJS-27	36.8	2.6	3.0	0.2	110.4	11.0
1-RJS-28A	33.8	2.4	2.5	0.2	84.4	8.4
1-RJS-32	35.6	2.5	2.5	0.2	89.1	8.9
1-RJS-33	31.8	2.2	1.9	0.2	60.2	6.0
1-RJS-36	32.2	2.3	2.9	0.2	93.5	9.3
1-RJS-38	35.5	2.5	2.7	0.2	95.8	9.6
1-RJS-41	33.9	2.4	2.8	0.2	94.8	9.5
4-RJS-42	32.5	2.3	2.5	0.2	81.2	8.1
1-RJS-43	30.2	2.1	2.3	0.2	69.6	7.0
1-RJS-44	37.9	2.7	2.4	0.2	91.0	9.1
1-RJS-45	33.4	2.3	2.5	0.2	83.4	8.3
1-RJS-46	32.3	2.3	2.9	0.2	93.6	9.4
1-RJS-47E	34.2	2.4	2.8	0.2	95.8	9.6
1-RJS-48	33.6	2.4	2.5	0.2	84.1	8.4
1-RJS-49	30.9	2.2	2.7	0.2	83.6	8.4
1-RJS-50	37.2	2.6	2.5	0.2	92.9	9.3
1-RJS-52	34.4	2.4	2.3	0.2	79.1	7.9
1-RJS-53	33.6	2.4	2.5	0.2	84.1	8.4
1-RJS-54	35.0	2.4	2.4	0.2	84.0	8.4
4-RJS-55	28.3	2.0	2.5	0.2	70.8	7.1
1-RJS-56	33.0	2.3	2.8	0.2	92.4	9.2
1-RJS-57	37.9	2.7	2.5	0.2	94.6	9.5
1-RJS-58	35.3	2.5	2.2	0.2	77.6	7.8
3-RJS-59	34.7	2.4	2.5	0.2	86.6	8.7
1-RJS-60	33.3	2.3	2.3	0.2	76.7	7.7
4-RJS-62A	30.6	2.1	2.5	0.2	76.5	7.7
1-RJS-63A	28.3	2.0	2.8	0.2	79.2	7.9
1-RJS-64A	28.8	2.0	2.9	0.2	83.6	8.4
1-RJS-65	31.5	2.2	2.5	0.2	78.9	7.9
1-RJS-66	33.4	2.3	2.6	0.2	86.9	8.7
1-RJS-68	32.3	2.3	2.5	0.2	80.8	8.1
1-RJS-69	33.7	2.4	2.4	0.2	80.9	8.1
1-RJS-70	35.2	2.5	2.4	0.2	82.7	8.3
1-RJS-71	33.1	2.3	2.5	0.2	82.7	8.3
1-RJS-73	32.2	2.3	2.2	0.2	70.8	7.1
1-RJS-74	34.3	2.4	2.3	0.2	78.9	7.9
1-RJS-75	34.0	2.4	2.9	0.2	98.5	9.8
1-RJS-76	30.9	2.2	2.2	0.2	68.0	6.8
1-RJS-78	33.7	2.4	2.5	0.2	84.2	8.4
1-RJS-79	38.0	2.7	2.8	0.2	106.4	10.6
3-RJS-80	36.0	2.5	2.5	0.2	90.0	9.0
1-RJS-82	36.9	2.6	2.4	0.2	88.5	8.9
1-RJS-83	32.7	2.3	2.3	0.2	75.1	7.5

TABLE 5: Continued.

Well ID	Γ (°C/km)		λ (w/m/K)		q (mW/m^2)	
	Mean	σ_Γ	Mean	σ_λ	Mean	σ_q
1-RJS-84	35.1	2.5	2.5	0.2	87.9	8.8
1-RJS-85	33.8	2.4	2.2	0.2	74.4	7.4
1-RJS-88	35.0	2.5	2.5	0.2	87.5	8.8
1-RJS-89	34.0	2.4	2.4	0.2	81.5	8.2
1-RJS-90	33.3	2.3	2.5	0.2	83.3	8.3
1-RJS-91	36.0	2.5	2.6	0.2	93.7	9.4
1-RJS-92	32.1	2.2	2.7	0.2	86.6	8.7
1-RJS-93	33.2	2.3	2.5	0.2	83.0	8.3
1-RJS-94	41.7	2.9	2.8	0.2	116.8	11.7
1-RJS-95	33.6	2.3	2.2	0.2	73.8	7.4
1-RJS-96A	39.0	2.7	2.2	0.2	85.9	8.6
1-RJS-97C	33.3	2.3	2.3	0.2	76.5	7.7
1-RJS-99	29.4	2.1	2.1	0.2	61.8	6.2
1-RJS-100	30.8	2.2	2.5	0.2	76.9	7.7
1-RJS-101	40.1	2.8	2.1	0.2	84.1	8.4
1-RJS-102A	30.4	2.1	2.1	0.2	63.8	6.4
1-RJS-105	34.6	2.4	2.6	0.2	90.0	9.0
1-RJS-106	38.7	2.7	2.5	0.2	96.8	9.7
1-RJS-107	34.9	2.4	2.7	0.2	94.2	9.4
1-RJS-108	32.3	2.3	2.1	0.2	67.8	6.8
1-RJS-111	34.0	2.4	2.5	0.2	84.9	8.5
1-RJS-113	31.3	2.2	2.6	0.2	81.4	8.1
1-RJS-114	26.9	1.9	2.5	0.2	67.2	6.7
1-RJS-115	26.7	1.9	2.4	0.2	64.1	6.4
1-RJS-116	32.4	2.3	2.5	0.2	80.9	8.1
1-RJS-117	28.4	2.0	2.2	0.2	62.4	6.2
3-RJS-120	32.9	2.3	2.5	0.2	82.3	8.2
4-RJS-121	30.7	2.1	2.3	0.2	70.6	7.1
1-RJS-125	32.3	2.3	2.5	0.2	80.8	8.1

the basis of the CS and VS models. In this figure, the curve in green color represents the thermal subsidence derived for the VS model, while that in red color represents the subsidence for the CS model. The subsidence values for the CS model are in general similar to those of the VS model. However, thermal subsidence values derived from CS model are slightly higher than those for the VS model, a consequence of the fact that in the VS model the values of crustal stretching are systematically lower than those of the subcrustal layer. The curve in blue color represents the sediment-loaded subsidence. As expected, the difference between sediment-loaded and thermal subsidence increases with elapsed time. Another notable feature in Figure 12 is the indication of relatively high values of extension during the initial periods.

A summary of the stretching factors derived for the sites of six wells of the Campos basin, indicated in Figure 4, is presented in Table 8. It is important to point out that the values of elapsed time in the second column are approximate, derived from [69]. In general, the results obtained point to values in the range of 1.1 to 1.6 for the CS model. For the VS model, the stretching factors for the crust are found to

TABLE 6: Evolutionary changes in thicknesses and porosities of sedimentary formations at six sites in the Campos basin, determined from results of the backstripping process.

Well ID	Formation	Thickness (m) at elapsed time				Porosity (%) at elapsed time			
		0 Ma	68 Ma	90 Ma	108 Ma	0 Ma	68 Ma	90 Ma	108 Ma
RJS-5b	Emborê	3135				20			
	Cretaceous	714	886			16	32		
RJS-13	Emborê	936				28			
	Mb. Siri	759	854			21	30		
	Macaé	801	845	903		14	19	24	
	Lagoa Feia	669	698	736	800	13	16	21	27
RJS-23	Emborê	972				28			
	Mb. Siri	698	769			21	31		
	Macaé	808	831	867		26	29	32	
	Lagoa Feia	426	448	471	550	14	18	22	28
RJS-33	Emborê	793				30			
	Mb. Siri	351	397			25	34		
	Lagoa Feia	1845	1981		2062	16	21		25
RJS-70	Emborê	1470				28			
	Lagoa Feia	831			929	17			26
RJS-99	Emborê	2054				22			
	Mb. Siri	1347	1586			12	26		
	Macaé	228	237	276		11	20	31	
	Lagoa Feia	714	763	830	831	10	15	22	24

TABLE 7: Values of parameters used in models of backstripping and thermal subsidence.

Parameter	Description	Value	Unit
ρ_w	Water density	1	g/cm^3
ρ_m	Mantle density	3.3	g/cm^3
ρ_c	Crustal density	2.8	g/cm^3
a	Initial thickness of lithosphere	125	km
t_c	Initial thickness of the crust	34	km
α	Linear thermal expansion coefficient	3.3×10^{-5}	°C^{-1}
τ	Time constant of the lithosphere	62.8	Ma
λ	Thermal conductivity of the lithosphere	4	W/m/K
k_w	Thermal conductivity of water	0.56	W/m/K
T_l	Temperature at the base of the lithosphere	1333	°C
κ	Thermal diffusivity of the lithosphere	8×10^{-7}	m^2/s

fall in the range of 1.1 to 1.7, while those for the subcrustal lithosphere fall in the range of 1.3 to 2. In the case of well RJS-5B there are no significant differences between the values of δ and β. On the other hand, values of δ are systematically lower than those of β for data sets of wells RJS-13, RJS-23, RJS-33, and RJS-70. The results also reveal that the values of the parameters δ and β are different for the main periods of stretching. Note that the differences in the values of δ and β can lead to significant differences in the evolutionary history.

4.3. Paleoheat Flux. According to the extensional models, the stretching factors derived from thermal subsidence analysis may be employed in determination of heat flow during the evolutionary history of the basin. The relation for variation of heat flow (q_p) during the period following the stretching episode is [67]

$$q_p(t) = \frac{\lambda T_m}{a}\left\{1 + 2\sum_{n=1}^{\infty}\left(\frac{\beta}{n\pi}\operatorname{sen}\frac{n\pi}{\beta}\right)e^{-n^2(t/\tau)}\right\}, \quad (13)$$

where λ is the thermal conductivity of basement rocks, a the thickness of the lithosphere, β the stretching factor, t the time elapsed after the initial stretching event, τ the thermal time constant of the lithosphere, and T_m its basal temperature.

Paleoheat flow values may also be calculated using the variable stretching model. In this case, the relation for heat flux is

$$q_p(t) = \frac{\lambda T_1}{a}\left\{1 + \sum_{n=1}^{\infty}(2x_n)e^{-n^2(t/\tau)}\right\}, \quad (14)$$

TABLE 8: Stretching parameters calculated for the sites of six wells, indicated in Figure 4. The error values refer to uncertainties in estimates of model fits.

Well ID	Elapsed time (Ma)	Extensional model		
		Constant stretching (CS)	Variable stretching (VS)	
		Crust and subcrust (β)	Subcrustal (β)	Crustal (δ)
RJS_5B	0–60	1.1 ± 0.1	2 ± 0.2	1.1 ± 0.1
	60–130	1.3 ± 0.1	1.6 ± 0.2	1.3 ± 0.1
RJS_13	0–40	1.6 ± 0.2	2 ± 0.2	1.5 ± 0.1
	40–60	1.4 ± 0.2	2 ± 0.2	1.3 ± 0.1
	60–130	1.2 ± 0.1	1.5 ± 0.1	1.2 ± 0.1
RJS_23	0–40	1.6 ± 0.2	1.6 ± 0.2	1.5 ± 0.1
	40–60	1.4 ± 0.1	1.6 ± 0.2	1.4 ± 0.1
	60–130	1.2 ± 0.1	1.5 ± 0.2	1.2 ± 0.1
RJS_33	0–60	1.6 ± 0.2	1.6 ± 0.2	1.5 ± 0.1
	60–100	1.4 ± 0.1	1.6 ± 0.2	1.3 ± 0.1
	100–130	1.2 ± 0.1	1.3 ± 0.1	1.1 ± 0.1
RJS_70	0–22	1.4 ± 0.1	1.6 ± 0.2	1.3 ± 0.1
	22–60	1.3 ± 0.1	1.5 ± 0.2	1.1 ± 0.1
	60–130	1.2 ± 0.1	1.3 ± 0.1	1.1 ± 0.1
RJS_99	0–40	1.6 ± 0.2	2 ± 0.2	1.7 ± 0.2
	40–60	1.5 ± 0.2	2 ± 0.2	1.5 ± 0.1
	60–130	1.4 ± 0.1	1.7 ± 0.2	1.1 ± 0.1

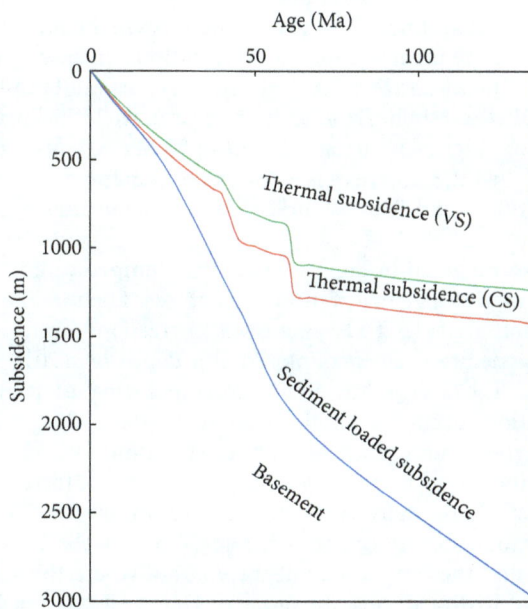

FIGURE 12: Thermal subsidence curves for models of uniform (CS) and nonuniform (VS) stretching. The blue curve indicates sediment-loaded subsidence.

where, according to [68], the term x_n is given by the relation

$$x_n = \gamma + \left\{ (1-\gamma) \left[(\delta - \beta)\, \mathrm{sen}\, n\pi \left(1 - \frac{y}{a\delta}\right) \right. \right.$$

$$\left. + \beta\, \mathrm{sen}\, n\pi \left(1 - \frac{y}{a\delta} - \frac{1-y/a}{\beta}\right) \right].$$

$$\times \left. \frac{(-1)^{n+1}}{n\pi} \right\}$$

$$(15)$$

A comparison with (13) revealed the presence of a discrepancy in (15). The correct expression may be written as [64]

$$x_n = \gamma + \left\{ (1-\gamma) \left[(\delta - \beta)\, \mathrm{sen}\, n\pi \left(1 - \frac{y}{a\delta}\right) \right. \right.$$

$$\left. + \beta\, \mathrm{sen}\, n\pi \left(\frac{y}{a\delta} + \frac{1-y/a}{\beta}\right) \right] \quad (16)$$

$$\times \left. \frac{(-1)^{n+1}}{n\pi} \right\}.$$

The stretching factors listed in Table 8 were used in deriving paleoheat flow values for the sites of six wells selected in the present work. Comparison of results obtained using (13) and (14) reveals that the overall trends of paleoheat flow derived from CS and VS models are similar. However, VS model leads to heat flow values that are in general higher, a consequence of the larger stretching factors in the subcrustal layer.

The variations of paleoheat flow values with elapsed time are illustrated in Figure 13, for the six sites considered in the present work. It reveals heat flow in the range of 100 to 130 mW/m^2 during the initial rift phase, which lasted from 130 to 100 Ma. During the transition stage which followed this initial phase, heat flow decreased systematically with

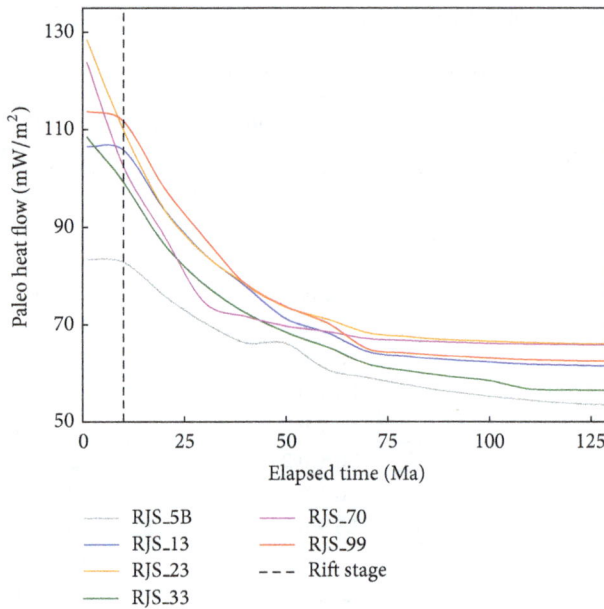

FIGURE 13: Evolution of heat flux as a function of elapsed time calculated on the basis of the variable stretching model [68] for sites of six wells indicated in the map of Figure 4.

age, reaching values of less than $70 \, \text{mW/m}^2$, at 60 Ma. No appreciable changes in heat flow occurred during the final stage, lasting from 60 Ma to present day. It is important to point out that heat flow variations illustrated in Figure 13 refer mainly to localities in the eastern parts of the Campos basin. As mentioned earlier, the present day heat flow in the central parts of this basin is relatively high, reaching values in excess of $90 \, \text{mW/m}^2$.

At this point, it is important to draw attention to a potential source of error in the calculation of paleoheat flow based on the model proposed by McKenzie [67]. As pointed out by Cardoso and Hamza [70], the definition of the time constant τ in the McKenzie model (see (11)) implies that the decay of the transient components of temperature and heat flow during the period immediately following episode of extension is determined by the initial thickness (L) of the lithosphere. This is obviously an inappropriate assumption as the lithospheric thickness immediately after extension is (L/β). It returns to the initial value L only after the dissipation of the thermal perturbation produced by the stretching event. In other words, the relevant parameter for heat dissipation from the underlying uplifted asthenosphere is variable.

Cardoso and Hamza [70] proposed that the postextensional growth in the thickness (a) of the lithosphere obeys a relation of the type

$$a(t) = \frac{L}{\beta} + \left(L - \frac{L}{\beta}\right) \text{erf}(\gamma t), \qquad (17)$$

where erf is the error function and γ is a suitable scaling constant. According to (17), the initial (i.e., at time $t = 0$) value for the thickness of the stretched lithosphere is (L/β), while, for large times (i.e., for $t \to \infty$), it tends to the initial value (L) before the stretching event. The error function term

arises from the underlying assumption that solidification process of the intruded asthenosphere during the period following extension has similarities with that of a stagnant, semi-infinite fluid. Cardoso and Hamza [70] demonstrated that the correction for thermal time constant leads to postrift heat flow values invariably lower than those predicted by the McKenzie model. However, results of numerical simulations indicate that departures from McKenzie model become significant only in cases where values of stretching parameter are greater than 2. It is therefore unlikely to be a significant source of error in the results of the present work.

4.4. Paleotemperatures. The temperatures during evolutionary history of the basin have been calculated making use of the relation

$$T(z,t) = T_0 + \int_0^z Q(t) R_t(z,t) \, dz, \qquad (18)$$

where T_0 is the surface temperature, Z the thickness of the layer under consideration, Q the heat flux at time t, and R_t the thermal resistance of the layer at depth z and time t. An example of paleotemperatures is presented in Figure 14 for well RJS-99. In this figure, the dotted lines in red color indicate the isotherms. The numbers beside the isotherms indicate values of paleotemperatures in degrees centigrade. The continuous lines indicate the sequences in the subsidence history of the main sedimentary formations. In case of wells RJS-13, RJS-23, and RJS-70, the isotherms are found mostly to be parallel and subhorizontal. In the case of well RJS-5B, there is a drop in temperature at the time of 30 Ma, whereas, in the case of well RJS-33, there is a rise in temperatures. In case of wells RJS-33 and RJS-99, the rise of temperatures occurs at age of 68 Ma.

Also included in Figure 14 are time-temperature indices (TTI) of thermal maturation, as per the Lopatin method [71]. Note that the TTI value of 15, corresponding to onset of oil generation, is encountered at a depth of nearly 3000 meters. The TTI value of 60, corresponding to peak oil generation, occurs at depth of approximately 4200 meters. The corresponding indices for the remaining five well sites are found to occur at slightly larger depths. Unfortunately, most of these wells (with the exception of RJS-99) are located outside the central rift zone. These wells have not penetrated the deeper sedimentary strata where the source rocks of hydrocarbons are situated. Hence, it has not been possible to estimate ages of peak oil generation in deep seated source rocks of Barremian, Albian, and Turonian periods. Results of numerical simulations using commercial software (e.g., PETROMOD [72]) indicate periods of peak oil generation in the range of approximately 80 to 60 Ma [73]. It is, however, important to point out that such calculations are based on paleoheat flow values derived from the McKenzie model [67]. On the other hand, results of model calculations that incorporate corrections for the variable thermal time constant of the lithosphere (Cardoso and Hamza [70]) point to a lower age range (of 40 to 20 Ma) for peak oil generation in the Campos basin.

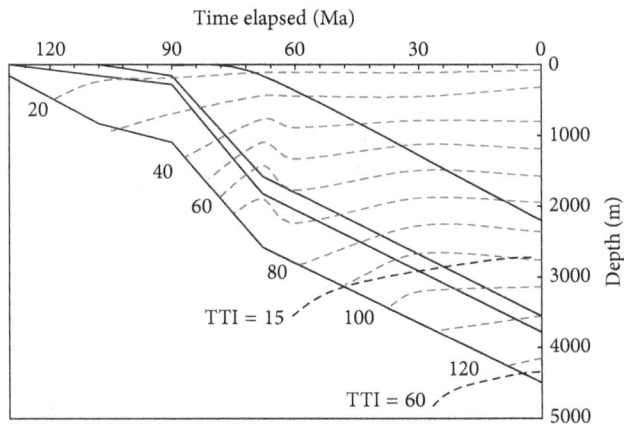

FIGURE 14: Evolution of temperatures at the site of well RJS-99 during the evolutionary history of the Campos basin. The dotted lines in red color indicate the isotherms. The numbers beside the isotherms indicate values of temperatures in degree centigrade. The dashed curves indicate TTI indices of 15 and 60, corresponding, respectively, to onset and peak of oil generation. The continuous lines indicate time-depth sequences in the subsidence history of the main sedimentary formations.

5. Conclusions

Analysis of data sets on bottom-hole temperatures, physical properties derived from geophysical well logs, and descriptions of lithologic sequences encountered in drilling operations of deep oil wells has contributed to revised assessments of the geothermal field of the Campos sedimentary basin. According to the results obtained, the present day geothermal gradients in the Campos basin vary from 24 to 41°C/km, while crustal heat flow values are in the range of 50 to 100 mW/m^2. The regional distribution of available data sets has allowed identification of a northsouth trending zone of relatively high geothermal gradients and heat flow in the central part of the Campos basin. This anomalous zone has the shape of an arcuate belt and is roughly coincident with areas of known occurrences of oil and gas deposits. There are indications that the high heat flow belt is not restricted to the Campos basin but extends to the south into the Santos basin and also to the north into the Espirito Santo basin. The existence of a major geothermal belt adjacent to the continental margin of southeast Brazil is an altogether surprising result as it departs from the usual trend of decreasing heat flow age of ocean crust [74, 75]. The width of the high heat flow zone is narrow implying that the heat source is located at shallow depths in the upper crust and is relatively recent. The mechanism responsible for high heat flow with such characteristics brings into question the nature of central rift zone of the Campos basin. It appears to have the characteristics of a unique "heat-leaking plate boundary," situated between the continental and oceanic segments of the South American lithosphere.

The present work also provides improved assessment of the subsidence history at six localities in the Campos basin. The results have allowed determination of paleothermal

conditions of the basement beneath the sediments. There are indications that heat flow was substantially higher (in excess of 100 mW/m^2) during the initial rifting episode, which lasted from approximately 130 to 110 Ma. The transition stage, which followed the initial rift stage, is characterized by systematic decrease in heat flow, reaching values of less than 70 mW/m^2 at 60 Ma. No appreciable changes in heat flow seem to have occurred during the final stage in most parts of the Campos basin. However, present heat flow is high within the central rift zone, pointing to recent thermal reactivation of the contact zone between the oceanic and continental segments of the South American lithosphere. Thermal maturation indices (TTI) calculated on the basis of Lopatin method indicate that significant oil and gas generation occur at depths greater than 3 km. The age of peak oil generation, estimated on the basis of McKenzie model of lithospheric extension, is found to fall in the range of 80 to 60 Ma. However, the age of peak oil generation is found to be less than 40 Ma, if we allow for variable thermal relaxation periods for the extended lithosphere.

Conflict of Interests

The authors declare that there is no conflict of interests regarding the publication of this paper.

Acknowledgments

The present work was carried out as part of M.S. Thesis Project of the first author, who has been a recipient of a scholarship granted by CAPES, during 2005–2008. Thanks are due to Mr. Silvio Zembruscki, late Geologist of PETROBRÁS, for exchange of information on bottom-hole temperature data of oil wells in sedimentary basins of Brazil. The National Agency for Petroleum (ANP) provided data on geophysical logs of six oil exploration wells in the Campos basin. Information on ocean water temperatures along the coastal area of southeast Brazil was provided by Directorate of Hydrograph and Navigation (DHN). The authors thank Engineer Fábio Vieira (National Observatory, Rio de Janeiro, Brazil) for complementary data analysis and Blair Bryce (Marketing Manager of Modus Medical Devices Inc., Canada) for help in reformatting the graphics. The second author is recipient of a research scholarship granted by Conselho Nacional de Desenvolvimento Cientifico e Tecnológico, CNPq (Project no. 301865/2008-6; Produtividade de Pesquisa, PQ). The authors thank Dr. Andrés Papa, Coordinator of the Department of Geophysics of Observatório Nacional, ON/MCT, for institutional support.

References

[1] C. H. L. Bruhn, C. Cainelli, and R. M. D. Matos, "Habitat do Petróleo e Fronteiras Exploratórias nos Rifts Brasileiros," *Boletim de Geociências da Petrobrás*, vol. 2, pp. 217–253, 1988.

[2] H. D. Rangel, F. A. L. Martins, F. R. Esteves, and F. J. Feijó, "Bacia de Campos," *Boletim de Geociências da Petrobrás*, vol. 8, no. 1, pp. 203–217, 1994.

[3] E. J. Milani, J. A. S. L. Brandão, P. V. Zalán, and L. A. P. Gamboa, "Petróleo na margem continental brasileira: geologia, exploração, resultados e perspectivas," *Revista Brasileira de Geofísica*, vol. 18, no. 3, pp. 351–396, 2000.

[4] E. M. Meister, "Gradientes geotérmicos nas bacias sedimentares Brasileiras," *Boletim Tecnico da Petrobrás*, vol. 16, no. 4, pp. 221–232, 1973.

[5] J. Rossi Filho, "Mapa de gradiente geotérmico na plataforma continental brasileira," Internal Report, Centro de Pesquisas da PETROBRÁS - CENPES/SUPEP/DIVEX/SEGEL, 1981.

[6] S. G. Zembruscki and C. H. Kiang, "Gradiente geotérmico das bacias sedimentares brasileiras," *Boletim de Geociências da Petrobrás*, vol. 3, no. 3, pp. 215–227, 1989.

[7] S. Ross and J. L. Pantoja, "Estudo geotérmico da bacia de Campos," Internal Report, PETROBRÁS, DEPEX, DISUD, Vitória (ES), Brazil, 1978.

[8] R. J. Jahnert, "Gradiente geotérmico da Bacia de Campos," *Boletim de Geociências da Petrobrás*, vol. 1, no. 2, pp. 183–189, 1987.

[9] R. A. Cardoso and V. M. Hamza, "Gradiente e fluxo geotérmico da plataforma continental da região sudeste do Brasil," in *Proceedings of the 8th International Congress of the Brazilian Geophysical Society*, Rio de Janeiro, Brazil, 2003.

[10] A. J. L. Gomes and V. M. Hamza, "Geothermal gradient and heat flow in the state of Rio de Janeiro," *Brazilian Journal of Geophysics*, vol. 23, no. 4, pp. 325–347, 2008.

[11] F. P. Vieira, R. R. Cardoso, and V. M. Hamza, "Global heat loss: new estimates using digital geophysical maps and GIS techniques," in *Proceedings of the 4th Brazilian Symposium on Geophysics*, pp. 14–17, Brasília, Brazil, November 2010.

[12] F. P. Vieira and V. M. Hamza, "Global heat flow: comparative analysis based on experimental data and theoretical values," in *Proceedings of the 12th International Congress of the Brazilian Geophysical Society*, Rio de Janeiro, Brazil, August 2011.

[13] F. F. M. de Almeida, "Origem e evolusão da plataforma brasileira," Tech. Rep. 241, Boletim da Divisão de Geologia e Mineralogia, Departamento Nacional da Produção Mineral, Rio de Janeiro, Brazil, 1967.

[14] G. O. Estrella, "O estagio, "rift" nas bacias marginais do leste brasileiro," in *Proceedings of the 26th Brazilian Geological Congress*, vol. 3, pp. 29–34, 1972.

[15] H. E. Asmus and F. C. Ponte, "The Brazilian marginal basins," in *The Ocean Basins and Margins, Volume 1: The South Atlantic*, A. E. M. Nairn and F. G. Stehli, Eds., pp. 87–133, Springer, New York, NY, USA, 1973.

[16] F. C. Ponte and H. E. Asmus, "The Brazilian marginal basins—current state of knowledge," *Brazilian Academy of Sciences*, vol. 48, supplement, pp. 215–240, 1976.

[17] F. F. M. de Almeida, "The system of continental rifts bordering the Santos Basin, Brazil," *Anais da Academia Brasileira de Ciências*, vol. 48, supplement, pp. 14–26, 1976.

[18] P. D. Rabinowitz and J. la Brecque, "The Mesozoic South Atlantic ocean and evolution of its continental margins.," *Journal of Geophysical Research B*, vol. 84, no. 11, pp. 5973–6002, 1979.

[19] T. P. Gladczenko, K. Hinz, O. Eldholm, H. Meyer, S. Neben, and J. Skogseid, "South Atlantic volcanic margins," *Journal of the Geological Society of London*, vol. 154, no. 3, pp. 465–470, 1997.

[20] H. E. Asmus and R. Porto, "Classificação das bacias sedimentares brasileiras segundo a tectônica de placas," in *Proceedings of the 26th Brazilian Geological Congress*, vol. 2, pp. 67–90, 1972.

[21] C. W. M. Campos, F. C. Ponte, and K. Miura, "Geology of the Brazilian continental margin," in *The Geology of Continental Margins*, C. A. Burk and C. L. Drake, Eds., pp. 447–461, Springer, New York, NY, USA, 1974.

[22] H. A. O. Ojeda, "Structural framework, stratigraphy, and evolution of Brazilian marginal basins," *American Association of Petroleum Geologists Bulletin*, vol. 66, no. 6, pp. 732–749, 1982.

[23] R. L. M. Azevedo, J. Gomide, and M. C. Viviers, "Geo-história da Bacia de Campos, Brasil: do Albiano ao Maastrichtiano," *Revista. Brasileira de Geociências*, vol. 17, no. 2, pp. 137–146, 1987.

[24] J. D. Milliman, "Morphology and structure of upper continental margin off southern Brazil," *American Association of Petroleum Geologists Bulletin*, vol. 62, no. 6, pp. 1029–1048, 1978.

[25] F. C. Ponte and H. E. Asmus, "Geological framework of the Brazilian continental margin," *Geologische Rundschau*, vol. 67, no. 1, pp. 201–235, 1978.

[26] W. U. Mohriak, M. R. Mello, G. D. Karner, J. F. Dewey, and J. R. Maxwell, "Structural and stratigraphic evolution of the Campos Basin, offshore Brazil," in *Extension Tectonics and Stratigraphy of the North Atlantic Margins*, AAPG Special Volume, chapter 38, Analogs, pp. 577–598, American Association of Petroleum Geologists Memoir, 1989.

[27] W. U. Mohriak, M. R. Mello, J. F. Dewey, and J. R. Maxwell, "Petroleum geology of the Campos Basin, offshore Brazil," *Geological Society of London, Special Publication*, vol. 50, no. 1, pp. 119–141, 1990.

[28] W. U. Mohriak, A. Z. N. de Barros, and A. Fujita, "Magmatismo e tectonismo Cenozóico na região de Cabo Frio, RJ," in *Proceedings of the 36th Brazilian Geological Congress*, vol. 6, pp. 2873–2885, 1990.

[29] A. M. P. Mizusaki and W. U. Mohriak, "Seqüências Vulcano-sedimentares na Região da Plataforma Continental de Cabo Frio, RJ," in *Proceedings of the 37th Brazilian Geological Congress*, vol. 2, pp. 468–469, 1992.

[30] J. L. P. Moreira, C. A. Esteves, J. J. G. Rodrigues, and C. S. Vasconcelos, "Magmatismo, sedimentação e estratigrafia da porção norte da Bacia de Santos," *Boletim de Geociências da Petrobrás*, vol. 14, no. 1, pp. 161–170, 2006.

[31] S. G. Oreiro, *Magmatismo e sedimentação em uma área na plataforma continental de Cabo Frio, Rio de Janeiro, Brasil, no intervalo Cretáceo Superior—Terciário [M.S. thesis]*, State Uiversity of Rio de Janeiro, Rio de Janeiro, Brazil, 2002.

[32] Diretoria de Hidrografia e navegação (DHN), *Banco Nacional de Dados Oceanográficos—BNDO*, Centro de Hidrografia da Marinha, Rio de Janeiro, Brazil, 1998.

[33] S. G. Zembruscki, "Gradiente geotérmico das bacias sedimentares brasileiras," Internal Report of CENPES (PETROBRÁS) 483, Boletim de Geociencias PETROBRAS, 1982.

[34] American Association of Petroleum Geologists, *Basic Data from AAPG Geothermal Survey of North America*, University of Oklahoma, Norman, Okla, USA, 1976.

[35] F. B. Ribeiro and V. M. Hamza, "Modelling thermal disturbances induced by drilling activity: advances in theory and practice," *Brazilian Journal of Geophysics*, vol. 4, pp. 91–106, 1986.

[36] A. G. Cavalcante, R. M. Argollo, and H. S. Carvalho, "Correção de Dados de Temperatura de Fundo de Poço (Tfp)," *Brazilian Journal of Geophysics*, vol. 22, pp. 233–243, 2004.

[37] J. Gallardo and D. D. Blackwell, "Thermal structure of the Anadarko basin," *AAPG Bulletin*, vol. 83, no. 2, pp. 333–361, 1999.

[38] D. D. Blackwell and M. Richards, "Geothermal Map of North America," American Association of Petroleum Geologists, scale 1:6, 500, 000, 2004.

[39] V. M. Hamza and F. P. Vieira, "Global distribution of the lithosphere-asthenosphere boundary: a new look," *Solid Earth*, vol. 3, pp. 1–13, 2012.

[40] Y. R. Marangoni and V. M. Hamza, "Condutividade térmica de sedimentos da plataforma continental sudeste do Brasil," *Brazilian Journal of Geophysics*, vol. 2, pp. 11–18, 1983.

[41] A. J. L. Gomes and V. M. Hamza, "Avaliação de Recursos Geotermais do Estado do Rio de Janeiro," in *Proceedings of the 8th International Congress of the Brazilian Geophysical Society*, Rio de Janeiro, Brazil, September 2003.

[42] A. C. del Rey and S. G. Zembruscki, "Hydrothermic study of the Espirito Santo and Mucuri Basins," *Boletim de Geociências da Petrobrás*, vol. 5, no. 1–4, pp. 25–38, 1991.

[43] S. M. Viana, *Fluxo Térmico em uma Bacia Sedimentar da Margem Continental Brasileira*, Monografia de Graduação—FGEL/UERJ, 1999.

[44] F. Birch and H. Clark, "The thermal conductivity of rocks and its dependence upon temperature and composition," *American Journal of Science*, vol. 238, pp. 529–558, 1940.

[45] W. Woodside and J. H. Messmer, "Thermal conductivity of porous media, I. Unconsolidated sands, II. Consolidated rocks," *Journal of Applied Physics*, vol. 32, no. 9, pp. 1688–1706, 1961.

[46] J. J. H. C. Houbolt and P. R. A. Wells, "Estimation of heat flow in oil wells based on a relation betweeen heat conductivity and sound velocity," *Geologie en Mijnbouw*, vol. 59, no. 3, pp. 215–224, 1980.

[47] G. R. Beardsmore and J. P. Cull, *Crustal Heat Flow: A Guide to Measurement and Modelling*, Cambridge University Press, 2001.

[48] N. Lu and S. Ge, "Effect of horizontal heat and fluid flow on the vertical temperature distribution in a semi-confining layer," *Water Resources Research*, vol. 32, no. 5, pp. 1449–1453, 1996.

[49] J. D. Bredehoeft and I. S. Papadopulos, "Rates of vertical groundwater movement estimated from the earth's thermal profile," *Water Resources Research*, vol. 1, pp. 325–328, 1965.

[50] M. Reiter, "Using precision temperature logs to estimate horizontal and vertical groundwater flow components," *Water Resources Research*, vol. 37, no. 3, pp. 663–674, 2001.

[51] E. T. Pimentel and V. M. Hamza, "Use of geothermal methods in outlining deep groundwater flow systems in Paleozoic interior basins of Brazil," *Hydrogeology Journal*, vol. 22, pp. 107–128, 2014.

[52] P. Szatmari, J. C. J. Conceição, M. C. Lana, E. J. Milani, and A. P. Lobo, "Mecanismo tectônico do rifteamento Sul-Atlântico," in *Proceedings of the 33rd Brazilian Geological Congress*, pp. 1589–1601, Rio de Janeiro, Brazil, 1984.

[53] P. Szatmari and W. U. Mohriak, "Plate model of post breakup tectono-magmatic activity in SE Brazil and the adjacent Atlantic," in *Proceedings of the 1995 National Symposium on Tectonic Studies*, vol. 5, pp. 213–214, Gramado (RS), Brazil, 1995.

[54] F. F. M. de Almeida, C. D. R. Carneiro, and A. M. P. Mizusaki, "Correlação do magmatismo das Bacias da Margem Continental Brasileira com o das areas emersas adjacentes," *Revista Brasileira de Geociências*, vol. 23, no. 3, pp. 125–138, 1996.

[55] W. U. Mohriak, J. M. Macedo, R. T. Castelani et al., "Salt tectonics and structural styles in the Deep-Water province of the Cabo Frio Region, Rio de Janeiro, Brazil," in *Salt Tectonics: A Global Perspective*, D. G. Jackson, D. G. Roberts, and S. Snelson, Eds., vol. 65, pp. 273–304, American Association of Petroleum Geologists, Memoir, 1996.

[56] W. R. Winter, R. J. Jahnert, and A. B. França, "Bacia de Campos," *Boletim de Geociências da Petrobrás*, vol. 15, no. 2, pp. 511–529, 2007.

[57] A. Thomaz Filho, A. M. P. Mizusaki, E. J. Milani, and P. Césero, "Rifting and magmatism associated with the South America and Africa Breakup," *Revista Brasileira de Geociências*, vol. 30, no. 1, pp. 17–19, 2000.

[58] A. T. Filho, P. de Cesero, A. M. Mizusaki, and J. G. Leão, "Hot spot volcanic tracks and their implications for south American plate motion, Campos basin (Rio de Janeiro state), Brazil," *Journal of South American Earth Sciences*, vol. 18, no. 3-4, pp. 383–389, 2005.

[59] A. A. Bender, U. T. Mello, and C. H. Kiang, "Reconstituição bidimensional da história geológica de bacias sedimentares—teoria e uma aplicação na Bacia de Campos," *Boletim de Geociências da Petrobrás*, vol. 3, no. 1-2, pp. 67–85, 1989.

[60] H. K. Chang, A. A. Bender, and U. T. Mello, "Origem e evolução termomecânica de bacias sedimentares," in *Origem e Evolução de Bacias Sedimentares*, G. P. Raja Gabaglia and E. J. Milani, Eds., pp. 49–71, Petrobras, 1990.

[61] F. G. Gonzaga, *Simulação Geoquímica 1d ao longo de uma seção geológica da bacia de Campos [M.S. thesis]*, Federal University of Rio de Janeiro, 2005.

[62] R. A. Cardoso, "Estudo da evolução tectônica e história térmica da plataforma continental do Estado do Rio de Janeiro," Monography of Graduation, State University of Rio de Janeiro, 2005.

[63] R. A. Cardoso and V. M. Hamza, "Evolução Termotectônica da Parte Oeste da Bacia de Campos," in *Proceedings of the 9th International Congress of the Brazilian Geophysical Society*, pp. 1–6, Salvador, Brazil, September 2005.

[64] R. A. Cardoso, *Evolução Termo-Tectônica da Plataforma Continental do Estado do Rio de Janeiro [M.S. thesis]*, Observatório Nacional, Rio de Janeiro, Brazil, 2007.

[65] M. S. Steckler and A. B. Watts, "Subsidence of the Atlantic-type continental margin off New York," *Earth and Planetary Science Letters*, vol. 42, pp. 1–3, 1978.

[66] J. G. Sclater and P. A. F. Christie, "Continental stretching: an explanation of the pre-mid-cretaceous subsidence of the Central North Sea Basin," *Journal of Geophysical Research*, vol. 85, pp. 3711–3739, 1980.

[67] D. McKenzie, "Some remarks on the development of sedimentary basins," *Earth and Planetary Science Letters*, vol. 40, no. 1, pp. 25–32, 1978.

[68] L. Royden and C. E. Keen, "Rifting process and thermal evolution of the continental margin of Eastern Canada determined from subsidence curves," *Earth and Planetary Science Letters*, vol. 51, no. 2, pp. 343–361, 1980.

[69] K. M. Cohen, S. Finney, and P. L. Gibbard, *International Chronostratigraphic Chart. Version 2013/01*, International Commission on Stratigraphy, 2013, http://www.stratigraphy.org/ICSchart/ChronostratChart2013-01.pdf.

[70] R. R. Cardoso and V. M. Hamza, "A source of error in McKenzie model of lithospheric extension and its implications for petroleum play in sedimentary basins," in *Proceedings of the 11th International Congress of the Brazilian Geophysical Society*, pp. 1–6, Salvador, Brazil, August 2009.

[71] N. V. Lopatin, "Temperature and geological time as factors of carbonifaction," *Izvestiya Akademii Nauk SSSR, Seriya Geologicheskaya*, vol. 3, pp. 95–106, 1971.

[72] PETROMOD, "IES GMBH Integrated Exploration Systems," Aachen, Germany, 2006.

[73] J. L. Dias, J. C. Scarton, F. R. Esteves, M. Carminatti, and L. R. Guardado, "Aspectos da Evolução Tectono-sedimentar e a ocorrência de hidrocarbonetos na Bacia de Campos," in *Origem e Evolução de Bacias Sedimentares*, G. P. Raja Gabaglia and E. J. Milani, Eds., pp. 333–360, Petrobrás, 1990.

[74] V. M. Hamza, R. R. Cardoso, and C. F. Ponte Neto, "Spherical harmonic analysis of earth's conductive heat flow," *International Journal of Earth Sciences*, vol. 97, no. 2, pp. 205–226, 2008.

[75] V. M. Hamza, R. R. Cardoso, and C. H. Alexandrino, "A magma accretion model for the formation of oceanic lithosphere: implications for global heat loss," *International Journal of Geophysics*, vol. 2010, Article ID 146496, 16 pages, 2010.

[76] M. J. Pereira and F. J. Feijó, "Bacia de Santos," *Boletim de Geociências da Petrobrás*, vol. 8, no. 1, pp. 219–234, 1994.

[77] V. M. Hamza, "Pesquisas geotérmicas na exploração de hidrocarbonetos na Bacia do Paraná," Internal Report of Instituto de Pesquisas Tecnologicas—IPT 17, Instituto de Pesquisas Tecnologicas, 1982.

12

Structural Interpretation of the Mamfe Sedimentary Basin of Southwestern Cameroon along the Manyu River Using Audiomagnetotellurics Survey

Jean Jacques Nguimbous-Kouoh,[1] Eric M. Takam Takougang,[2] Robert Nouayou,[1] Charles Tabod Tabod,[1] and Eliezer Manguelle-Dicoum[1]

[1] *Department of Physics, Faculty of Science, University of Yaoundé I, P.O. Box 812, Yaoundé, Cameroon*
[2] *Department of Earth Sciences, Simon Fraser University, 8888 University Drive, Burnaby, BC, Canada V5A1S6*

Correspondence should be addressed to Jean Jacques Nguimbous-Kouoh, nguimbouskouoh@hotmail.fr

Academic Editor: F. Monteiro Santos

Five audiofrequency magnetotelluric AMT soundings were collected northwest-southeast along the Manyu river in the Mamfe sedimentary basin of southwestern Cameroon. The soundings were performed with frequencies in the range 3 to 2500 Hz and covered a distance of approximately 28 km. Sounding curves and geoelectric and geological sections were processed, and the results were compared with rocks' resistivity to characterize the lithostratigraphy of the eastern part of the basin. The results show above 1000 m depth, sedimentary layers with resistivities in the range of 1 to 100 Ohm-m, which decrease with depth. We identified three types of sedimentary rocks: laterite-clay mixture, shale, and sandstones. Various faults were also identified, illustrating the structural complexity of the Mamfe basin, along the Manyu River.

1. Introduction

The geological studies of the Mamfe basin (southwestern Cameroon) were carried out for the first time by Le Fur [1], Dumort [2], and Paterson et al. [3]. These initial studies provided the framework for the first geological map of the area. The geophysical studies in the area integrate gravity works by Collignon [4], Fairhead and Okereke [5], Fairhead and Okereke [6], Fairhead et al. [7], Ndougsa-Mbarga [8], and Ndougsa-Mbarga et al. [9]. The geophysical studies also include audio-magnetotelluric (AMT) works by Manguelle-Dicoum et al. [10], Nguimbous-Kouoh [11], Nouayou [12], and Tabod et al., [13] to map subsurface resistivities.

The aim of this paper is to use AMT field data to characterize the shallow structure of the Mamfe sedimentary basin along the Manyu River. To achieve this, various interpretation techniques were employed. The sounding curves were interpreted to derive the stratigraphy under each AMT station. A pseudosection and geoelectric and geological resistivity sections were then derived along the AMT profile, to deduce the continuity of the subsurface layers and the distribution of associated electrical resistivities. The combination of geoelectrical sections and pseudosections enables a more thorough interpretation [11, 14–18]. The AMT profile has five stations: Ndwap (M1), Abonando (M2), Esagem1 (M3), Esagem2 (M4), and Baku (M5).

2. Geology of the Study Area

The Mamfe sedimentary basin is a rifting basin formed in response to the Gondwana break-up and subsequent separation of the South American and African plates. It lies on an NW-SE trending trough with a length of 130 km and a width of 60 km and constitutes a small prolongation of the Benue trough where important oil fields have been discovered (Figure 1(a)). The basin is favorable for the exploration of ores like lignite, lead, and zinc [2, 8, 11, 12, 19–23]. It is located between the latitudes 5°30′ N and 6°00′ N, and longitudes 8°50′ E and 9°40′ E (Figure 1(b)), with average altitude ranging between 90 and 300 m above sea level.

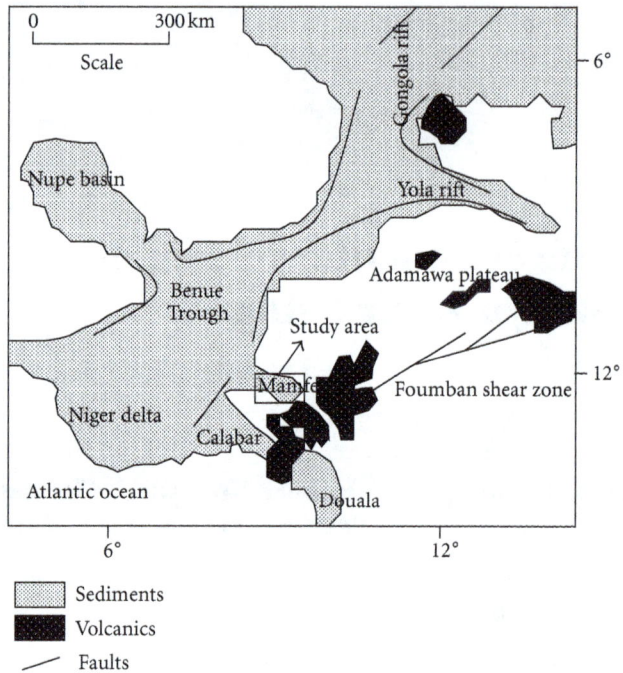

Sediments

Volcanics

Faults

(a)

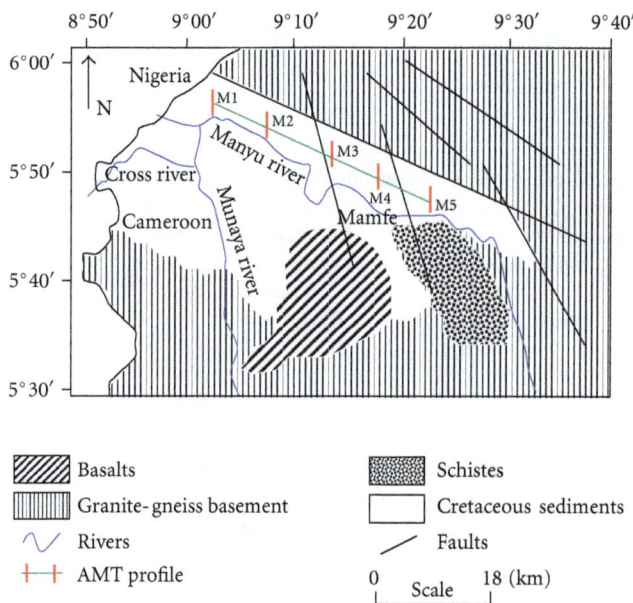

Basalts   Schistes

Granite-gneiss basement   Cretaceous sediments

Rivers   Faults

AMT profile

(b)

FIGURE 1: Geologic map of the study area. (a): Map showing the location of the study area and the location of rifts in the Benue Trough (from Fairhead et al. [7]). (b): Geologic map of the Mamfe basin and AMT stations from [20].

Figure 1(b) shows the available geological map. Some geological features have been extrapolated or withdraw. This is a preliminary geological map that was updated following several studies [2, 9, 20]. The geomorphology of the area is characterized by a succession of horst and grabens [2, 7–9, 20]. The basin is bordered by faults and rivers such as Manyu and Munaya which extend from Cameroon to Nigeria.

The lithology of the basin consists of a thick layer of sediments that impedes the identification and mapping of some major discontinuities at shallow depths. The sedimentary package lies on the Precambrian granite-gneiss basement. The sequence, from bottom to top, presents a succession of granites, schists, shales, sandstones, clays, and laterite (Figure 2). The stratification forms a sigmoid structure typical of a synclinal which is oriented E-W and its axes plunges 10° to 20° W [2, 7, 20, 22–24].

Le Fur [25] recognized five series of sedimentary rocks in the eastern part of the Mamfe basin (Figure 2), and these include from top to bottom [23]:

(i) cross River sandstone series,

(ii) clayey sandstone series,

(iii) upper conglomeratic sandstone series,

(iv) Manyu sandy clay series (This series is mineralized with galena, blende, and pyrite, as well as lignite or brown coal products),

(v) lower conglomeratic sandstone series.

3. Review of the AMT Method

The audiomagnetotelluric method (AMT) is based on the calculation of the transfer functions between the telluric and magnetic fields measured on the surface of the ground. These transfer functions define an impedance tensor Z_{xy} that is calculated assuming a linear relationship between the geomagnetic field Hy (in gamma) and the telluric field Ex (in mV/km) resulting from the interaction between the solar wind and the earth magnetic field in the upper atmosphere [26, 27]. The subsurface apparent resistivity is related to this impedance by the relation $\rho = 0.2/f \cdot |Z_{xy}|^2$. The apparent depth at a given frequency f (Hz) and for an apparent resistivity ρ (Ohm-m) is given by the relation ρ (Km) = $0.503(\rho/f)^{1/2}$ [28–30]. Electromagnetic waves at a given frequency penetrate deeper in resistive rocks than in conductive rocks. Therefore, by changing the frequency, different depth of the subsurface can be imaged. This constitutes a fundamental basis for the AMT method.

Processing of observed data (sounding curves) is based on best-fitting the observed data with the computed data. In order to mitigate the nonuniqueness of solution, a priori geological information of the study area is usually necessary. If the subsurface is isotropic, the telluric and magnetic fields are perpendicularly to one another, and the telluric field shows a shift of 45° compared to magnetic field (advance in phase). When the subsurface is anisotropic, the telluric and magnetic fields are no more orthogonal, and the shift deviates from 45°. The apparent resistivity and the phase shift can be estimated in the direction of the telluric field (ρaE and $\Delta\varphi$aE, resp.) and in the direction of the magnetic field (ρaH and $\Delta\varphi$aH, resp.) [31, 32].

4. Data Acquisition

The data were collected with an audio-frequency resistivimeter [1], which measures the values of the electrical resistivity

Structural Interpretation of the Mamfe Sedimentary Basin of Southwestern Cameroon along the Manyu River Using Audiomagnetotellurics Survey

111

Lithology	Description	Age
	Basalts, syenites, and trachytes	Tertiary
	Sandstones, arkosik, and conglomeratic	Cenomanian
	Oolitic, shales with centimetric veinlets of lignite, and bituminous shales	Albian
	Conglomerates overlain by arkosic sandstones showing cross bedding	
	Siliceous and calcareous shales intercalated with arkosic sandstones	
	Conglomeratic sandstones with arkosic cements containing basement fragments up to 15 cm in diameter	
	Basement made up of granites schists and gneisses	Precambrian

FIGURE 2: Generalised stratigraphic column of the Mamfe Basin showing ages of units, lithology, and probable source rocks from Le Fur [25] and Eseme et al. [23].

and the apparent phase for twelve different frequencies (2500 to 3 Hz). The stations were laid out with an equidistance of approximately 5 km, due to irregularities of the relief and ease to equipment accessibility. The electrodes were positioned at each station on a surface area of about 10.000 m² with two electrodes aligned in the N-S magnetic axis (Hy) and two in the E-W telluric axis (Ex). The location of the stations, and the directions of the telluric line, which was assumed to be parallel to the basin, were obtained with a compass and a GPS. Measurements were carried at five successive stations (M1–M5) between the localities of Ndwap (M1) and Baku (M5) (Figure 1(b)). The geographical coordinates that help locating these stations are shown in Table 1. The average-apparent resistivity and the average phase for each frequency were calculated to obtain a good image of the pseudosection and the geoelectric section. The formulas $\rho_{am} = \sqrt{\rho_{N-S} \cdot \rho_{E-W}}$ for the average apparent resistivity and $\varphi_{am} = 1/2 \, (\varphi_{N-S} + \varphi_{E-W})$ for the average phase were used [11, 12, 32]. We applied the one-dimensional AMT principle: when two sounding curves measured in both telluric and magnetic directions along the main axes show similar apparent resistivity trends, 1D interpretation of average sounding curves is justified and the pseudosection and geoelectric section can be derived without any other errors beside those due to data quality [11, 12, 14–16]. The static shift was removed by compensating for site anisotropy [33–37]. The data inversion was done using IPI 2WIN and AMTINV software [17, 18].

5. Results and Interpretation

5.1. Interpretation of the Sounding Curves. The 1D models in Figure 3 show the sounding and phase curves at each station. The number of layers, thickness, and resistivity was derived

from the best fit between experimental and theoretical curves. The data were interpreted with a maximum number of four layers, considering the sedimentation sequence and the lower resistivity of rocks in the area [7, 11, 12]. The number of layers at each station was controlled by the use of the root means square (RMS) misfit (0.006 to 0.20) between the field and modeled data and by the use of a damping-mean (0.71 to 0.87).

The interpreted models show the subsurface resistivity of the basin, along the Manyu River, to a depth of about 800 m. Overall, the first layer is relatively resistive (4–79 Ohm-m) with thickness of about 7 to 31 m. The underlying second layer is less resistive (2 to 48 Ohm-m) and extends between 66 to 250 m depth. The third layer has a thickness ranging between 250 and 600 m with electrical resistivities of about 1 to 3 Ohm-m and overlies the fourth layer, which has a resistivity of about 4 to 20 Ohm-m.

5.2. Interpretation of the Phase Curves. An important property of the phase is that it reacts slowly to variation in subsurface resistivities. Another application of the phase is that it is often used to measure the quality of the data [38, 39]. In all the models of Figure 3, the phase curves are plotted as a function of the frequency. At high frequencies, the phase curves at Abonando (M2), Esagem1 (M3), and Baku (M5) stations are slightly below 45 degree, showing that the first layer is less resistive than the second. At low frequencies, the phase curves are slightly above 45 degree and imply that the third layer is increasingly conductive. However, the phase curves are greater than 45 degree or converge asymptotically towards 45 degree, in Ndwap (M1) and Esagem2 (M4). This shows that in these areas, the first layer is more resistive than the whole subsurface layers at all the frequencies.

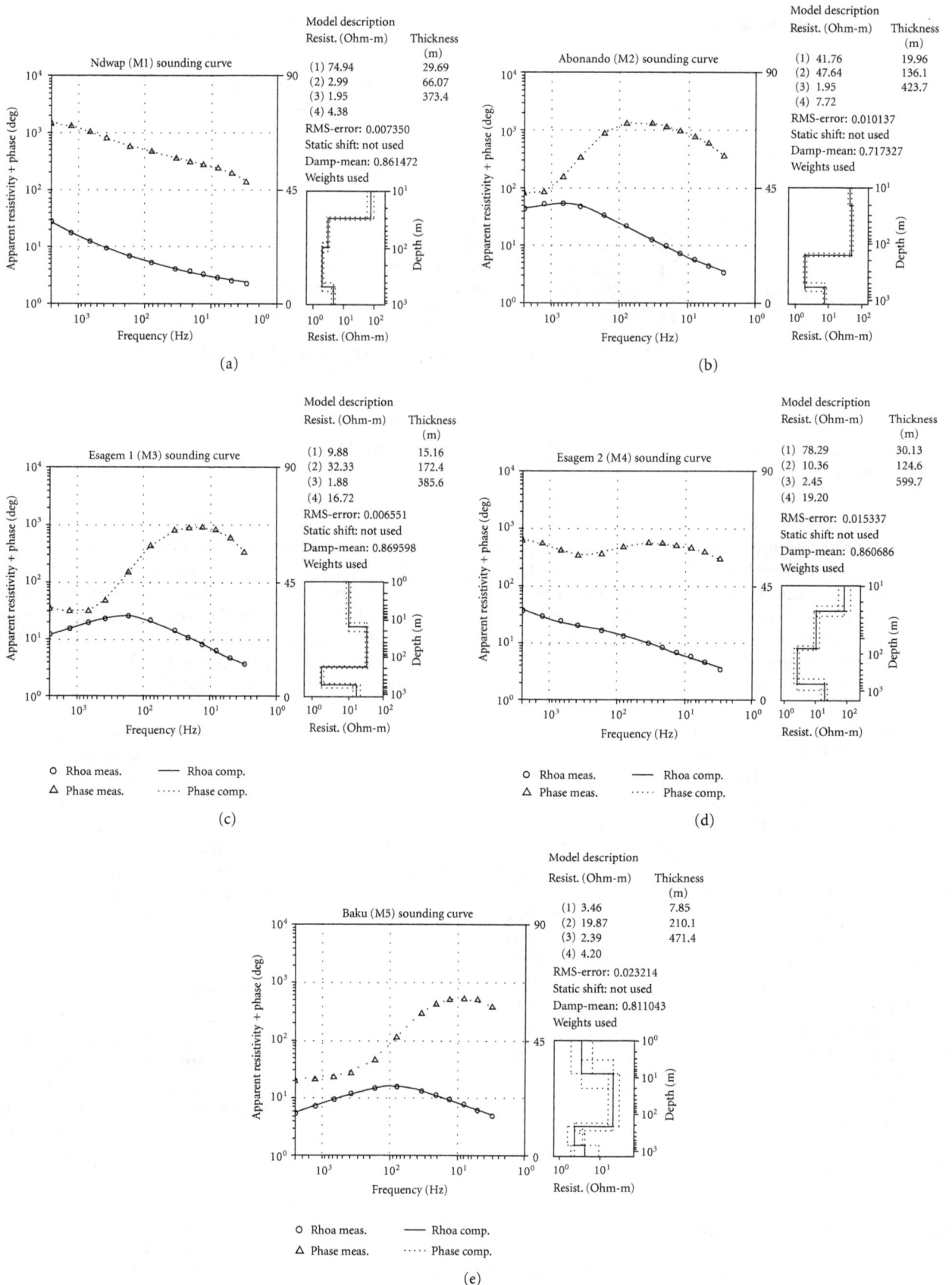

FIGURE 3: 1D modelling results of five stations.

Structural Interpretation of the Mamfe Sedimentary Basin of Southwestern Cameroon along the Manyu River Using
Audiomagnetotellurics Survey

113

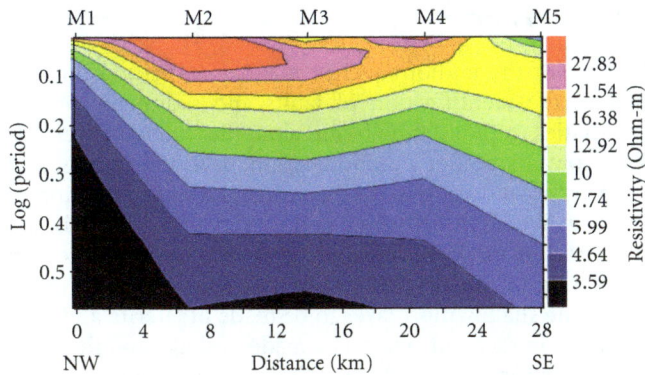

FIGURE 4: Derived pseudosection along the profile.

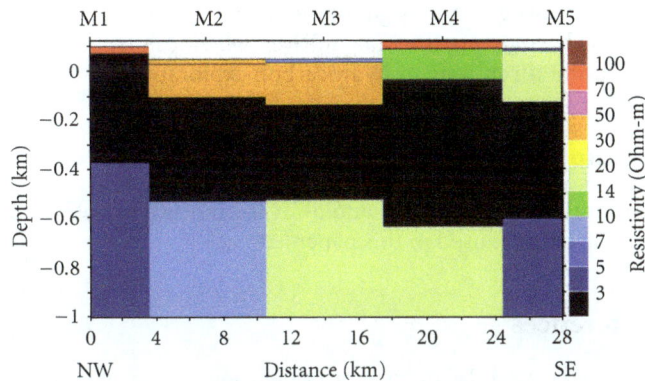

FIGURE 5: Geoelectric model derived from correlation of sounding curves along the profile.

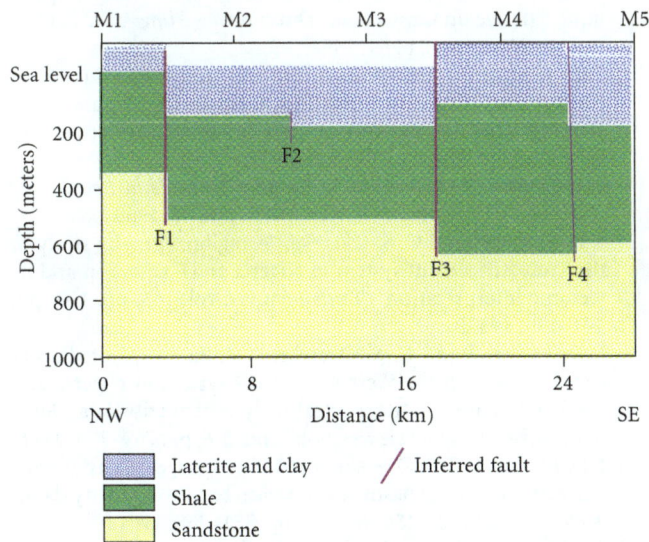

FIGURE 6: Geological section along Manyu River, showing probable stratigraphy and formation.

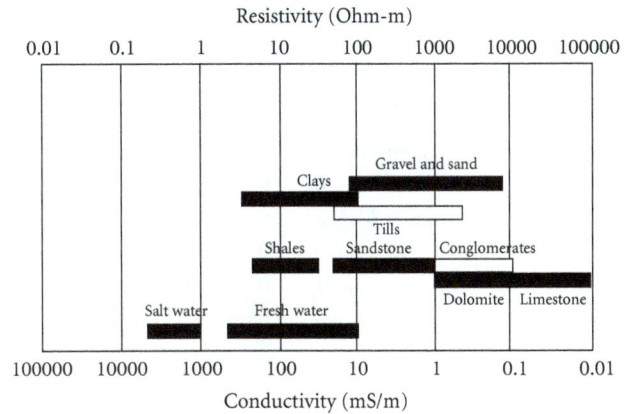

FIGURE 7: Resistivity of geological formations from [42].

structures and formations. In this type of representation, the distances between stations are displayed in the abscissa (linear scale), and the square roots of the periods are displayed in the ordinate (logarithmic scale) [11, 12, 17]. It can be observed that the variations in resistivities are slightly contrasted at the surface. These variations highlight three major superficial discontinuities in M2, M3, and M4. The pseudosection also shows a regular decrease of the resistivity with the period, thus suggesting the occurrence of conductive rocks at greater depths. These relatively conductive layers are characterized by resistivities that are lower than 17 Ohm-m. The conductive layers stretch across the pseudosection.

The geoelectric section (Figure 5) shows the electrical resistivity model which bestfits AMT responses along the profile. This model displays a relatively conductive zone with electrical resistivities ranging from 1 to 100 Ohm-m, located down to a depth of about 1000 m. The model highlights a layer whose electrical resistivities vary between 1 and 3 Ohm-m (in black). This layer has a depth, which varies between 100 and 200 m and can be related to conductive layers previously mentioned in the interpretation of sounding curves. This layer is sandwiched between two layers with electrical resistivity 10 to 100 Ohm-m and thickness 100 to 200 m, and with deeper layers characterized by lower electrical resistivities (20 Ohm-m).

Although the model does not allow delineation of the granite-gneiss basement geometry, it helps to locate the limits of the geological formations and the fractured zones cutting the profile, either by a reduction or an increase in the electrical resistivity. As a whole, this geoelectric section helped to conceive the geologic section of the first 1000 m of this part of the basin.

In the absence of borehole data, which are generally necessary for this kind of interpretation [36, 40, 41], the geoelectrical section (Figure 5), which was obtained by correlating 1D models at each station, was compared with rocks' resistivity (Figure 7) and the lithostratigraphic column of the eastern part of the Mamfe basin proposed by Le Fur [25] and Eseme et al. [23] (Figure 2) to deduce a geological section (Figure 6) along the profile. The geological section in Figure 6 highlights three different resistivity layers. The first layer, which is the combination of the first two layers

5.3. Structural Interpretation. Our structural interpretation is based on the pseudosection (Figure 4) and the geoelectric section (Figure 5) that enables us to derive a geological section (Figure 6).

The apparent resistivity pseudosection (Figure 4) helps to constrain the underground continuity of geological

TABLE 1: Geographic coordinates of the measuring sites. The geodetic system used to locate the stations is WGS 84, and the system of projection used is UTM (Universal Transverse Mercator). The latitudes and longitudes are in degree.

Stations	Longitude °E	Latitude °N	Elevation (m)
NDWAP (M1)	009°03.920	05°56.151	100
ABONANDO (M2)	009°06.985	05°53.820	50
ESAGEM1 (M3)	009°11.846	05°52.574	50
ESAGEM2 (M4)	009°15.091	05°47.664	120
BAKU (M5)	009°18.360	05°48.007	89

identified on the 1D models and geoelectric section (Figures 3 and 5), has an average thickness of approximately 200 m with resistivities in the range 20–100 Ohm-m. This layer can be associated with mixture of laterite and clay of Tertiary age [11, 12, 23]. The second layer has an average thickness of 400 m with resistivities in the range of approximately 1 to 3 Ohm-m. It can be due to the black shale of Manyu [11, 12, 23]. The third layer has an average thickness of 500 m with resistivities that range from 3 to 20 Ohm-m; it can be associated with sandstones of Albian age [11, 12, 23]. We also identified four discontinuities between stations M1 and M5, which appear to correlate with faults of geological map (Figure 1(b)) or a change in altitude (Table 1). The geological map shows a fault between stations M2 and M3 and between M4 and M5. We interpret the fault between M2 and M3 (F2) to be located at about 150 m depth below the surface. The fault between M4 and M5 (F4) is characterized in our model by a discontinuity. Another discontinuity is observed between stations M3 and M4, and we postulate that this discontinuity may also be due to a fault (F3). It should be noted that, due to the relatively large spacing between stations (5 km), the precise delimitation of these faults is not possible.

6. Discussion

We identified three major resistivity layers covering a depth of 7 to 1000 m, with low resistivities in the range 1 to 100 Ohm-m, indicating the conductive nature of the subsurface in this part of the basin. The models in Figures 3 and 5 show a progressive decrease in resistivity with depth. The granite-gneiss basement characterize by greater resistivity values seems sufficiently far from the surface in our models. The gravity studies of Fairhead et al. [7] and Ndougsa-Mbarga [8] showed that the depth of the basin can vary between 700 m to 3000 m at different sites of the basin. Based on the geoelectrical section, we can conclude that sedimentary layer can exceed 1000 m thickness.

The results of Nouayou [12] and Tabod et al. [13] predict the occurrence of volcanic rocks in the southern part of the basin with resistivities that vary between 250 and 1000 Ohm-m and depths 30 to 100 m. We did not see evidence for shallow volcanic rocks in our profile because the resistivity values are well below 250 Ohm-m.

7. Conclusion

The use of the audiofrequency magnetotellurics technique enables the identification of faults and sediments of low resistivities in the Mamfe basin, along the Manyu River. We use a combination of a pseudosection, sounding curves, and a geoelectrical section to derive a geological interpretation of the area. We identified three main layers of low resistivities (1–100 Ohm-m), covering a depth of 800–1000 m. Based on correlation of our derived resistivities with rocks, we interpreted the shallowmost layer to be made of laterite and clay, the following second layer to be made of shale, and the third layer to be made of sandstone. Our interpretation of faults, which is based on discontinuities in the derived geoelectric section, illustrates the structural complexity of the basin. These results give an insight of the geological structure of the Mamfe basin along the Manyu River, which can be useful in the future for mining or hydrocarbon exploration.

Acknowledgment

The authors thank IRGM and SNH for the authorization to use the data set used in this paper.

References

[1] Lagas, 1993, Laboratoire de Géophysique Applique et structurale du CNRS.

[2] J. C. Dumort, "Carte géologique de reconnaissance à l'échelle 1/500000. Note explicative sur la feuille Douala-Ouest. République fédérale du Cameroun," Direction des Mines et de la Géologie du Cameroun, p. 69, 1968.

[3] Paterson, Grant and Watson, "Etudes aéromagnétiques sur certaines régions de la République Unie du Cameroun," Rapport d'interprétation, Agence Canadienne de Développement International, Toronto, Canada, 1976.

[4] F. Collignon, Gravimétrie de reconnaissance de la République Fédérale du Cameroun, Ostrom, Paris, France, 1968.

[5] J. D. Fairhead and C. S. Okereke, "A regional gravity study of the West African rift system in Nigeria and Cameroon and its tectonic interpretation," Tectonophysics, vol. 143, no. 1–3, pp. 141–159, 1987.

[6] J. D. Fairhead and C. S. Okereke, "Depths to major density contrats beneath the West African rift system in Nigeria and Cameroon based on the spectral analysis of gravity data," Journal of African Earth Sciences, vol. 7, no. 5-6, pp. 769–777, 1988.

[7] J. D. Fairhead, C. S. Okereke, and J. M. Nnange, "Crustal structure of the Mamfe basin, West Africa, based on gravity data," Tectonophysics, vol. 186, no. 3-4, pp. 351–358, 1991.

[8] T. Ndougsa-Mbarga, Etude géophysique, par méthode gravimétrique des structures profondes et superficielles de la région de Mamfé, Thèse de Doctorat, Faculté des Sciences, Université de Yaoundé I, Yaoundé, Cameroun, 2004.

[9] T. Ndougsa-Mbarga, E. Manguelle-Dicoum, J. O. Campos-Enriquez, and Q. Y. Atangana, "Gravity anomalies, sub-surface structure and oil and gas migration in the Mamfe, Cameroon-Nigeria, sedimentary basin," Geofisica Internacional, vol. 46, no. 2, pp. 129–139, 2007.

[10] E. Manguelle-Dicoum, R. Nouayou, C. Tabod, and T. E. Kwende-Mbanwi, "Audio and helio magnetotelluric study of the Mamfe sedimentary basin," Tech. Rep., 1999.

[11] J. J. Nguimbous-Kouoh, *Apport de l'audio-magnétotellurique (AMT) pour l'Etude des couches superficielles le Long du Fleuve Manyu*, Mémoire de DEA, Faculté des Sciences, Université de Yaoundé I, Yaoundé, Cameroun, 2003.

[12] R. Nouayou, *Contribution à l'étude géophysique du bassin sédimentaire de Mamfe par prospections audio et hélio magnétotelluriques*, M.S. thesis, spécialité Géophysique Interne, Université de Yaoundé I, Yaoundé, Cameroun, 2005.

[13] C. T. Tabod, A. P. Tokam Kamga, E. Manguelle-Dicoum, R. Nouayou, and S. Nguiya, *An Audio-Magnetotelluric Investigation of the Eastern Margin of the Mamfe Basin, Cameroon*, The Abdus Salam International Centre for Theoretical Physics, Trieste, Italy, 2008.

[14] A. Dupis, *Première Application de la Magnétotellurique à la Prospection Pétrolière*, Doctorat d'état ès-sciences, Géologique ou Minière de diverses régions Métropolitaines, 1970.

[15] Y. Benderritter, *Interprétation des mesures magnétotelluriques à l'aide d'un résistivimètre ECA*, Centre de recherches géophysiques, Nièvre, France, 1982.

[16] P. Andrieux, 1987, Application des Sondages Magnétotellurique à l'exploration à moyennes et à grandes profondeurs. Dossier présenté en appui à une demande d'habilitation à diriger des recherches (U.P.M.C., Paris VI).

[17] A. Bobachev, *IPI2WIN in (MT) V.2.0 Is Designed for Automated and Interactive Semi-Automated Interpreting of Magnetotelluric Sounding Data Using Amplitude and/or Phase Curves*, Geoscan-M, Moscow, Russia, 2001.

[18] Pitisharvi, 2004, AMTINV for automated and interactive 1-D interpretation of audio-magnetotelluric EM soundings Version 1.3 (c).

[19] M. M. Eben, "Report of the geological expedition in the gulf of Mamfe: archives of the department of mines & geology," Tech. Rep., Ministry of Mines & Power, Douala, Cameroon, 1984.

[20] J. V. Hell, V. Ngako, V. Bea, J. B. Olinga, and J. T. Eyong, "Rapport des travaux sur l'étude de reconnaissance géologique du bassin sédimentaire de Mamfé: IRGM-SNH," Tech. Rep., 2000.

[21] J. T. Eyong, *Litho-Biostratigraphy of the Mamfe Cretaceous Basin, Communication of the Department of Earth Sciences*, Faculty of Science University of Yaoundé I, Yaoundé, Cameroon, 2001.

[22] R. Kangkolo, "Aeromagnetic study of the Mamfe basalts of southwestern Cameroon," *Journal of the Cameroon Academy of Sciences*, vol. 2, no. 3, pp. 173–180, 2002.

[23] E. Eseme, C. M. Agyingi, and J. Foba-Tendo, "Geochemistry and genesis of brine emanations from Cretaceous strata of the Mamfe Basin, Cameroon," *Journal of African Earth Sciences*, vol. 35, no. 4, pp. 467–476, 2002.

[24] R. Kangkolo and S. B. Ojo, "Integration of aeromagnetic data over the Mamfe basin of Nigeria and Cameroon," *Nigeria Journal of Physics*, vol. 7, pp. 53–56, 1995.

[25] Y. le Fur, "Mission socle-Crétacé. Rapport 1964–1965 sur les indices de plomb et zinc du golfe de Mamfe," Tech. Rep., Rapport B.R.G.M., Cameroun, 1965.

[26] A. N. Tikhonov, "On determining electrical characteristics of the deep layers of the earth's crust," *Doklady Akademii Nauk, USSR*, vol. 73, no. 2, pp. 295–297, 1950.

[27] L. Cagniard, "Basic theory of the magnetotelluric method in geophysical prospecting," *Geophysics*, vol. 18, no. 3, pp. 605–635, 1953.

[28] K. Vozoff, "The magnetotelluric method," in *Electromagnetic Methods in Applied Geophysics*, M. N. Nabighian, Ed., vol. 2, part B, pp. 641–711, Society of Exploration Geophysicists, Tulsa, Okla, USA, 1991.

[29] M. Bastani, *EnviroMT New Controlled Source/Radio Magnetotelluric*, Ph.D. thesis, Uppsala University, Uppsala, Sweden, 2001.

[30] M. Bastani and L. B. Pedersen, "Estimation of magnetotelluric transfer functions from radio transmitters," *Geophysics*, vol. 66, no. 4, pp. 1038–1051, 2001.

[31] J. M. Travassos and P. T. L. Menezes, "Geoelectric structure beneath limestones of the Sao Francisco Basin, Brazil," *Earth, Planets and Space*, vol. 51, no. 10, pp. 1047–1058, 1999.

[32] L. B. Pedersen and M. Engels, "Routine 2D inversion of magnetotelluric data using the determinant of the impedance tensor," *Geophysics*, vol. 70, no. 2, pp. G33–G41, 2005.

[33] A. G. Jones, "Static shift of magnetotelluric data and its removal in a sedimentary basin environment," *Geophysics*, vol. 53, no. 7, pp. 967–978, 1988.

[34] L. Pellerin and G. W. Hohmann, "Transient electromagnetic inversion: a remedy for magnetotelluric static shifts," *Geophysics*, vol. 55, no. 9, pp. 1242–1250, 1990.

[35] E. Manguelle-Dicoum, R. Nouayou, A. S. Bokosah, and T. E. Kwende-Mbanwi, "Audiomagnetotelluric soundings on the basement-sedimentary transition zone around the eastern margin of the Douala Basin in Cameroon," *Journal of African Earth Sciences*, vol. 17, no. 4, pp. 487–496, 1993.

[36] M. Chouteau, P. Zhang, D. J. Dion, B. Giroux, R. Morin, and S. Krivochieva, "Delineating mineralization and imaging the regional structure with magnetotellurics in the region of Chibougamau (Canada)," *Geophysics*, vol. 62, no. 3, pp. 730–748, 1997.

[37] X. Garcia and A. G. Jones, "A new methodology for the acquisition and processing of audio-magnetotelluric (AMT) data in the dead band," *Geophysics*, vol. 70, no. 5, pp. 119–126, 2005.

[38] P. Zhang, R. G. Roberts, and L. B. Pedersen, "Magnetotelluric strike rules," *Geophysics*, vol. 52, no. 3, pp. 267–278, 1987.

[39] M. Hjärten, *Master thesis in interpretation of controlled-source radiomagnetotelluric data from Hallandsåsen*, M.S. thesis, Uppsala universitet Institutionen för Geovetenskaper—Geofysik, Uppsala, Sweden, 2007.

[40] T. Dahlin, "2D resistivity surveying for environmental and engineering applications," *First Break*, vol. 14, no. 7, pp. 275–283, 1996.

[41] J. Pratt and J. Craven, 2010, magnetotelluric imaging of the Nachako basin, Bristish Colombia. Geological survey of Canada, Current research 2010-3, 9p.

[42] Bemex Consulting International, *Interpretation of Apparent Resistivity Maps and Resistivity cross Sections from the Kotcho Region, N.E. British Columbia*, British Columbia Ministry of Energy and Mines, British Columbia, Canada, 2004.

13

Streaming Potential and Electroosmosis Measurements to Characterize Porous Materials

D. T. Luong and R. Sprik

Van der Waals-Zeeman Institute, University of Amsterdam, 1098 XH Amsterdam, The Netherlands

Correspondence should be addressed to R. Sprik; r.sprik@uva.nl

Academic Editors: E. Liu, A. Streltsov, and A. Tzanis

Characterizing the streaming potential and electroosmosis properties of porous media is essential in applying seismoelectric and electroseismic phenomena for oil exploration. Some parameters such as porosity, permeability, formation factor, pore size, the number of pores, and the zeta potential of the samples can be obtained from elementary measurements. We performed streaming potential and electro-osmosis measurements for 6 unconsolidated samples made of spherical polymer particles. To check the validity of the measurements, we also used alternative analysis to determine the average pore size of the samples and, moreover, used a sample made of sand particles to determine the zeta potential.

1. Introduction

Recently, seismoelectric and electroseismic conversions which arise due to the coupling of seismic waves and electromagnetic waves have been studied in order to investigate oil and gas reservoirs [1] or hydraulic reservoirs [2–4]. These phenomena have been used to deduce the depth and the geometry of the reservoir [5]. The coupling coefficients of conversion between electric wave and flow depend strongly on the fluid conductivity, porosity, permeability, formation factor, pore size, zeta potential of porous media, and other properties of the rock formation [6]. Therefore, determining these parameters is very important in studying electrokinetics in general and to model seismoelectric and electroseismic conversions. Li et al. [7] used two reciprocal electrokinetic phenomena known as streaming potential and electro-osmosis by Ac measurement to determine the effective pore size and permeability of porous media. In [8], the authors used image analysis to determine the number of pores per cross-sectional area of porous samples (see Figure 1 for the schematic of the porous medium with different length scales). This parameter is especially important in processes of contaminant removal from low-permeability porous media under a Dc electric field [8], and in building electro-osmosis micropumps [9].

However, the method used in [8] did not work for porous media with very small pores such as Bentonite clay soils or tight-gas sandstones (the pore radius is smaller than $1\,\mu m$) that are relevant for application in the oil and gas industry. In oil exploration and production, the typical pore sizes in rocks are necessary information for considering the location of oil and fluid flow through the rocks. The characteristics of porous media also determine differential gas pressures needed to overcome capillary resistance of tight-gas sandstones in gas production

Alternative methods such as nuclear magnetic resonance (NMR) or magnetic resonance imaging (MRI) [10] can also be used to determine characteristics of porous media such as the porosity and pore size distribution, the permeability, and the water saturation. But this technique is quite expensive and is not able to determine the zeta potential—one of the most important parameters in electrokinetic phenomena.

Here we used Dc measurements of streaming potential and electro-osmosis in porous samples and other simple measurements to fully characterize porous media and determine parameters needed for the experimental study of seismoelectric and electroseismic conversions. Our approach works well for very small pores in particular.

This work includes five sections. Section 2 describes the theoretical background of electrokinetics. Section 3 presents

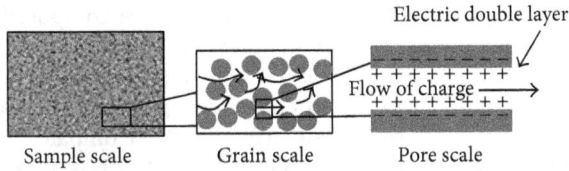

FIGURE 1: Schematic of the porous medium with different length scales: sample scale, grain scale, and pore scale.

the investigated samples and the experimental methods. Section 4 contains the experimental results and their interpretation using the model proposed by [11]. Conclusions are provided in Section 5.

2. Theoretical Background of Electrokinetics

2.1. Electric Double Layer. Electrokinetic phenomena are induced by the relative motion between the fluid and a wall, and they are directly related to the existence of an electric double layer (EDL) between the fluid and the solid surface. When a solid surface is in contact with a liquid, an electric field is generated perpendicular to the surface which attracts counterions (usually cations) and repulses anions in the vicinity of the liquid-solid interface. This leads to the charge distribution known as the EDL. The EDL is made up of the Stern layer, where cations are adsorbed on the surface and are immobile due to the strong electrostatic attraction, and the Gouy diffuse layer, where the ions are mobile. In the bulk liquid, the number of cations and anions is equal so that it is electrically neutral. The closest plane to the solid surface in the diffuse layer at which flow occurs is termed the shear plane or the slipping plane, and the electrical potential at this plane is called the zeta potential. The characteristic length over which the EDL strongly exponentially decays is known as the Debye length, and it is on the order of a few nanometers for typical grain electrolyte combinations [6] (for more detail, see [12–14]).

2.2. Streaming Potential. The streaming current is created by the motion of the diffuse layer with respect to the solid surface induced by a fluid pressure drop over the channel. This streaming current is then balanced by a conduction current, leading to the streaming potential. In a porous medium (see Figure 1), the electric current density and the fluid flux are coupled, so fluids moving through porous media generate a streaming potential [15]. The streaming potential increases linearly with the fluid pressure difference that drives the fluid flow, provided that the flow remains laminar [16]. The steady-state streaming potential coupling coefficient, C_S, is defined when the total current density is zero as follows:

$$C_S = \frac{\Delta V}{\Delta P} = \frac{\epsilon \zeta}{\eta \sigma_{\text{eff}}}, \tag{1}$$

where ΔV is the streaming potential, ΔP is the fluid pressure difference, ϵ is the dielectric permittivity of the fluid, η is the dynamic viscosity of the fluid, ζ is the zeta potential, and σ_{eff} is the effective conductivity which includes the intrinsic fluid

conductivity and the surface conductivity (that is due to the electric double layer and the surface itself). The streaming potential is independent of the sample geometry.

According to [15], C_S can be written as

$$C_S = \frac{\epsilon \zeta}{\eta \sigma_{\text{eff}}} = \frac{\epsilon \zeta}{\eta F \sigma_S}, \tag{2}$$

where σ_S is the electrical conductivity of the sample saturated with a fluid with a conductivity σ_f and F is the formation factor. If surface conductivity is negligible, then $\sigma_{\text{eff}} = \sigma_f$ and the coupling coefficient becomes the equation

$$C_S = \frac{\epsilon \zeta}{\eta \sigma_f}. \tag{3}$$

2.3. Electroosmosis. Electroosmosis was first observed by Reuss in 1809 in an experiment where a direct current was applied to a clay-sand-water mixture in a U-tube [17]. When an electric field is applied parallel to the wall of a capillary, ions in the diffuse layers experience a Coulomb force and move toward the electrode of opposite polarity, which creates a motion of the fluid near the wall and transfers momentum via viscous forces into the bulk liquid. So a net motion of bulk liquid along the wall is created and is called electroosmosis flow.

A complex porous medium (see Figure 1) with the physical length L and cross-sectional area A can be approximated as an array of N parallel capillaries with inner radius equal to the average pore radius a of the medium and an equal value of zeta potential ζ. For each of these idealized capillaries, the solution for electro-osmosis flow in a single tube can be analyzed to estimate the behavior of the total flow in a porous medium by integrating over all pores [18].

In a U-tube experiment, when potential difference is applied across the fluid-saturated porous medium, the liquid rises on one side (the cathode compartment for our experiment) and lowers on the other side (the anode compartment). This height difference increases with the time, and this process stops when the hydraulic pressure caused by the height difference equals the electro-osmosis pressure (see Figure 2). At that time, the height difference is maximum.

The expression for the height difference as function of time is given by [8]

$$\Delta h = \frac{\Delta P_{\text{eq}}}{\rho_f g} \left[1 - \exp\left(-\frac{N \rho_f g a^4}{4 \mu R^2 L} t \right) \right]$$
$$= \frac{\Delta P_{\text{eq}}}{\rho_f g} \left[1 - \exp\left(-\frac{t}{\tau} \right) \right], \tag{4}$$

with

$$\Delta P_{\text{eq}} = \frac{8\epsilon |\zeta| V}{a^2} \left[1 - \frac{2\lambda I_1 (a/\lambda)}{a I_0 (a/\lambda)} \right], \tag{5}$$

$$\tau = \frac{4 \mu R^2 L}{N \rho_f g a^4}, \tag{6}$$

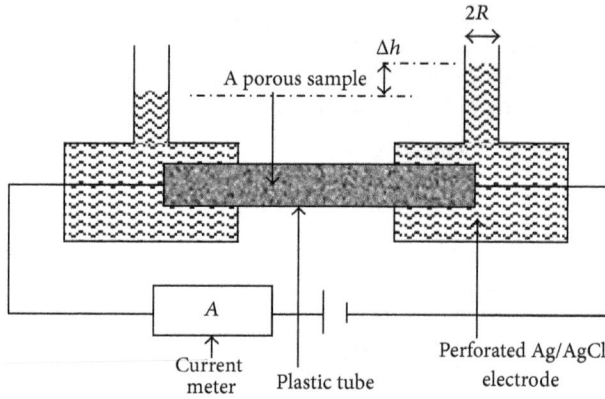

FIGURE 2: Experimental setup for electro-osmosis measurements in which Δh and R are height difference of liquid and the radius of the tubes in both sides, respectively.

where τ is response time, ΔP_{eq} is the pressure difference caused by the electro-osmosis flow at equilibrium which corresponds to maximum height difference, V is the applied voltage across the porous medium, ρ_f is the fluid density, g is the acceleration due to gravity, λ is the Debye length, R is the radius of the tubes in both sides, and I_0 and I_1 are the zero-order and the first-order modified Bessel functions of the first kind, respectively.

For a conductive liquid such as distilled water, the Debye length λ is about 2 nm [8], and a typical pore radius of our samples a (see below) is around 3 μm in this case; the ratio $I_1(a/\lambda)/I_0(a/\lambda)$ can be neglected [19]. Under these conditions, (5) may be simplified as

$$\Delta P_{eq} = \frac{8\epsilon |\zeta| V}{a^2}, \tag{7}$$

and (4) can be rewritten as follows:

$$\Delta h = \frac{8\epsilon |\zeta| V}{\rho_f g a^2} \left[1 - \exp\left(-\frac{t}{\tau}\right) \right] = \Delta h_{max} \left[1 - \exp\left(-\frac{t}{\tau}\right) \right], \tag{8}$$

with

$$\Delta h_{max} = \frac{8\epsilon |\zeta| V}{\rho_f g a^2}. \tag{9}$$

3. Experiment

To demonstrate that one can characterize porous media by obtaining parameters such as porosity, permeability, formation factor, pore size, the number of pores, and zeta potential of the liquid fully saturated porous media through electrokinetics, streaming potential and electro-osmosis measurements have been performed on 7 unconsolidated samples. Six of them are spherical monodisperse particle packs with different diameters (10 μm, 20 μm, 40 μm, 140 μm, 250 μm, and 500 μm) of the particles. These are obtained

from Microbeads AS Company, and they are composed of polystyrene polymers. Those samples are designated as TS10, TS20, TS40, TS140, TS250, and TS500, respectively. We also used an unconsolidated sample made up of blasting sand particles obtained from Unicorn ICS BV Company with diameter in the range of 200–300 μm, and this is designated as S_{sand}.

When using low electrical conductivity solutions such as deionized water, the magnitude of the coupling coefficient is large. The electrical conductivity of the saturated samples slowly stabilizes in about 24 h for our samples. Perhaps due to CO_2 uptake from the air, that changes the conductivity. We, therefore, use a 10^{-3} M NaCl solution of low enough conductivity of 10×10^{-3} S/m measured by the conductivity meter (Consort C861) for the measurements. All measurements were carried out at room temperature (20°C).

3.1. Sample Assembly. Samples were constructed by filling polycarbonate plastic tubes (1 cm in inner diameter and 7.5 cm in length) successively with 2 cm thick layers of particles that were gently tamped down, and they were then shaken by a shaker (TIRA-model TV52110). Filter paper was used in both ends of the tube to retain the particles and is permeable enough to let the fluid pass through. The samples were flushed with deionized water to remove any powder or dust.

3.2. Porosity, Permeability, and Formation Factor Measurements. The porosity was measured by a simple method [9]. The sample was first dried in oven for 24 hours, then cooled to room temperature, and finally fully saturated with deionized water under vacuum. The sample was weighed before (m_{dry}) and after saturation (m_{wet}), and the porosity was determined as

$$\phi = \frac{\left(m_{wet} - m_{dry}\right)/\rho}{AL}, \tag{10}$$

where ρ is density of the deionized water and A and L are the inner cross-sectional area and the physical length of the tubing, respectively.

The permeability k was measured by constant flow-rate method. A high-pressure pump (LabHut, Series III-Pump) ensures a constant flow through the sample, and a high-precision differential pressure transducer (Endress and Hauser Deltabar S PMD75) is used to measure the pressure drop. At different flow rates the pressure drop was measured to determine the permeability of the sample.

Method of determining the tortuosity was proposed in [20]. They defined the formation factor F as

$$F = \frac{\alpha_\infty}{\phi} = \frac{\sigma_f}{\sigma_S}, \tag{11}$$

where α_∞ is the tortuosity, σ_S is the electrical conductivity of the saturated sample, σ_f is the intrinsic fluid conductivity, and ϕ is the porosity of the sample.

Our experimental setup is similar to the one used in [21] and is shown in Figure 3. The electrodes, Ag/AgCl mesh discs,

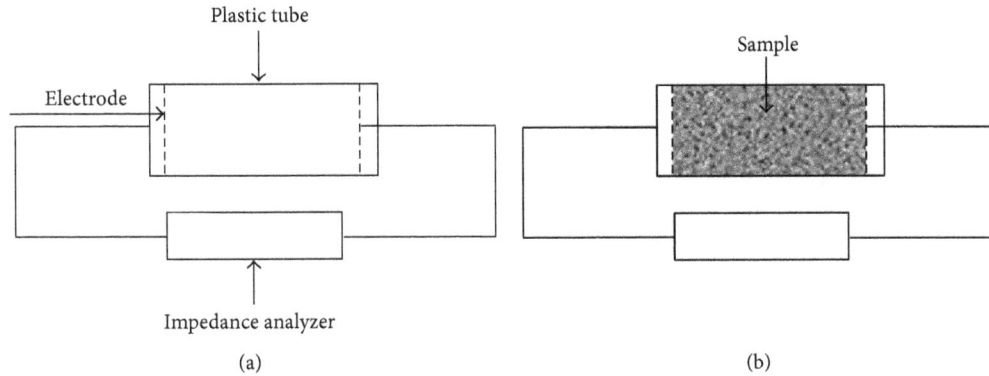

FIGURE 3: Setup for measuring the electrical conductivity of a porous medium saturated with an electrolyte on the right. On the left-hand side is the identical setup without the porous medium for measuring the fluid conductivity.

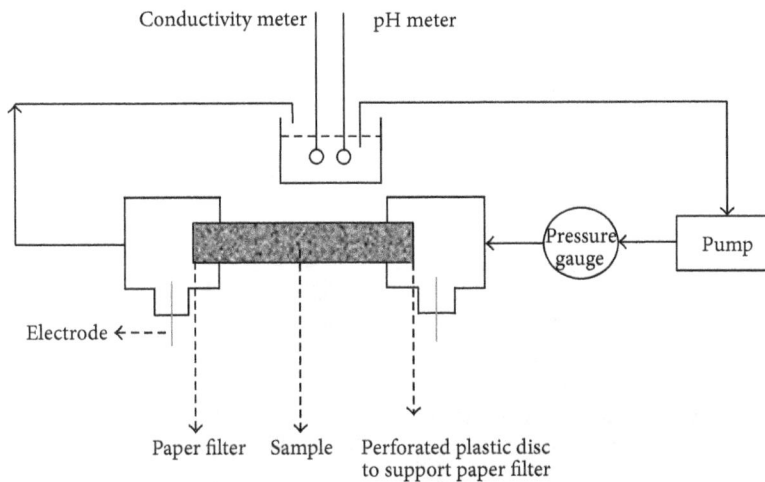

FIGURE 4: Experimental setup for streaming potential measurements.

were placed on both sides against the porous sample that was saturated successively with the following set of aqueous NaCl solutions with different conductivities (0.13, 0.47, 0.81, 1.23, 1.51, and 1.98 S/m). For consolidated sandstone cores, when the electrical conductivity of solution is higher than 0.60 S/m, the surface conductivity is negligible [22]. According to [16], the surface conductivity is around 0.43 mS/m for a 50–60 μm sand pack and is inversely proportional to the grain diameter. For the materials we used, it is likely that the surface conductivity is smaller than that of traditional materials such as sand or sandstone, so that the surface conductivity can also be neglected for our experiments. The electrical conductivity was measured using a Hioki IM3570 impedance analyzer at different frequencies (varying from 100 Hz to 100 kHz).

3.3. Streaming Potential Measurement.
The experimental setup for the measurement of the streaming potential is shown in Figure 4. The pressure differences across the sample were created by the high-pressure pump and were measured by the pressure transducer. The electrical potential was measured by two Ag/AgCl wire electrodes (A-M systems). The electrodes were put in the vicinity of the end faces of

the sample but not within the liquid circulation to avoid the electrical noise from liquid movement around the electrodes [15].

The tubing circuit is shown in Figure 4; the electrolyte from the outlet tube is not in contact with the electrolyte used to pump liquid through the samples, preventing an electric current leakage through the liquid in the tube. The solution was circulated through the samples until the electrical conductivity and pH of the solution reached a stable value. Electrical potentials across the samples were then measured by a high-input impedance multimeter (Agilent 34401A) connected to a computer and controlled by a Labview program (National Instruments). The electrical potentials at a given pressure difference fluctuate around a specific value (see Figure 7); the Labview program averages the value of electrical potentials. The pH values of equilibrium solutions, measured with the pH meter (Consort C861), are in the range 7.1 to 7.6, and the solutions were also used for electro-osmosis measurements.

3.4. Electroosmosis Measurement.
The experimental setup for the electro-osmosis measurement is shown in Figure 2. The

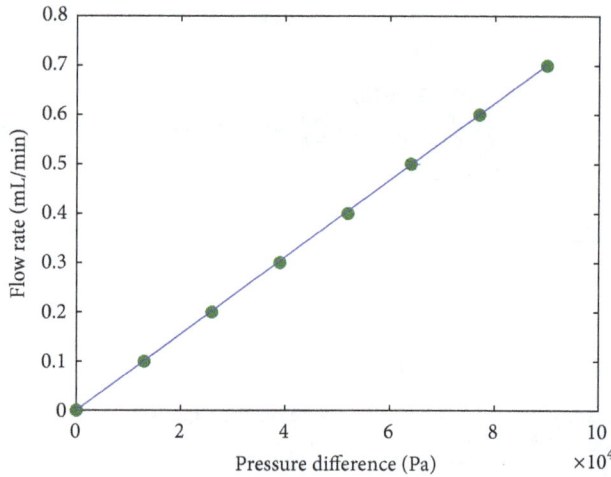

FIGURE 5: The flow rate against pressure difference. Two runs are shown for sample TS10.

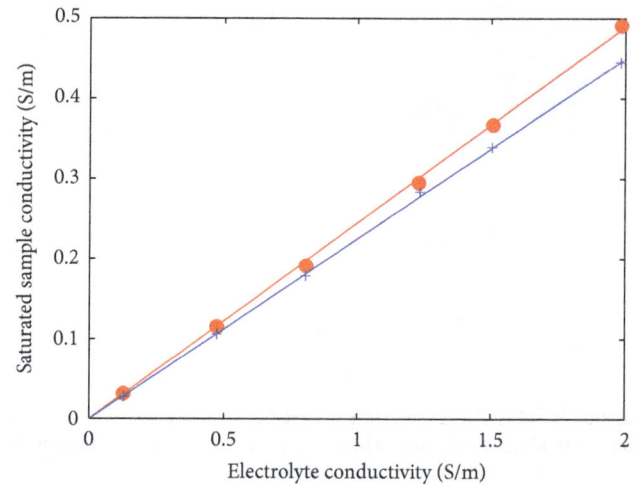

FIGURE 6: Saturated sample conductivity versus electrolyte conductivity for 2 samples (red dots for TS10 and blue cross symbols for TS500). The slopes of the straight lines yield the formation factors.

same solution used in the streaming potential measurement was also used for this measurement. The zeta potential is consequently the same for both kinds of measurements. The electrodes used to apply a Dc voltage across the samples are perforated Ag/AgCl electrodes (MedCaT). To measure the maximum height difference, $\Delta h_{\max}$, and the height difference as a function of time at a given voltage, cameras (Philips SPC 900NC PC) with the assistance of HandiAVI software were used to take pictures of the heights of the liquid columns in time. For each new measurement (new applied voltage, new sample), the samples were dried, mounted in the setup, evacuated by a vacuum pump, and fully saturated by the same solutions.

It should be noted that, when an applied voltage exceeds a critical value (1.48 V for water [23]), there will be electrolysis at the anode and the cathode. These electrode reactions produce ions and gas in both electrodes. If these ions are not removed, then these reactions induce a low pH at the anode, a high pH at the cathode, and a change in electrical conductivity. The rate of electrolysis reaction is largely determined by the current. If the current density is smaller than $<35\ \mu A$ per cm^2 cross-sectional area, then the effects due to the electrolysis can be ignored [24]. The resistances of the fully saturated samples that we used in this paper are about 400 kΩ, so applied voltages were limited below 10 V to avoid unwanted electrolysis effects.

4. Results and Discussion

4.1. Porosity, Permeability, and Formation Factor. The measured porosity of the packs as mentioned in Section 3.2 is 0.39 independently of the size of the particles with an error of 5%. Figure 5 shows the typical graph of flow rate as a function of applied pressure difference for sample TS10.

The graph shows that there is a linear relationship between flow rate and pressure difference, and Darcy's law is obeyed. So the flow is laminar, and (1) and (2) are valid. This behavior is identical for all samples. Two measurements

TABLE 1: Measured properties of the samples in which d, k_o, F, and σ_S are diameter, permeability, formation factor, and electrical conductivity of the fully saturated sample at equilibrium for all samples, respectively.

Sample	d (in μm)	k_o (in m^2)	F	σ_S (in S/m)
TS10	10	0.15×10^{-12}	4.0	4.1×10^{-3}
TS20	20	0.30×10^{-12}	4.2	3.6×10^{-3}
TS40	40	0.85×10^{-12}	4.2	3.0×10^{-3}
TS140	140	1.36×10^{-12}	4.3	3.1×10^{-3}
TS250	250	1.71×10^{-12}	4.0	3.5×10^{-3}
TS500	500	2.36×10^{-12}	4.3	2.9×10^{-3}
S_{sand}	200–300	1.22×10^{-12}	4.0	3.0×10^{-3}

were performed for all samples to find the graphs of flow rate versus pressure difference. From the slope of the graph and Darcy's law (the viscosity of the fluid was taken as 10^{-3} Pa.s), the permeability of the sample was calculated. We obtained permeabilities of all samples (see Table 1) with an uncertainty of 15% in the reported values.

An example of the electrical conductivity of the samples versus the electrical conductivity of the electrolyte is shown in Figure 6 for the 2 samples with the largest differences in the formation factors. We calculated the formation factor F as the reciprocal of the slope of a linear regression through the data points. Values of the formation factors for all samples are also reported in Table 1 with the 5% error.

The measured formation factor of the samples is the range from 4.0 to 4.3 (see Table 1). According to Archie's law, $F = \phi^{-m}$ (F is the formation factor, ϕ is the porosity of the sample, and m is the socalled cementation exponent), m was found to be in the range 1.47–1.55. For unconsolidated samples made of perfect spheres, the exponent m should be 1.5 [25]. So the measured formation factors of the samples are in good agreement with Archie's law. The electrical conductivities of the samples fully saturated by the solutions are also shown in Table 1.

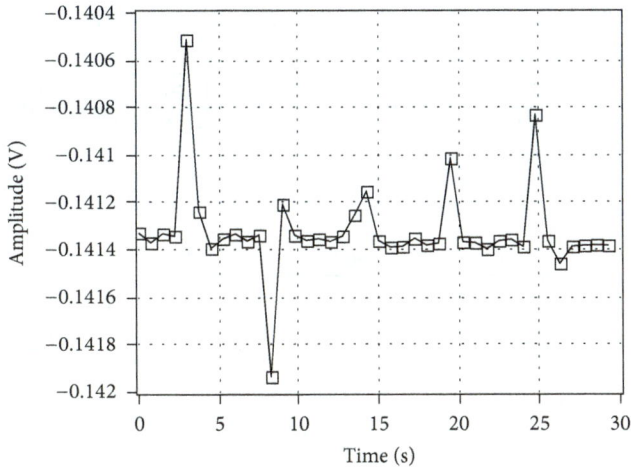

FIGURE 7: The electrical potential (V) fluctuating with time (s) at a given pressure drop for sample TS10 was taken by Labview.

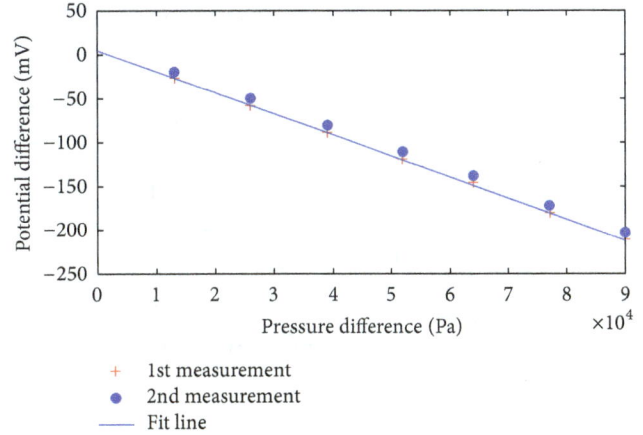

FIGURE 8: Measured streaming potential versus applied pressure difference for sample TS10 at two different times for equilibrium electrical conductivity of liquid of $11 \cdot 10^{-3}$ S/m.

4.2. Streaming Potential.

The typical fluctuation of electrical potentials at a given pressure difference is shown in Figure 7. The final value of electric potential for each pressure difference was taken as the average value of all datapoints.

The streaming potential as a function of pressure difference was measured twice for each sample. Figure 8 shows two typical sets of measurements for sample TS10 in which the second measurement was carried out 10 h after the first one. The graph shows that there was a very small variation of streaming potentials with time (the drift is about 1 mV/h), and the straight lines fitting the data points do not go through the origin. This may be due to the electrode polarizations. However, this variation has no influence on the coupling coefficient because the slopes of straight lines are almost the same for two separate measurements.

Streaming potential coupling coefficients at 3 different electrical conductivities for sample TS10 are shown in Figure 9. From Figures 8 and 9, we see that the magnitude of the streaming potential is proportional to the driving pressure difference and is inversely proportional to the liquid electrical conductivity, as expected from (1) and (2).

From the coupling coefficients, the formation factors, and the electrical conductivities of the samples, the zeta potential of the samples can be obtained by using (2) as shown in Table 2. Because the measured coupling coefficient was always negative, the zeta potential obtained from the measurement was also negative.

4.3. Electroosmosis.

Figure 10 shows the measured maximum height difference versus applied voltage for sample TS10. We observe that there is a linear relationship between the maximum height difference and the applied voltage as expected from (9) except for the last datapoint (when the applied voltage was 7.5 V). That last datapoint which deviated from the linear trend could be due to the electrolysis happening on both electrodes as mentioned in Section 3.4. Using the slope of the graph and the zeta potential obtained from streaming

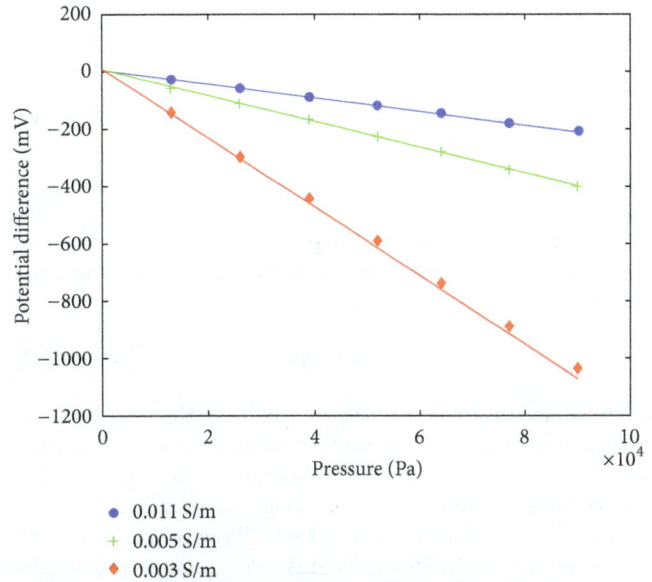

FIGURE 9: Example of typical runs for sample TS10 at three different liquid conductivities. We observe linear relationships between measured streaming potentials and applied pressure differences. At each electrical conductivity, the streaming potential coupling coefficient is equal to the slope of the linear trend.

potential measurement, we can estimate the average pore size of the samples from (9) (see Table 2).

Because of the limitation of applied voltage, the electroosmosis measurements were only performed for sample TS10 and TS20. The height difference as a function of time carried out for sample TS10 at possible maximum applied voltage of 6 V is shown in Figure 11. The graph has an exponential curve as expected from (8). By using the exponential part of the graph, the response time in (6) was obtained (see Figure 12). From the calculated response time and parameters of the samples, the number of pores on average can be determined by electro-osmosis measurements (see Table 2).

TABLE 2: Calculated parameters of the samples in which ζ is the zeta potential in mV, a is average pore radius in μm, N is the average number of pores per cross-sectional area of the samples, and k_o is the permeability of the samples in m^2.

Sample	TS10	TS20	TS40	TS140	TS250	TS500	S_{sand}
ζ	−32.4	−5.2	−6.3	−12.5	−7.2	−9.1	13.7
a	2.3	3.2					
N	775×10^3	482×10^3					
k_o	0.16×10^{-12}	0.31×10^{-12}					
a from [11]	1.5	2.9	5.9	20.6	36.8	73.5	

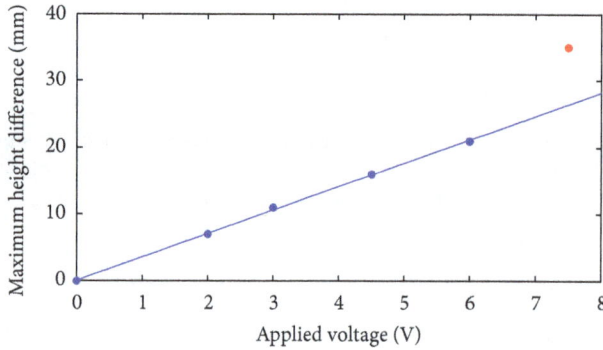

FIGURE 10: Maximum height difference as a function of applied voltage for sample TS10.

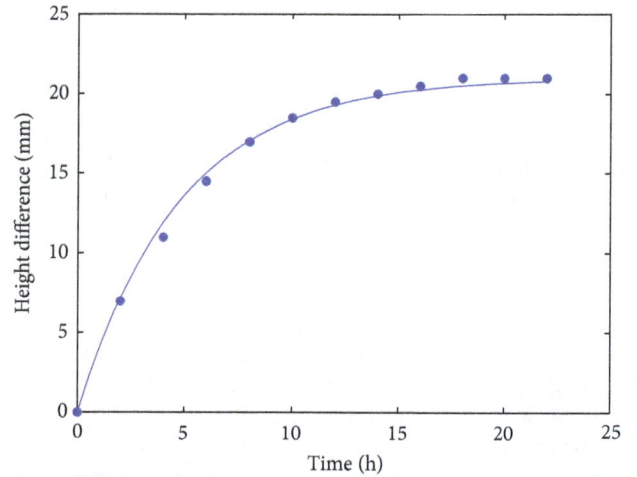

FIGURE 11: The dotted line is the experimental measurement of height difference as a function of time for sample TS10 at voltage of 6 V (dotted line). The solid line is fit through the datapoints.

To check the validity of the pore size estimation, we used the relationship between grain diameter and effective pore radius given by [11]

$$d = 2\theta a, \qquad (12)$$

where θ is the theta transform function that depends on parameters of the porous samples such as porosity, cementation exponent, and formation factor of the samples. For the samples made of the monodisperse spherical particles arranged randomly, θ is taken to be 3.4. From pore sizes determined by the electro-osmosis measurements, permeabilities of the samples were also calculated by the model of [11] (see Table 2) as

$$k_o = \frac{a^2 \phi^{3/2}}{8}, \qquad (13)$$

where ϕ is the porosity of samples. From Table 2, we see that the pore size estimated from the electro-osmosis measurement is in good agreement with that estimated from the model of [11] and that, in addition, the calculated permeabilities are in good agreement with the measured ones in Table 1.

5. Conclusions

Streaming potential measurements have been performed for 7 unconsolidated samples fully saturated with a 10^{-3} M NaCl solution to determine the zeta potentials. Because of the limitation of the voltage one can apply, we carried out the electro-osmosis measurements for two smallest particle

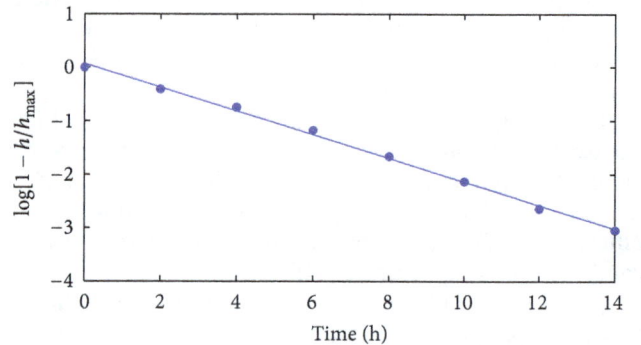

FIGURE 12: The slope of the straight line is equal to the reciprocal of the response time.

samples. This allows us to estimate average pore sizes of the samples as well as the number of pores these cannot be determined by a method such as the image analysis performed in [8] for very smallpore porous media. The estimated pore sizes and the measured permeabilities have been compared to those calculated from the model of [11] to check the validity of the measurements.

The comparison shows that the measured pore sizes are in good agreement with the model of [11]. If the electro-osmosis measurements can be improved, for example, by using ion-exchange membranes or a large crosssectional area of the samples to reduce the experimental time, then this method can be applied for any kind of porous media, and it will be especially useful for very small pores.

According to [26, 27], an average value of zeta potential is about −17 mV for sands or sandstones so that the value of −13.7 mV from our measurements is within the range expected. Therefore, our approach can be effectively used to characterize porous media by using simple measurements, in particular, for very smallpore porous media that are relevant for application in the oil and gas industry.

Moreover, our streaming potential and electro-osmosis measurements also worked for the interface between a liquid and a polymeric material. The polymer may be a more promising material for electro-osmosis micropumps besides traditional materials like silica particles. The zeta potential of the polymer material that we used in this paper is a little bit smaller than that of sands or sandstones.

Acknowledgments

The authors would like to thank Professor Dr. Daniel Bonn for his very helpful suggestions and comments which have helped them to improve this paper. They also would like to thank Karel Heller for his help in improving the setups as well as measurements.

References

[1] A. Thompson, S. Hornbostel, J. Burns et al., "Field tests of electroseismic hydrocarbon detection," *SEG Technical Program Expanded Abstracts*, vol. 24, p. 565, 2005.

[2] S. S. Haines, A. Guitton, and B. Biondi, "Seismoelectric data processing for surface surveys of shallow targets," *Geophysics*, vol. 72, no. 2, pp. G1–G8, 2007.

[3] M. Strahser, L. Jouniaux, P. Sailhac, P.-D. Matthey, and M. Zillmer, "Dependence of seismoelectric amplitudes on water content," *Geophysical Journal International*, vol. 187, no. 3, pp. 1378–1392, 2011.

[4] S. Garambois and M. Dietrich, "Seismoelectric wave conversions in porous media: field measurements and transfer function analysis," *Geophysics*, vol. 66, no. 5, pp. 1417–1430, 2001.

[5] A. H. Thompson, J. R. Sumner, and S. C. Hornbostel, "Electromagnetic-to-seismic conversion: a new direct hydrocarbon indicator," *Leading Edge*, vol. 26, no. 4, pp. 428–435, 2007.

[6] S. Pride, "Governing equations for the coupled electromagnetics and acoustics of porous media," *Physical Review B*, vol. 50, no. 21, pp. 15678–15696, 1994.

[7] S. X. Li, D. B. Pengra, and P.-Z. Wong, "Onsager's reciprocal relation and the hydraulic permeability of porous media," *Physical Review E*, vol. 51, no. 6, pp. 5748–5751, 1995.

[8] T. Paillat, E. Moreau, P. O. Grimaud, and G. Touchard, "Electrokinetic phenomena in porous media applied to soil decontamination," *IEEE Transactions on Dielectrics and Electrical Insulation*, vol. 7, no. 5, pp. 693–704, 2000.

[9] S. Zeng, C.-H. Chen, J. C. Mikkelsen Jr., and J. G. Santiago, "Fabrication and characterization of electroosmotic micropumps," *Sensors and Actuators B*, vol. 79, no. 2-3, pp. 107–114, 2001.

[10] D. Bonn, S. Rodts, M. Groenink, S. Rafaï, N. Shahidzadeh-Bonn, and P. Coussot, "Some applications of magnetic resonance imaging in fluid mechanics: complex flows and complex fluids," *Annual Review of Fluid Mechanics*, vol. 40, pp. 209–233, 2008.

[11] P. W. J. Glover and E. Walker, "Grain-size to effective pore-size transformation derived from electrokinetic theory," *Geophysics*, vol. 74, no. 1, pp. E17–E29, 2009.

[12] R. J. Hunter, *Zeta Potential in Colloid Science*, Academic Press, New York, NY, USA, 1981.

[13] P. W. J. Glover and M. D. Jackson, "Borehole electrokinetics," *The Leading Edge*, vol. 29, no. 6, pp. 724–728, 2010.

[14] F. C. Schoemaker, N. Grobbe, M. D. Schakel, S. A. L. de Ridder, E. C. Slob, and D. M. J. Smeulders, "Experimental validation of the electrokinetic theory and development of seismoelectric interferometry by cross-correlation," *International Journal of Geophysics*, vol. 2012, Article ID 514242, 23 pages, 2012.

[15] L. Jouniaux, M. L. Bernard, M. Zamora, and J. P. Pozzi, "Streaming potential in volcanic rocks from Mount Pelée," *Journal of Geophysical Research B*, vol. 105, no. 4, pp. 8391–8401, 2000.

[16] A. Boleve, A. Crespy, A. Revil, F. Janod, and J. L. Mattiuzzo, "Streaming potentials of granular media: influence of the Dukhin and Reynolds numbers," *Journal of Geophysical Research*, vol. 112, no. 8, 2007.

[17] F. Reuss, "Sur un nouvel effet de l'électricité galvanique," *Mémoires de la Societé Imperiale de Naturalistes de Moscou*, vol. 2, pp. 327–336, 1809.

[18] S. Yao and J. G. Santiago, "Porous glass electroosmotic pumps: theory," *Journal of Colloid and Interface Science*, vol. 268, no. 1, pp. 133–142, 2003.

[19] C. L. Rice and R. Whitehead, "Electrokinetic flow in a narrow cylindrical capillary," *Journal of Physical Chemistry*, vol. 69, no. 11, pp. 4017–4024, 1965.

[20] R. J. S. Brown, "Connection between formation factor for electrical resistivity and fluid-solid coupling factor in Biot's equations for acoustic waves in fluid-filled porous media," *Geophysics*, vol. 45, no. 8, pp. 1269–1275, 1980.

[21] J. G. M. van der Grinten, "Tortuosity measurement and its spin-off to a new electrical conductivity standard," in *Learned and Applied Soil Mechanics*, F. B. J. Barends and P. M. P. C. Steijger, Eds., pp. 143–148, Taylor & Francis, New York, NY, USA, 2002.

[22] S. F. Alkafeef and A. F. Alajmi, "Streaming potentials and conductivities of reservoir rock cores in aqueous and non-aqueous liquids," *Colloids and Surfaces A*, vol. 289, no. 1, pp. 141–148, 2006.

[23] D. R. Lide, *CRC Handbook of Chemistry and Physics*, CRC Press, New York, NY, USA, 2004.

[24] D. H. Gray and J. K. Mitchell, "Fundamental aspects of electro-osmosis in soils," *Journal of the Soil Mechanics and Foundations Division*, vol. 93, no. 6, pp. 209–236, 1967.

[25] P. N. Sen, C. Scala, and M. H. Cohen, "A self-similar model for sedimentary rocks with application to the dielectric constant of fused glass beads," *Geophysics*, vol. 46, no. 5, pp. 781–795, 1981.

[26] L. Jouniaux, "Electrokinetic techniques for the determination of hydraulic conductivity," in *Hydraulic Conductivity—Issues, Determination and Applications*, L. Elango, Ed., InTech, 2011.

[27] L. Jouniaux and T. Ishido, "Electrokinetics in earth sciences: a tutorial," *International Journal of Geophysics*, vol. 2012, Article ID 286107, 16 pages, 2012.

14

Assessment of Economically Accessible Groundwater Reserve and Its Protective Capacity in Eastern Obolo Local Government Area of Akwa Ibom State, Nigeria, Using Electrical Resistivity Method

N. J. George,[1] E. U. Nathaniel,[1] and S. E. Etuk[2]

[1] Department of Physics, Akwa Ibom State University, Ikot Akpaden 53001, Nigeria
[2] Department of Physics, University of Uyo, Uyo 53001, Nigeria

Correspondence should be addressed to N. J. George; nyaknojimmyg@gmail.com

Academic Editors: E. Liu, G. Mele, F. Monteiro Santos, and B. Tezkan

The application of geophysical method employing vertical electrical sounding (VES) method in combination with laboratory analysis of aquifer sediments has been used to access the economically accessible groundwater reserve and its protective capacity in some parts of Eastern Obolo Local Government area, the eastern region of the Nigerian Niger Delta. Schlumberger electrode configuration was used to sound twelve VES to occupy the areas that have borehole locations and accessibility for the spread of current electrodes to at least 1000 m. Based on the results, the safe and economic aquifer potential has groundwater reserve of about 168480558 ± 18532861 m^3. The desired aquifer thickness and its depth of burial have average value of 52.02 m and 73.14 m, respectively. The area has a fair protective capacity. This is indicated by 58.33% weak, 16.67% moderate, and 25% good protective capacity for the area. This study was done in one of the oil cities, where contaminated Salt River water is used as the major source of water for domestic uses and it is believed that the settlers will appropriate this result and sue for safe groundwater at the indicated depths.

1. Introduction

The need to assess safe groundwater repositories is increasingly important in Eastern Obolo Local Government Area (EOLGA), located in Eastern Niger Delta region of Nigeria, because of the extant health hazards posed by surface and underground contaminations which are associated with chemical, microbial, and physical contaminant plumes. The focus of groundwater quality protection is on the prevention of groundwater pollution; however, where groundwater has become polluted, it must be cleaned up and managed to ensure the ongoing protection of human health and the environment. Access to clean water is a human right and basic requirement for economic development. The worldwide development of past civilisations as well as the recent socioeconomic evolution of nations is based on and strongly

controlled by the availability of water which can be obtained either as surface or subsurface water.

The Niger Delta region (the site for which the study area is located) is one of the most industrialised parts in the entire Gulf of Guinea. These industries have contributed immensely to economic growth and development of the states within the region. The region which is one of the ten most important wetland and marine ecosystems in the world has suffered immensely from these unsustainable industrial activities [1]. Some of the human activities that have impacted negatively on the Niger Delta environment include industrial (hydrocarbon exploration and exploitation, noise pollution, gas flaring, oil spillage, and waste disposal), removal of backshore vegetation, construction of barges and other coastal control works, river dredging, agricultural (excessive and uncontrolled application of inorganic fertilizers, pesticides,

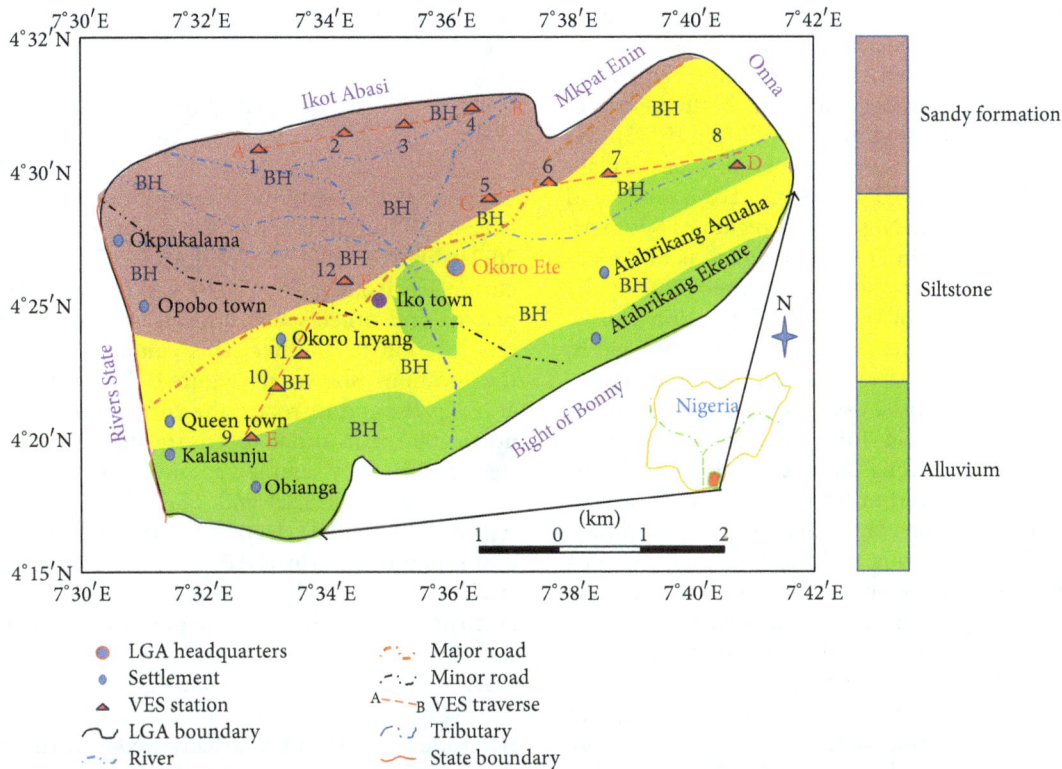

FIGURE 1: Location map showing the study location, VES points and VES traverses.

and herbicides), municipal waste disposal, urbanisation, and mining activities [2, 3]. Environmental problems like loss of biodiversity, coastal and riverbank erosion, flooding, land degradation, loss of soil fertility, and deforestation are now common [1, 4]. Most of the numerous surface water resource endowments of the area which the entire population used to be solely dependent on for their domestic, agricultural, industrial, and social needs are currently seriously polluted by both natural and anthropogenic sources [4]. Recently, the activities of criminals who are involved in illegal pipeline vandalism and crude oil theft and refining have compounded the environmental problems in the Niger Delta region. The quality of the groundwater resources in many parts of the region is gradually being degraded [5]. There is urgent need to understand the extent of natural protection, the distributions of the economically safe aquifer repository, and possibly the pollutant flow path in order to design appropriate mitigation, remediation, and protection strategies especially now that such plans are in the drawing board by governmental, environmental, and other international partners. Design of appropriate groundwater management strategies in any geologic environment depends to a reasonable extent on the nature of subsurface materials whose properties (physical and chemical) and spatial distribution constitute the goal of all hydrogeological and hydrogeophysical investigations. The surface water is found to be grossly degraded in quality because of its exposure to physical, biological, or chemical contaminants [6]. Groundwater on its own has less degree of contaminations when relatively compared with surface

water. Inadequate public water supply has led to increased demand for alternative sources of water supply in EOLGA due to rapidity in human growth in recent times [7]. Today, we are witnessing the increasing number of boreholes drilled by the government, nongovernmental organizations, and individuals. This clearly shows that groundwater is effectively complementing other sources of water supply in the area. This is due to the rate of contamination of surface water through the effect of inordinate quest for development. Wildcat drilling, which does not support any scientific search for the location of the water bearing sediments in the area, has exposed many groundwater resources to contaminations. In view of this, the area was mapped in order to access the quantitative reserve, distribution of the safe and economic water bearing units, its depth of burial, and the protective capacity of the groundwater repositories in the study area.

2. Location and Geology of the Study Area

The study area shown in Figure 1 lies between longitudes 7°30′ and 7° 42′ E and latitudes 4°15′ and 4°32′ N in the Niger Delta region of Southern Nigeria. The study was designed to cover some areas of Eastern Obolo Local Government Areas of Akwa Ibom State of Nigeria, one of the oil rich regions of the Arcuate Niger Delta, surrounded by both fresh and saline water. The study area is located in an equatorial climatic region that is characterised by two major seasons. The seasons are the rainy season (March–October) and dry season (November–February) [7, 8]. The dry season is a period of extreme aridity characterized by excruciating high temperatures that do reach 35°C. The area has been severely affected by the current global climatic changes in such a way that there have been shifts in both the upper and lower boundaries of these climatic conditions [9–13]. The study area is located in the tertiary to quaternary coastal plain sands (CPS) (otherwise called the Benin Formation) and alluvial environments of the Niger Delta region of Southern Nigeria (Figure 1). The Benin Formation which is underlain by the paralic Agbada Formation covers over 80% of the study area. The sediments of the Benin Formation consist of interfringing units of lacustrine and fluvial loose sands, pebbles, clays, and lignite streaks of varying thicknesses, while the alluvial units comprise tidal and lagoonal sediments and beach sands and soils [14–16] which are mostly found in the southern parts and along the river banks. The CPS is covered by thin lateritic overburden materials with varying thicknesses at some locations but is massively exposed near the shorelines. The CPS forms the major aquiferous units in the area. It comprises poorly sorted continental (fine, medium, and coarse) sands and gravels that alternate with lignite streaks, thin clay horizons, and lenses at some locations. The thin clay/shale horizons truncate the vertical and lateral extents of the sandy aquifers thereby building up multiaquifer systems in the area [17]. Thus both confined and partially confined aquifers can be found in the area. Southward flowing rivers like the Kwa River and their tributaries that empty directly into the Bight of Bonny drain the area.

3. Data Acquisition and Analysis

Surface electrical methods have been used in investigating different types of geological, geotechnical, and environmental problems for many years due to the dependence of earth resistivity on some geologic parameters. Generally, the electrical resistivity method involves injecting electrical current into the ground through a pair of electrodes (called current electrodes) and monitoring the potential difference created by the passage of the electrical current through the earth materials using another pair of electrodes technically called potential electrodes. Details of this can be found in [18–20]. Many electrode configurations and field procedures exist that can be used to perform geoelectrical investigations [18], but the Schlumberger array which was performed using the vertical electrical sounding field procedure was adopted to assess the subsurface electrical resistivity and the depth of the economically accessible aquifers. The Schlumberger electrode configuration used in measuring apparent resistivity ρ_a is shown in the following equation:

$$\rho_a = \pi \cdot \left(\frac{(AB/2)^2 - (MN/2)^2}{(MN)} \right) \cdot R_a, \qquad (1)$$

where AB is the distance between two current electrodes, MN is the distance between two potential electrodes, and R_a is the apparent electrical earth's resistance measured from the

equipment. The term in the equation shown below is called the geometric factor (G):

$$G = \pi \cdot \left(\frac{(AB/2)^2 - (MN/2)^2}{MN} \right). \qquad (2)$$

The electrical resistivity investigations were conducted in twelve (12) locations across the study area between 2011 and 2012 using a SAS1000 model of ABEM Terrameter. Maximum current electrode spacing (AB) was constrained by settlement pattern and other space limiting conditions to vary from one location to another. This confined the VES points to the locations shown in Figure 1. In locations with good access paths and/or roads, the current cables were extended up to 1000 m in order to ensure that depths above 200 m were comfortably sampled assuming that penetration depth varies between $0.25AB$ and $0.5AB$ [21, 22]. Corresponding receiving (potential) electrode separation (MN) varied from a minimum of 0.5 m at $AB = 2$ m to a maximum of 50 m at $AB = 1000$ m. At all the electrode positions, great care was taken to ensure that the separation between the potential electrodes did not exceed one-fifth of the separation of the current electrodes [23]. VES data quality was generally good especially in the wet season, but, in the dry season, the electrode positions were usually wetted with water and salt solution (where necessary) in order to lower the contact resistance and consequently ensure good electrical contact between the ground and the steel electrodes.

Information generated from the analyses of geophysical data was used to constrain drilling. The drilling phase started as soon as the geophysical reports were submitted to the Akwa Ibom State Millennium Development Goal [24] that funded the borehole projects. Manual drilling technique was adopted in drilling 6-inch borehole in all locations since the subsurface condition in the entire area was favourable to such drilling method. Some of the boreholes were cited adjacent to the VES stations while some that are located at where there was no access path to spread the cables at the vicinity of the boreholes were separated by more than 150 m (see Figures 2, 3, and 4). The drill cuttings were logged geologically in all 12 locations and from the desired aquifer samples were collected for laboratory measurement of porosity. The drilled boreholes were cased using 75 mm high pressure PVC casing materials. The PVC casings were slotted at various depths with sizable thicknesses and the slotted region of the well annulus was gravel packed to ensure good delivery of water to the borehole. Gravel packing is also important in checking the ingress of sediments into the borehole. A mixture of sand and cement was used to grout the boreholes to prevent backflow of water at the surface into the well [25]. The wells were developed.

The core samples were prewashed with distilled water to remove traces of clay and other argillaceous materials that might have originated from the coring operation [26]. The samples were later put into a vacuum desiccator and evacuated at a pressure of 0.3 mBar for a period of 1 hour (see [27]). Deaerated distilled water was gently poured into the desiccator until the water completely covers all the samples. All the samples were later soaked for a period of 24 hours

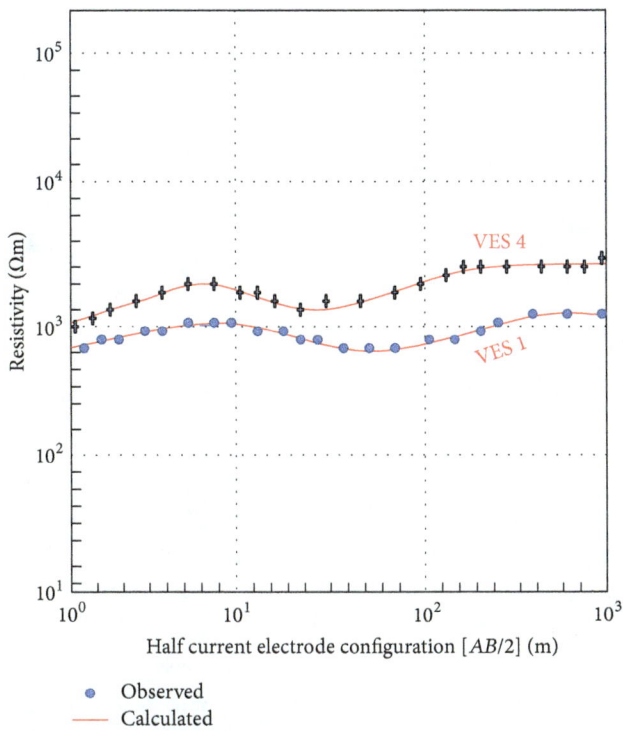

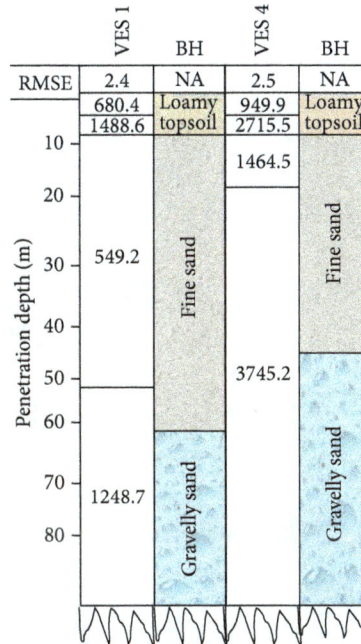

(a)

(b)

FIGURE 2: Samples of modelled VES curves along *AB* profile showing correlations between VES derived 1-D subsurface models and borehole lithologs.

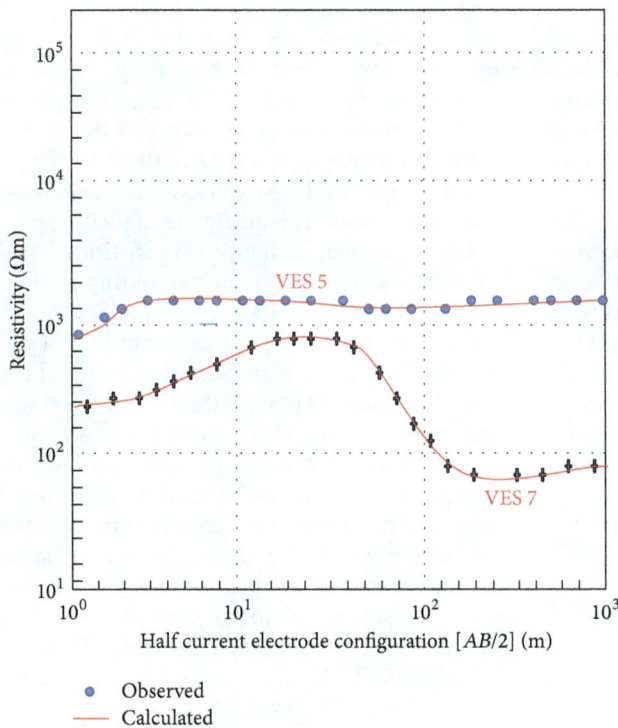

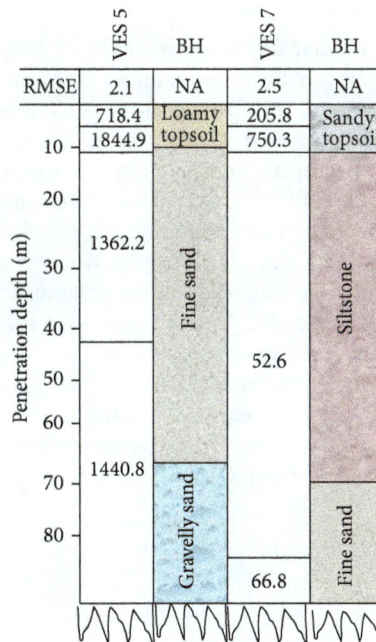

(a)

(b)

FIGURE 3: Samples of modelled VES curves along CD profile showing correlations between VES derived 1-D subsurface models and borehole lithologs.

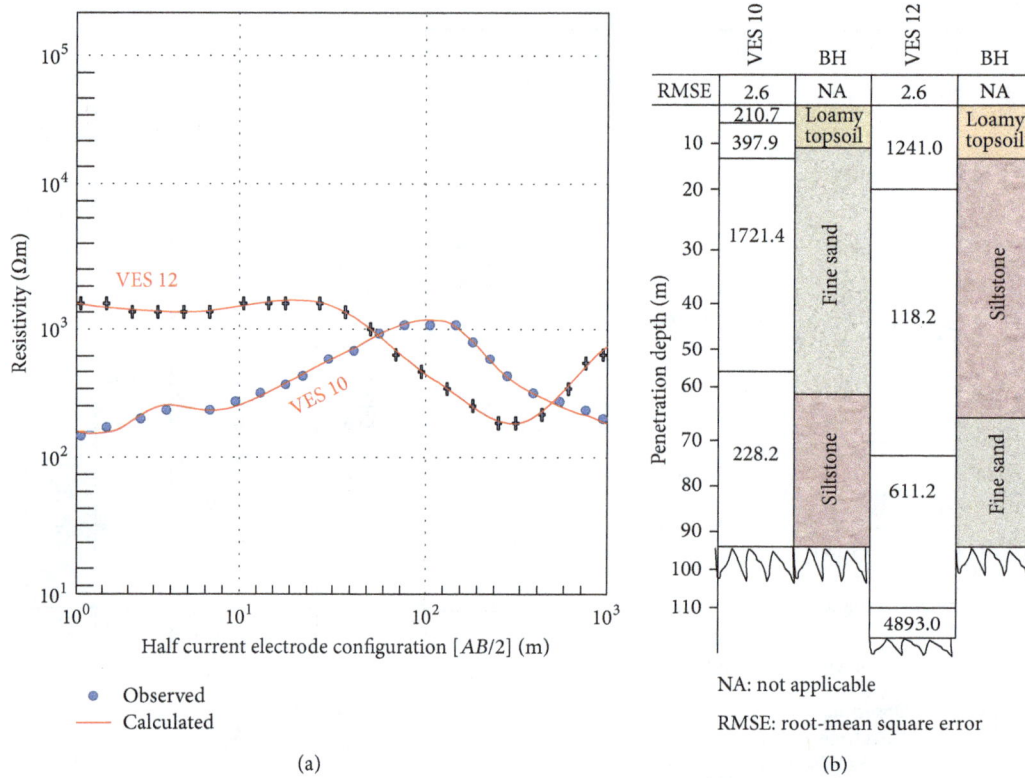

FIGURE 4: Samples of modelled VES curves along EF profile showing correlations between VES derived 1-D subsurface models and borehole lithologs.

in order to ensure that any trace of salt and other related soluble contaminants within the samples diffused out into the surrounding water. The cleaned samples were later dried in a temperature controlled oven at 105°C for 16 hours in order to check any irreversible change in the composition of the samples (see [27, 28]). The oven dried core samples were allowed to cool to normal air temperature in a desiccator. The weight of the cool and dry core samples W_d was measured using an electronic weighing balance five times and the mean was computed and recorded. The samples were soaked with distilled water that has been boiled for 30 minutes in a vacuum pressure of 0.3 mBar for 18 hours. The weight of the wet samples W_w was also measured five times and the mean was computed and recorded. Effective porosity (ϕ) of the samples was calculated using (6) as

$$\phi = 100 \cdot \left(\frac{W_w - W_d}{V} \right) \%, \tag{3}$$

where V is volume of the samples. Details of the experimental procedure can be found in [26–28].

4. Results and Discussion

The field data consisting of the apparent resistivity (ρ_a) and the current electrode spacing ($AB/2$) were partially curve matched, smoothened, and manually plotted against each other on a bilogarithmic scale with ρ_a on the ordinate and

current electrode $AB/2$ on the abscissa. It was performed by either averaging the two readings at the crossover points or deleting any outlier at the crossover points that did not conform to the dominant trend of the curve. Also deleted were data that stood out as outliers in the prevalent curve trend which could have caused serious increase in root mean square error (RMSE) during the modelling phase of the work. Where observed, such outliers constitute less than 2% of the total data generated in each sounding station and since we measured over ten data per decade, deleting such noisy data did not alter the trend of the sounding curve. Some of the deleted data might have been the electrical signatures of the thin clay materials that might have suffered suppression from the over- and underlying thick sandy aquifers [29, 30]. Any discontinuity observed in the smoothened curves was exclusively attributed to vertical variation of electrical resistivity with depth. Preliminary interpretation of the smoothened curves was made using the traditional partial curve marching technique to estimate primary layer parameters. The partial curve matching technique uses master curves and charts developed by [31]. A computer based VES modelling software called RESIST [32] that can perform automated approximation of the initial resistivity model from the observed data was later used to improve upon the preliminary interpreted results using the inversion technique. The RESIST software uses the initial layer parameters to perform some calculations and at the end generates a theoretical curve in the process. It then compares the theoretical curve with the field data curve.

TABLE 1: Summary of location coordinates, geoelectrical layer properties and protective strength of the study area.

VES number	Latitude (degree) y (m)	Longitude (degree) x (m)	Bulk resistivity of layers from top to bottom of the economic aquifer (Ωm)				Thickness of layers from top to bottom of economic aquifer (m)			Depth to economic aquifer (m)	Longitudinal conductance S (Ωm)	Economic aquifer fractional porosity	Protective strength	Curve type
			ρ_1	ρ_2	ρ_3	ρ_4	h_1	h_2	h_3					
1	2982	176	660.4	1488.6	549.2	1248.7	1.3	3.6	47.8	52.7	0.09	0.305	Weak	KH
2	2828	357	418.6	1913.0	392.1	—	1.7	112.3	—	114.0	0.06	0.350	Weak	K
3	2836	607	1304.5	720.2	2049.4	—	1.2	15	51.2	67.4	0.05	0.278	Weak	H
4	2882	794	949.9	2715.5	1464.5	3745.2	0.5	3.9	16.1	20.6	0.01	0.301	Weak	KH
5	1850	1009	718.4	1844.9	1362.2	1440.8	0.5	1.6	38.9	82.9	0.03	0.280	Weak	KH
6	1423	210	1433.0	210.0	860.5	8367.3	19.4	43.1	49.6	112.4	0.28	0.260	Moderate	HA
7	1950	1328	205.8	750.3	52.6	66.8	2.9	11.8	68.2	82.9	1.33	0.195	Good	KH
8	1363	1586	1826.8	107.5	40.1	64.7	3.6	24.6	45.2	73.4	1.36	0.312	Good	AH
9	581	104	3455.5	1825.6	393.9	2030.4	0.6	2.0	89.2	91.8	0.23	0.276	Good	QH
10	9562	150	210.7	397.9	1721.4	228.2	0.7	12.8	55.2	68.7	0.07	0.256	Weak	AK
11	9740	209	1335.1	3372.5	924.2	1681.6	4.4	55.6	50.9	110.9	0.07	0.289	Weak	KH
12	1867	356	1721.4	118.2	611.2	4893.0	91.6	56.7	36.8	115.1	0.59	0.406	Moderate	HA

TABLE 2: Protective capacity ratings of longitudinal conductance.

Range	Strength
$S > 10$	Excellent
$5 < S < 10$	Very good
$0.7 < S < 4.9$	Good
$0.2 < S < 0.69$	Moderate
$0.1 < S < 0.19$	Weak
$S < 0.1$	Poor

Since quantitative interpretation of geoelectrical sounding data is usually difficult due to the inherent problem of equivalence [33], borehole data were used to constrain all depth and minimise the choice of equivalent models by fixing layer thicknesses and depths while allowing the resistivities to vary [34]. The total number of observed minima and maxima on the smoothened VES curves was usually used as the starting number of layers (or models) over a half-space for the data inversion exercise. The software works iteratively by calculating at the end of each step updated parameters of the model and calculates the extent of fit between the calculated and the observed data using the root mean square error (RMSE) technique in which 5% was preset as the maximum acceptable value. Figures 2, 3, and 4 show some of the modelled VES curves observed and their correlation between the nearby borehole lithology and interpreted results. A good correlation was observed between the borehole lithology log data and the inverted results over half-space in many locations. These noticed distortions were suspected to have possibly originated from the failure of the 1D assumption of the shallow subsurface of the half-space [29]. The result of the computer iterations gave the layers of the subsurface penetrated by current, true resistivity of each layer, the thickness (h), depth (d) of each layer, and the total depth of overburden to the safe and economically accessible auriferous layers as shown in Table 1. Three to four layers with different curve types were delineated. The availability of H and K curve types indicates the high and low values of resistivities in sediments which translate from unsaturated zones into saturated zones. Specifically, the curve types in the areas occupied by VES are KH which takes 41.7% K, H, AH, AK, and QH which, respectively, take 8.3% of the total curve distributions. The other curve type also noticed is the AH curve type which takes 16.7% of the curve distributions (see Table 1 and Figure 5 for curve distributions). From the inferred layer resistivities and thicknesses, the longitudinal conductance known as one of the Dar Zarrouk parameters was used to classify the degree of protection of each of VES sites according to rating in Table 2 into weak, moderate, good, and excellent aquifers based on the numerical values assigned to each point. Based on the longitudinal conductance, the protective capacity of the overburden showed that 58.33% of the mapped area has weak protective capacity, 16.67% moderate protective capacity, and 25% good protective capacity. This is an indication that the chosen economic aquifer repository apart from having a sizeable thickness of average value of 52.05 m and average depth of 73.14 m also has a fair protection

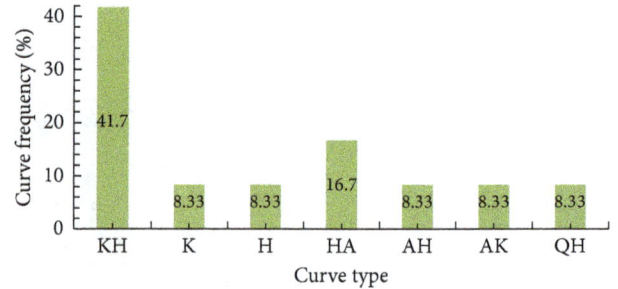

FIGURE 5: Bar chart showing frequency of curve type distribution in the study area.

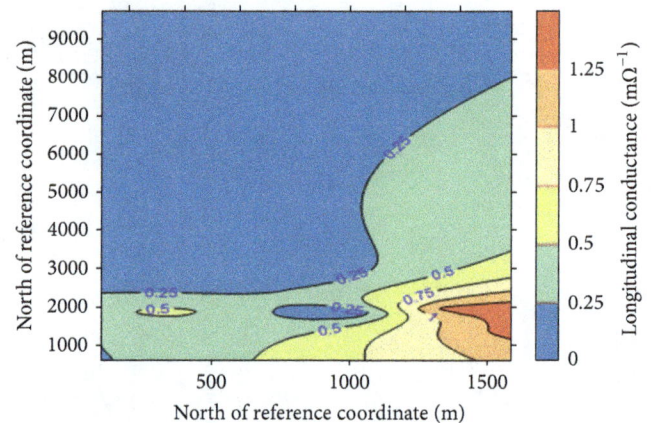

FIGURE 6: Distribution of longitudinal conductance map showing high values of S in VES 6, 7, and 8 in the north-eastern zone of the mapped area.

from the surface flow despite its open nature. The sizeable thickness and the average depth of the aquifer located in aquifer units ranging from fine sand to coarse or in siltstone can be economically exploited for domestic and industrial uses by the settlers of the region.

The longitudinal conductance (S) is given by

$$S = \sum_{i=1}^{n} \frac{h_i}{\rho_i} = \frac{h_1}{\rho_1} + \frac{h_2}{\rho_2} + \frac{h_3}{\rho_3} + \cdots + \frac{h_n}{\rho_n}, \quad (4)$$

where h_i and ρ_i are the saturated thickness of each of the layers and their corresponding true resistivities, respectively. The earth's medium acts as a natural filter to percolating fluid. The ability of the earth to retard or accelerate and filter percolating fluid is a measure of its protective capacity [35, 36]. The total longitudinal conductance is a parameter used to define the target areas of groundwater potential. High S values indicate relatively thick geologic succession and should be accorded with the highest priority in terms of groundwater potential while low S reflects thin geologic successions with low groundwater potential [35]. The contour map showing S distribution is given in Figure 6.

The results of the thickness and total depth of groundwater repository in Table 1 were, respectively, used in generating the 2D and 3D representation of the subsurface economic groundwater repository and 3D map showing the total depth

Assessment of Economically Accessible Groundwater Reserve and Its Protective Capacity in Eastern Obolo Local Government
Area of Akwa Ibom State, Nigeria, Using Electrical Resistivity Method

131

(a)

(b)

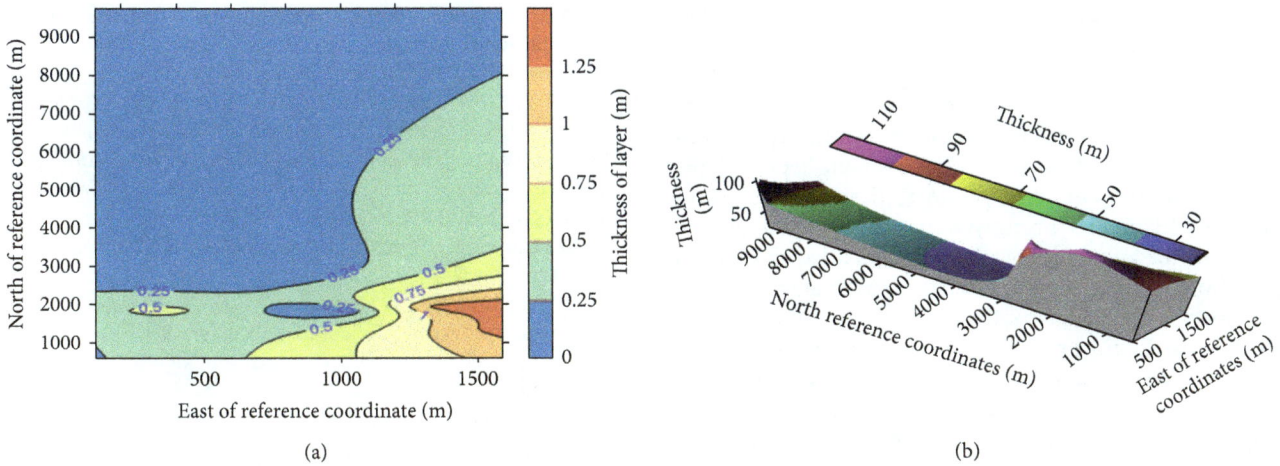

FIGURE 7: (a) 2D distribution of aquifer thickness in the mapped area. (b) 3D distribution map of the preferred aquifer thickness in the mapped area used in computing the groundwater reserve.

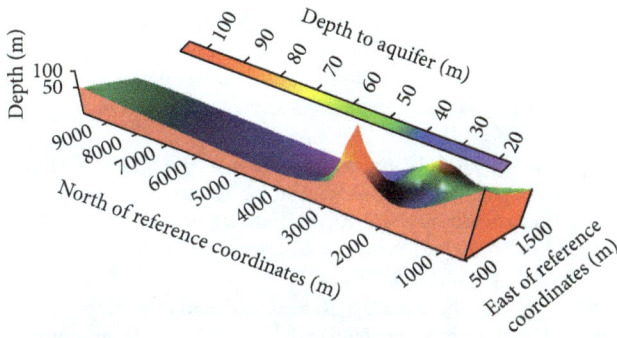

FIGURE 8: 3D distribution map of depth to aquifer in the mapped area.

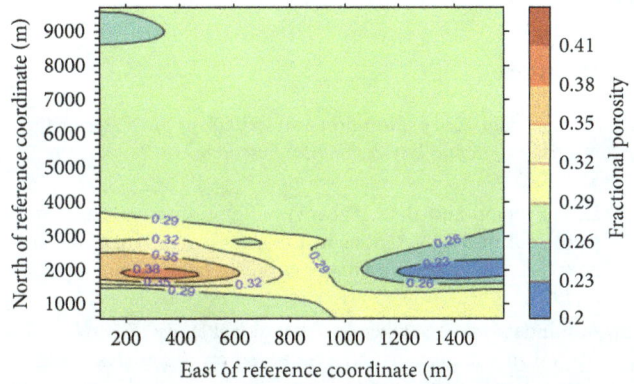

FIGURE 9: 3D distribution map of porosity of the preferred aquifer in the mapped area.

in a continuum (see Figures 7(a), 7(b), and 8). Applying SURFER programme to the thickness geometry, the total volume of the unconfined aquifer geometry was computed as 576396025 m^3 and the porous volume based on the average porosity of the unconfined aquifer of 29.23% (0.2923) was 168480558 m^3 using the latitude of each of the survey lines as the y-axis and the longitude as x-axis. The porosity distribution was also produced using fractional porosities in Table 1 and the porosity variation is shown in Figure 9. These maps describe the variations of the aquifer and can explain the flow pattern of groundwater. The thickness of the economic and safe aquifer repository to the estimated depth was considered as the z-axis. The error in porous volume was computed using the equation below:

$$\frac{\Delta V}{V} = \frac{\Delta l}{l} + \frac{\Delta w}{w} + \frac{\Delta h}{h} + \frac{\Delta \phi}{\phi}, \tag{5}$$

where ΔV, Δl, Δh, Δw, ϕ, and $\Delta \phi$ are error in volume V, error in length $l(0)$, error in thickness $h(0)$, error in width w, and

error in porosity. The error in length and width is negligible. Hence,

$$\frac{\Delta V}{V} = 0 + 0 + 0.1 + .001 = 0.11 \tag{6}$$

$$\Delta V = V \cdot 0.11 = 168480558 \cdot 0.11 = 18532861 \, \text{m}^3. \tag{7}$$

Therefore, the approximate volume of porous zone is 29.23% of the considered aquifer total volume. Hence, there exists 168480558 ± 18532861 m^3 of water residing in the study area.

5. Conclusion

The inferred safe and economic aquifer repository located in fine to gravelly sands does not show any interaction in the reference location based on the resistivity, the converse of conductivity. The estimated reserve shows that more boreholes can be drilled to exploit the described aquifer so that the reasonable population who relies on the Salt River water in the area can avoid the biological, chemical, and physical adverse consequences associated with river water.

The protection of the aquifer based on the overburden materials is commendable as the area has, on the average, good aquifers mostly within the vicinity of VES 6–12. Laboratory checks can be conducted from time to time in order to access the protective capacity of aquifers within the regions of VES 1–5 described as weak. The average depth of the overburden to the aquifer repository, the highly resistive topsoil, and estimated protective capacity on the average indicate that the safe economically recommended aquifer can be free from surface or near-surface contaminant plumes. The porosity, longitudinal conductance, thickness, and depth to aquifer distribution maps drawn could be used to derive input parameters for contaminant migration modelling and to improve the quality of model. The method is unique as the calculated aquifer parameters are well defined within the range of observed aquifer parameters.

Conflict of Interests

The authors declare that there is no conflict of interests regarding the publication of this paper.

References

[1] A. A. Kadafa, "Oil exploration and spillage in the Niger Delta of Nigeria," *Civil and Environmental Research*, vol. 2, no. 3, pp. 38–51, 2012.

[2] H. I. Ezeigbo and B. N. Ezeanyim, "Environmental pollution from coal mining activities in the Enugu area Anambka state Nigeria," *Mine Water and the Environment*, vol. 12, no. 1, pp. 53–61, 1993.

[3] A. E. Edet and C. S. Okereke, "A regional study of saltwater intrusion in southeastern Nigeria based on the analysis of geoelectrical and hydrochemical data," *Environmental Geology*, vol. 40, no. 10, pp. 1278–1289, 2001.

[4] United Nations Environmental Programme, "Environmental assessment of Ogoniland," Executive Summary, UNEP, 2011.

[5] K. S. Okiongbo and E. Akpofure, "Determination of aquifer properties and groundwater vulnerability mapping using geoelectric method in Yenagoa city and its environs in Bayelsa state, South Nigeria," *Journal of Water Resource and Protection*, vol. 4, no. 6, pp. 354–362.

[6] A. Edet and R. H. Worden, "Monitoring of the physical parameters and evaluation of the chemical composition of river and groundwater in Calabar (Southeastern Nigeria)," *Environmental Monitoring and Assessment*, vol. 157, no. 1–4, pp. 243–258, 2009.

[7] N. J. George, A. E. Akpan, and I. B. Obot, "Resistivity study of shallow aquifers in the parts of Southern Ukanafun Local Government Area, Akwa Ibom state, Nigeria," *E-Journal of Chemistry*, vol. 7, no. 3, pp. 693–700, 2010.

[8] U. F. Evans, N. J. George, A. E. Akpan, I. B. Obot, and A. N. Ikot, "A study of superficial sediments and aquifers in parts of Uyo Local Government Area, Akwa Ibom state, Southern Nigeria, using electrical sounding method," *E-Journal of Chemistry*, vol. 7, no. 3, pp. 1018–1022, 2010.

[9] A. G. Martínez, K. Takahashi, E. Núñez et al., "A multi-institutional and interdisciplinary approach to the assessment of vulnerability and adaptation to climate change in the Peruvian Central Andes: problems and prospects," *Advances in Geosciences*, vol. 14, pp. 257–260, 2008.

[10] D. Rapti-Caputo, "Influence of climatic changes and human activities on the salinization process of coastal aquifer systems," *Itaian Journal of Agronomy*, vol. 3, pp. 67–79, 2010.

[11] E. S. Riddell, S. A. Lorentz, and D. C. Kotze, "A geophysical analysis of hydro-geomorphic controls within a headwater wetland in a granitic landscape, through ERI and IP," *Hydrology and Earth System Sciences*, vol. 14, no. 8, pp. 1697–1713, 2010.

[12] G. Wagner and R. J. Zeckhauser, "Climate policy: hard problem, soft thinking," *Climatic Change*, vol. 110, no. 3-4, pp. 507–521, 2012.

[13] B. K. Farauta, C. L. Egbule, A. E. Agwu, Y. L. Idrisa, and N. A. Onyekuru, "Farmers' adaptation initiatives to the impact of climate change on agriculturein Northern Nigeria," *Journal of Agricultural Extension*, vol. 16, no. 1, pp. 132–144, 2012.

[14] T. J. A. Reijers, S. W. Petters, and C. S. Nwajide, "The Niger delta basin," in *African Basins*, R. C. Selley, Ed., vol. 3 of *Sedimentary Basin of the World*, pp. 151–172, Elsevier Science, Amsterdam, The Netherlands, 1997.

[15] T. J. A. Reijers and S. W. Petters, "Depositional environments and diagenesis of Albian carbonates on the Calabar Flank, SE Nigeria," *Journal of Petroleum Geology*, vol. 10, no. 3, pp. 283–294, 1987.

[16] T. N. Nganje, A. E. Edet, and S. J. Ekwere, "Concentrations of heavy metals and hydrocarbons in groundwater near petrol stations and mechanic workshops in Calabar metropolis, Southeastern Nigeria," *Environmental Geosciences*, vol. 14, no. 1, pp. 15–29, 2007.

[17] N. J. George, A. I. Ubom, and J. I. Ibanga, "Integrated approach to investigate the effect of leachate on groundwater around the Ikot Ekpene dumpsite in Akwa Ibom state, Southeastern Nigeria," *International Journal of Geophysics*, vol. 2014, Article ID 174589, 12 pages, 2014.

[18] W. M. Telford, L. P. Geldart, R. E. Sheriff, and D. A. Keys, *Applied Geophysics*, Cambridge University Press, 2nd edition, 1990.

[19] G. V. Keller and F. C. Frischknecht, *Electrical Methods in Geophysical Prospecting*, Pergamon, London, UK, 1966.

[20] M. S. Zhdanov and G. Keller, *The Geoelectrical Methods in Geophysical Exploration*, Methods in Geochemistry and Geophysics, Elsevier, 1994.

[21] K. K. Roy and H. M. Elliot, "Some observations regarding depth of exploration in DC electrical methods," *Geoexploration*, vol. 19, no. 1, pp. 1–13, 1981.

[22] K. P. Singh, "Nonlinear estimation of aquifer parameters from surficial resistivity measurements," *Hydrology and Earth System Sciences*, vol. 2, pp. 917–938, 2005.

[23] S. S. Gowd, "Electrical resistivity surveys to delineate groundwater potential aquifers in Peddavanka watershed, Anantapur District, Andhra Pradesh, India," *Environmental Geology*, vol. 46, no. 1, pp. 118–131, 2004.

[24] Akwa Ibom State Millennium Development Goal (AKSMDG), "Borehole completion report from the thirty one Local Government Areas of Akwa Ibom State," Tech. Rep., 2011.

[25] A. C. Ibuot, G. T. Akpabio, and N. J. George, "A survey of the repository of groundwater potential and distribution using geoelectrical resistivity method in Itu Local Government Area (L.G.A), Akwa Ibom State, Southern Nigeria," *Central European Journal of Geosciences*, vol. 5, no. 4, pp. 538–547, 2013.

[26] American Petroleum Institute (API), "Recommended practice for core analysis procedure," Tech. Rep. no. 40, 1990.

[27] D. W. Emerson, "Laboratory electrical resistivity," *Proceedings of the Australian Institute of Mining and Metallurgy*, vol. 230, no. 51, 1969.

Assessment of Economically Accessible Groundwater Reserve and Its Protective Capacity in Eastern Obolo Local Government Area of Akwa Ibom State, Nigeria, Using Electrical Resistivity Method

133

[28] J. S. Galehouse, "Sedimentation analysis," in *Procedures in Sedimentary Petrology*, R. F. Carver, Ed., p. 653, Wiley-Interscience, New York, NY, USA.

[29] V. V. S. G. Rao, G. T. Rao, L. Surinaidu, R. Rajesh, and J. Mahesh, "Geophysical and geochemical approach for seawater intrusion assessment in the Godavari Delta Basin, A.P., India," *Water, Air, & Soil Pollution*, vol. 217, no. 1–4, pp. 503–514, 2011.

[30] M. A. Sabet, "Vertical electrical resistivity sounding locate groundwater resources: a feasibility study," *Water Resources Bulletin*, vol. 73, p. 63, 1975.

[31] E. Orellana and A. M. Mooney, *Master Curve and Tables for Vertical Electrical Sounding over Layered Structures*, Interciencia, Escuela, Spain, 1966.

[32] B. P. A. V. Velpen, "A computer processing package for DC resistivity interpretation for an IBM compatibles," *ITC Journal*, vol. 4, 1988.

[33] R. A. van Overmeeren, "Aquifer boundaries explored by geoelectrical measurements in the coastal plain of Yemen: a case of equivalence," *Geophysics*, vol. 54, no. 1, pp. 38–48, 1989.

[34] A. T. Batayneh, "A hydrogeophysical model of the relationship between geoelectric and hydraulic parameters, Central Jordan," *Journal of Water Resource and Protection*, vol. 1, pp. 400–407, 2009.

[35] R. Barker, T. V. Rao, and M. Thangarajan, "Delineation of contaminant zone through electrical imaging technique," *Current Science*, vol. 81, no. 3, pp. 277–283, 2001.

[36] A. O. Olawepo, A. A. Fatoyinbo, I. Ali, and T. O. Lawal, "Evaluation of groundwater potential and subsurface lithologies in Unilorin quarters using resistivity method," *The African Review of Physics*, vol. 8, pp. 317–323, 2013.

15

Free Field Surface Motion at Different Site Types due to Near-Fault Ground Motions

Jagabandhu Dixit, D. M. Dewaikar, and R. S. Jangid

Department of Civil Engineering, Indian Institute of Technology Bombay, Mumbai 400076, PIN, India

Correspondence should be addressed to Jagabandhu Dixit, jagabandhu@iitb.ac.in

Academic Editors: A. Streltsov and P. Tosi

Seismic hazards during many disastrous earthquakes are observed to be aggravating at the sites with the soft soil deposits due to amplification of ground motion. The characteristics of strong ground motion, the site category, depth of the soil column, type of rock strata, and the dynamic soil properties at a particular site significantly influence the free field motion during an earthquake. In this paper, free field surface motion is evaluated via seismic site response analysis that involves the propagation of earthquake ground motions from the bedrock through the overlying soil layers to the ground surface. These analyses are carried out for multiple near-fault seismic ground motions at 142 locations in Mumbai city categorized into different site classes. The free field surface motion is quantified in terms of amplification ratio, spectral relative velocity, and spectral acceleration. Seismic site coefficients at different time periods are also evaluated for each site category due to near-fault ground motions from the acceleration response spectra of free field surface motion at each site and the corresponding acceleration response spectra at a reference rock outcrop site.

1. Introduction

Seismic response of a structure is dependent upon the nature of supporting soil. Severe structural damages to houses and manmade structures during many past earthquakes are observed to be concentrated in an area where the ground consisted of local alluvial deposits. Local soil deposits are found to have paramount influence on the characteristics of earthquake ground shaking and have played a major role in the damage and loss of life during many disastrous earthquakes such as the 1976 Tangshang, 1985 Mexico, 1989 Loma Prieta, 1994 Northridge, 1995 Kobe earthquakes, 2001 Bhuj earthquake, and 2005 Kashmir earthquake. The profound importance of the nature of the subsoil on the structural response of different types of structures has also been confirmed through several theoretical and experimental studies. The motion at the base of a structure founded on rock is identical to that occurring at the same point before the structure is built, but they are quite different if the structure is founded on soil. The motion that occurs in the soil or rock layers at some depth from the ground surface in the absence of any structure or excavation is defined as free-field motion. The motion at the base of a structure and the free field motion that would occur at the same point in the absence of the structure are different.

The study of wave propagation in horizontal layered media is an integral part of dynamic soil-structure interaction (SSI) analysis and it is the first stage of seismic SSI analysis [1]. Local soil stratigraphy, material heterogeneity, predominant excitation period, and the number of significant cycles have important roles on the characteristics of free field motion. Free field motions can be evaluated by treating the visco elastic soil column as a structure overlying an elastic rock half space with known excitations at the bedrock level. One-dimensional wave propagation theory is employed to simulate the propagation of seismic wave through given soil profiles at 142 sites in Mumbai city using the Standard Penetration Test (SPT) data and 100 selected near-fault acceleration time histories corresponding to several earthquake magnitudes of different fault types. The input seismic ground motion is applied at an assumed rock outcrop below the soil column. The control motion can be defined either at the free surface of the site, or at an assumed rock outcrop that is on the level of the rock under the assumption that there is no soil on top or at a point in middle soil layer.

Equivalent shear wave velocity of the soil layers overlying the bedrock is considered to be a fundamental parameter that represents the in situ properties of soil layers. Shear wave velocities of soil profiles can be calculated using the correlation developed in terms of penetration tests if the measured shear wave velocities are unavailable. The contrast in shear wave velocities and damping prevailing at the interface of two layers plays an important role on the amplification of ground motion. The influence of the soft ground at a site on the seismic response of structures is sometimes higher due to a resonance effect when the predominant periods of the structures and site periods are close to each other. The site period and duration of an earthquake increases usually with the softness of the ground. The amplification at sites with soft soil deposits is larger and longer than that compared with the shaking experienced at a hard rock site [2]. The effects of soft soil conditions are generally expressed in terms of fundamental vibration periods of the soil and the bedrock, aggravation factor, spectral acceleration, peak ground acceleration at the bedrock level, and ground surfaces.

2. Effects of Local Soil Deposits

The shaking caused by seismic waves can cause damage or collapse of the buildings. Local soil deposits significantly modify the amplitude, duration, and frequency content of free field motion [3]. Nonlinear site response is often demonstrated during the transmission of high intensity seismic waves through the horizontal soil layers. Acceleration records obtained in the near-field region during earthquakes at relatively short distances from the site have demonstrated the significant influence of certain source factors such as fault type, fault orientation, rupture mechanism, and rupture directivity as well as geotechnical site conditions such as properties of soil layers, soil stratification, and depth of ground water table on strong motion characteristics at the ground surface.

The source mechanism and fault rupture are very complex phenomenon. The coupled effect of source, path, and site conditions significantly modifies the characteristics of earthquake ground motion at the ground surface. The source characteristics are very effective and dominant especially in the near-field zones and affect the directional properties of the ground motion [4]. These ground motions with forward-directivity impose high deformation demands and also induce high energy demands to structures. These could have significant energy in low-frequency ranges that can seriously affect long-period structures [5]. Therefore, the effect of near-fault ground motions at the soft soil deposits is imperative to be studied to understand the potential seismic risk. The seismic site response analyses using one- or two-dimensional numerical model should consider the variability and uncertainties of the source and site conditions using

TABLE 1: Correlations between V_s and SPT N value for all types of soil.

Imai [6]	$V_s = 91.0 \, N^{0.337}$
Ohta and Goto [7]	$V_s = 85.35 \, N^{0.348}$
Imai and Tonouchi [8]	$V_s = 97.0 \, N^{0.314}$

selected input acceleration time histories to quantify the site effects accurately.

3. Data Collection

The characteristics of strong motion near the ground surface are much more affected by site conditions. In an attempt to evaluate the free field motion for an ensemble of near fault earthquakes, SPT soil data with bore log characteristics containing the details of the soil profiles along the depth, namely, number of blow counts in standard penetration test (N), dry density, wet density, specific gravity, and groundwater depth, among others have been collected from 142 representative borehole locations in Mumbai city. The depths of these boreholes are in the range of 7.5 m to 30 m. The SPT N values are measured at every 1.5 m depth intervals until rock strata is encountered. Majority of the soil profile data used in this analysis are contributed by DBM Geotechnics and Constructions Pvt. Ltd. and IIT Bombay. The soil profile data used in this study belong to a wide range of soil deposits comprising of sand, silt, clay, clayey sand, silty sand, silty gravel, and so forth, in different layers. The SPT blow counts at some places are in the order of 2 to 10 indicating very soft deposits of clay, whereas at some other places, it is up to 40 showing dense silty sand. The SPT N value at a particular site represents most of the important mechanical properties of the soil required for the site response analysis.

Site conditions play a major role in the amplification of soil sites during seismic events. To evaluate the free field surface motion for different site classes due to near-fault earthquakes, the site conditions have been categorized into different classes according to the mean shear wave velocity of the upper 30 m below the ground surface ($\overline{V}_{s-30}$) as per the provisions in National Earthquake Hazards Reduction Program [9]. $\overline{V}_{s-30}$ is defined as the ratio of 30 m to the time for vertically propagating shear waves to travel from 30 m depth to the surface. Estimation of free field surface motion by one dimensional approach requires shear wave velocity variation with depth as an important input parameter. For this purpose the SPT blow count of soil layers at each borehole is used for the determination of shear wave velocity from the empirical relationship between SPT N value and shear wave velocity (V_s) by Imai [6], Ohta and Goto [7], and Imai and Tonouchi [8] as given in Table 1. These relationships are reported to be valid for a wide range of soil type. The shear wave velocity for the soil layers at each site used in the present study are taken as the average of these three values so obtained.

Using the shear wave velocities of the soil layers, $\overline{V}_{s-30}$ is calculated as,

$$\overline{V}_{s-30} = \frac{\sum_{i=1}^{n} d_i}{\sum_{i=1}^{n} (d_i/V_{si})}, \qquad (1)$$

where d_i and V_{si} are the thickness and the shear wave velocity (V_s) of ith soil or rock layer to depth of 30 m, respectively, ($\Sigma\, d_i = 30$ m) and n is number of layers. This site classification approach suggests that 39 sites of 142 available sites correspond to C-type (360 m/s $\leq \overline{V}_{s-30} \leq 760$ m/s), 94 sites correspond to D-type (180 m/s $\leq \overline{V}_{s-30} \leq 360$ m/s), and 9 sites correspond to E-type ($\overline{V}_{s-30} < 180$ m/s) (Figure 7).

4. Seismic Input Excitation

Multiple near-fault strong ground motions comprising of 100 recorded acceleration time histories with peak ground acceleration (PGA) 0.1 g to 1.3 g of earthquakes of magnitudes M_w 5.5 to M_w 7.8 are downloaded from the strong motion database of Pacific Earthquake Engineering Research [10] center to excite at the bottom of the soil column consisting of horizontal layers overlying the elastic rock half space. The distance of recording station from the location of fault rupture is less than or equal to 10 km. These long-period pulse-type ground motions correspond to different source mechanisms and fault orientations. The seismic ground motions are nonstationary with respect to both amplitude and frequency. The near-fault acceleration time histories are characterized by long-period high-velocity pulse, higher frequencies, and short-duration strong-impulsive motion with considerable energy in the long-period regions due to short travel distance [11]. Figure 1 shows acceleration, velocity, and displacement time history of a typical near-fault ground motion for 1992 Landers earthquake recorded at Lucerne. The magnitude (M_w), epicentral distance (R), and PGA of near-fault acceleration time histories used in this study are shown in Figure 2.

The PEER [10] ground motion database includes the set of the most important ground motions recorded worldwide and includes corresponding earthquake magnitude, style of faulting, epicentral distance, site type at the recording stations, and so forth. This database is considered to be a reliable source of ground, motion and these ground motions are used for assessing seismic performance of structures and for analyzing the responses at sites due to a suite of strong ground motions. The response at a site is greatly influenced by source characteristics, path characteristics, local site conditions epicentral distance, and other relevant engineering parameters. The main purpose of this paper is to carry out a typical study to observe the responses at sites of different site types categorized based on average shear wave velocity due to near-fault ground motion. Due to the availability of the details of soil profiles at several sites of different categories in Mumbai city, those are chosen for this typical analysis.

The acceleration response spectra have been calculated for each of the acceleration-time histories used in this study. The mean (μ) and standard deviation (σ) of all

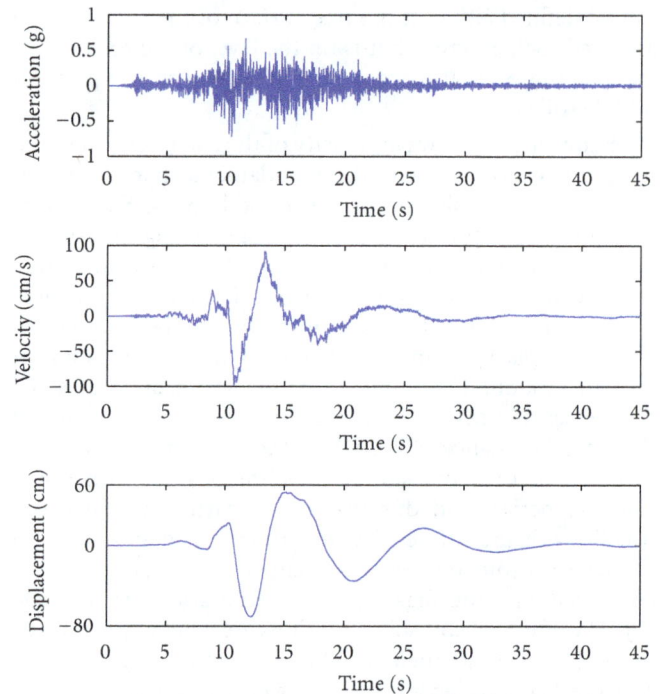

FIGURE 1: Acceleration-, velocity-, displacement-time histories of 1992 Landers earthquake recorded at Lucerne ($M_w = 7.3$, $R = 1.1$ km).

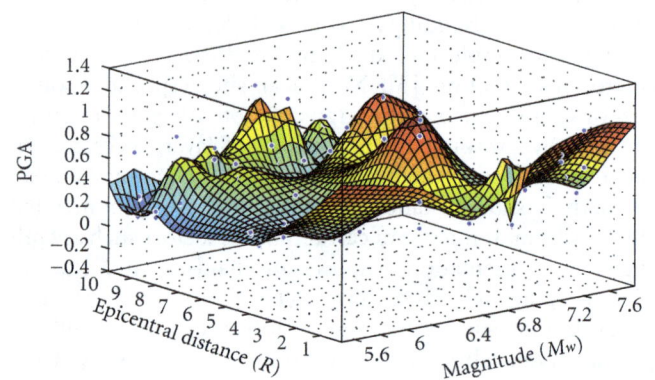

FIGURE 2: Magnitude (M_w), epicentral distance (R), and PGA of near-fault ground motions selected for the present study.

the acceleration-time histories taken together are shown in Figure 3.

A huge amount of energy is released in a short time period in case of the pulse-type motions with forward directivity. Forward directivity is caused due to propagation of the fault rupture towards the site. The fault-normal ground motions are significantly larger than the fault-parallel motions at higher periods. In this study, both fault-normal and fault-parallel components are considered to address their effects on the free field motion.

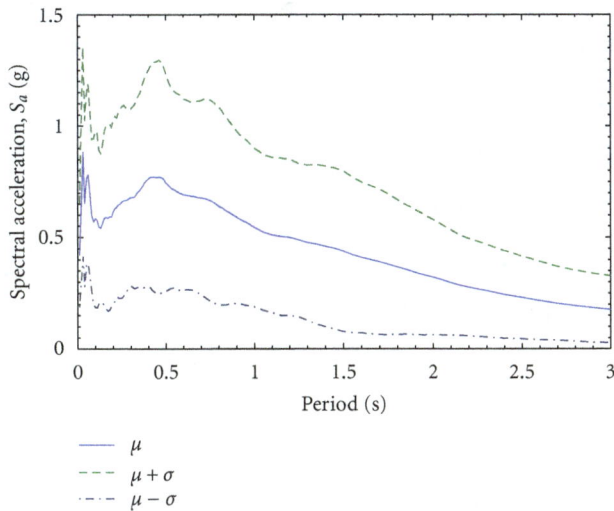

FIGURE 3: Mean and standard deviation of acceleration response spectra of selected near-fault ground motions.

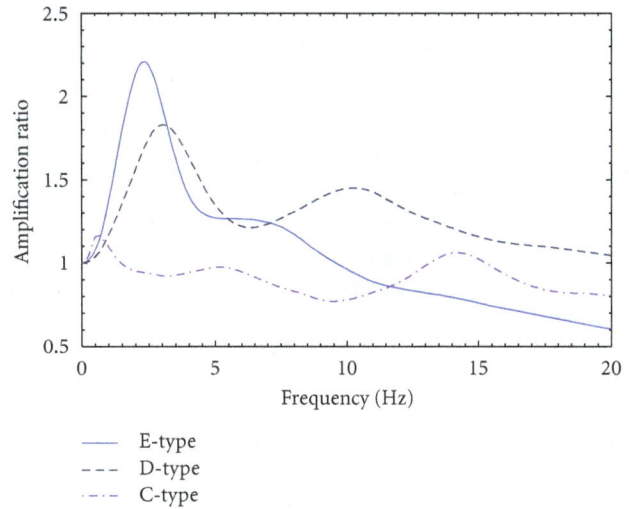

FIGURE 4: Amplification ratio at 5% damping at C-, D-, and E-type sites.

5. Evaluation of Free Field Motion

A comprehensive assessment of free field motion should consider source characteristics as well as geotechnical and geological site conditions. The free field surface motions are the result of wave reflection, refraction, and surface wave propagation. The major factors modifying the free field surface motions are depth of the soil column, the dynamic properties of soil layers, the type of rock strata, and properties of ground motion. The input ground motions are excited at the bottom of the soil column. Each soil layer in the soil column overlying the bedrock is assumed to be horizontal and homogeneous. One-dimensional wave propagation equation for vertically propagating shear waves through the linear viscoelastic system consisting of n layers is expressed as the following damped wave equation in the time domain:

$$\rho_j \frac{\partial^2 u}{\partial t^2} = G_j \frac{\partial^2 u}{\partial z^2} + \eta_j \frac{\partial^3 u}{\partial z^2 \partial t}, \tag{2}$$

where u is the horizontal displacement, t is time, z is the depth within jth layer, ρ_j, G_j, and η_j are mass density, shear modulus, and viscosity, respectively, of jth layer.

Equivalent linear one-dimensional wave propagation analyses are performed at 142 soil sites in Mumbai city for 100 near-fault strong motion using computer program SHAKE91 [12]. The site responses are computed in the frequency domain using equivalent shear modulus and equivalent damping ratios. Figure 4 shows a typical site response analysis at a site in Mumbai city. It shows the soil profile, shear wave profile, input excitation at the bedrock level, and the free field motion at the ground surface. The acceleration response spectra for both rock level and surface level motions are evaluated. The characteristics of free field motions can be quantified by amplification spectra, relative spectral velocity, acceleration response spectra, site coefficients, and so forth. Amplification spectra are defined in the frequency domain as the ratio of amplitude of free field motion at the surface level at different frequencies to the amplitude at the bedrock level at corresponding frequencies [13].

One-dimensional site response analysis is justified for horizontal grounds. The surface level free field motion at any particular site type in this study is obtained by taking the mean of the free field motions at all the sites of same category due to all the near-fault ground motion. Soil nonlinearity has a limited influence on mean free field motion at a site and the equivalent linear approach suitably takes care of the nonlinear aspects. SHAKE91 [12] is most widely used computer programs for site response analysis due to its simplicity and relative accuracy. Results obtained from SHAKE91 [12] compare well with field measurements. The sensitivity of this program to parameter variations is acceptable for most practical applications. Equivalent linear approach used in SHAKE91 [12] is based on viscoelastic constitutive model. The equivalent linear approach is computationally efficient and easy to implement. The shear modulus and damping ratio of the various soil layers at different strain levels are iteratively updated during the analysis through modulus degradation and damping curves to obtain revised values of shear modulus and damping of the materials constituting the different layers and the revised shear modulus and damping ratio are compatible with the deformation level induced in each soil layer by the seismic force. The algorithm of fast Fourier transform implemented in SHAKE91 [12] program allows performing the analysis in the frequency domain in an iterative procedure till a converged solution is arrived. SHAKE91 [12] is considered appropriate if the number of runs is significantly high. The number of runs of the program in the present study is 14200. However, for the sites involving slopes or earth retaining structures one-dimensional analysis has to be used judiciously. Two-dimensional equivalent linear or nonlinear site response analysis should be performed when one-dimensional analysis is not adequate.

SHAKE 91 [12] captures nonlinear cyclic response of soil through equivalent linear approximation by modifying the linear elastic properties of the soil based on the induced strain level. An effective shear strain is computed as 65% of peak strain for each soil layer for a given acceleration time series and an initial estimate of modulus and damping values. The strains induced in soil layers depend on the soil properties. Shear modulus and damping ratios vary with Fourier amplitude of shear strain. The strain compatible shear modulus and damping ratio values are iteratively calculated.

Similarly, site coefficients are the ratio of spectral acceleration of free field surface motion at a site at different periods to that of the assumed reference outcrop rock site at those corresponding periods [14]. Site coefficients can be expressed as

$$F_s(T) = \frac{S_a^s(T)}{S_a^r(T)}, \qquad (3)$$

where $S_a^s(T)$ and $S_a^r(T)$ are the acceleration response spectra at different periods at the surface level and reference outcrop rock site respectively. $F_s(T)$ are site coefficients at corresponding periods.

6. Results

Amplification spectra are evaluated individually at each site for 100 input excitations at 5% damping, followed by the evaluation of mean amplification spectrum at each site. Amplification spectrum for C-, D-, and E-type sites for near-fault ground motions are evaluated from mean amplification spectrum at different sites as depicted in Figure 4. The amplification is found to be more at low frequency for soft soil sites and deamplification occurs at higher frequencies. Soil profiles at D- and E-type sites being softer than that at C-type sites show appreciable amplification up to a factor of 2.5 at certain frequency. The majority of C-type sites considered in this study consist of alternate layers of dense sand and clay underlain by thick layers of soft rock.

The spectral relative velocity and spectral acceleration response of ground motion at a reference rock outcrop and the corresponding spectral acceleration response at the surface level for 5% damping are estimated considering all of these near-fault input motions. Mean spectral relative velocity for different site categories is evaluated from velocity response at each site as shown in Figure 5. The mean acceleration response spectra of free field surface motion and its corresponding standard deviation are evaluated for C-, D-, and E-type sites as shown in Figure 6 through Figure 8. A significant difference in frequency, duration, and amplitude of the free field surface motions are observed for different site categories. Site coefficients for C-, D-, and E-type sites are shown in Figure 9. These coefficients are highly dependent upon the site category. There is no significant variation in the values of site coefficients with respect to periods in case of dense soil or soft rock sites. A shift in peak response towards higher periods is also observed in the surface level time history. These site coefficients are smaller at higher values of spectral acceleration values at the reference rock site.

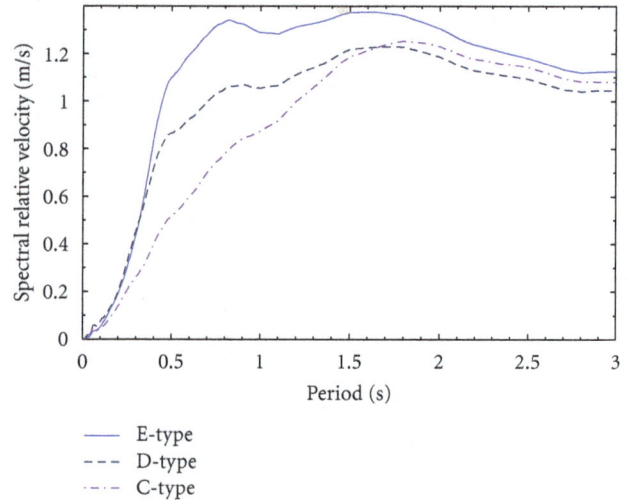

FIGURE 5: Spectral relative velocity at 5% damping of at C-, D-, and E-type sites.

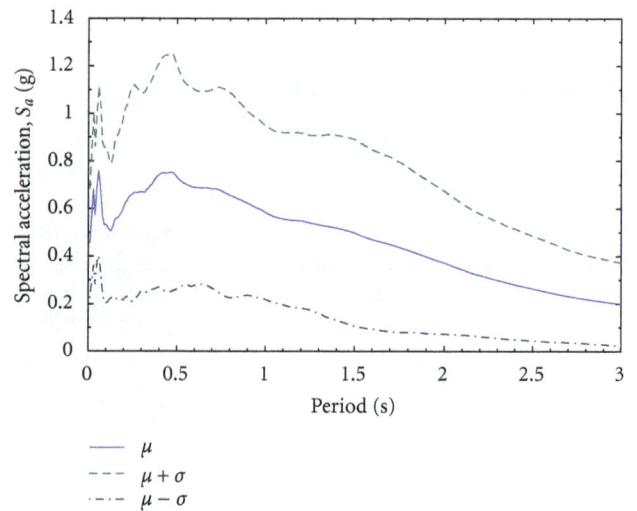

FIGURE 6: Spectral acceleration at C-type sites for near-fault motion.

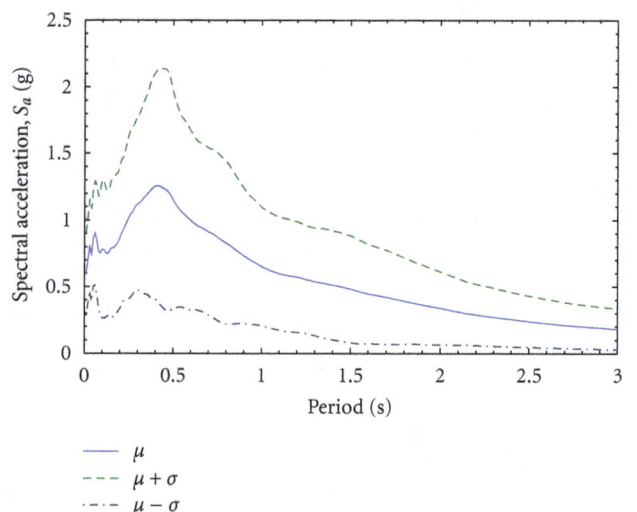

FIGURE 7: Spectral acceleration at D-type sites for near-fault motion.

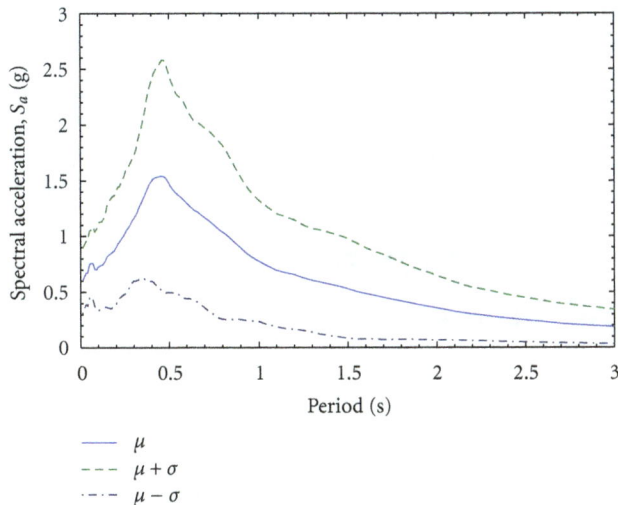

FIGURE 8: Spectral acceleration at E-type sites for near-fault motion.

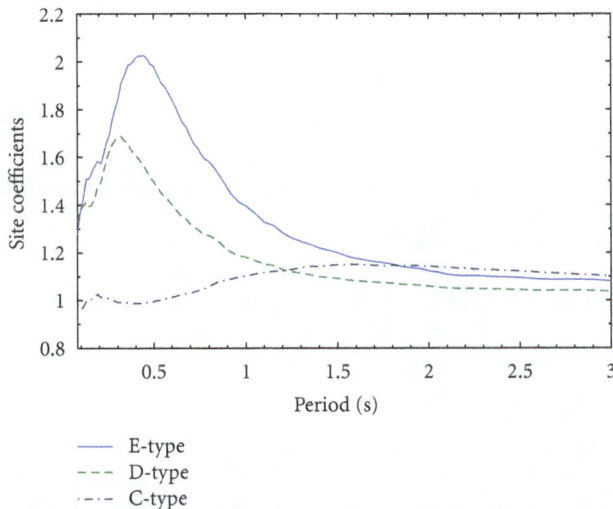

FIGURE 9: Site coefficients for near-fault ground motion.

7. Conclusions

Free field surface motion is evaluated for 100 near-fault strong ground motions of different magnitudes. Dynamic properties of soil column and ground motion characteristics are found to have a significant influence on free field surface motion. Free field surface motions are quantified in terms of amplification ratios, acceleration response spectra, and site coefficients. The free field surface motions and the corresponding response spectra are found to be significantly affected by the geological and geotechnical characteristics of local soil deposits. The effect of soil nonlinearity is also shown at higher peak ground accelerations.

Acknowledgments

The authors extend their acknowledgments to M/S DBM Geotechnics and Constructions Pvt. Ltd., Mumbai, for providing the borehole data at different sites at Mumbai city.

References

[1] J. Wolf, *Dynamic Soil-Structure Interaction*, Prentice Hall, Englewood Cliffs, NJ, USA, 1985.

[2] R. D. Borcherdt, "Estimates of site-dependent response spectra for design (methodology and justification)," *Earthquake Spectra*, vol. 10, pp. 617–653, 1994.

[3] K. Aki, "Local site effects on weak and strong ground motion," *Tectonophysics*, vol. 218, no. 1-3, pp. 93–111, 1993.

[4] G. P. Mavroeidis and A. S. Papageorgiou, "A mathematical representation of near-fault ground motions," *Bulletin of the Seismological Society of America*, vol. 93, no. 3, pp. 1099–1131, 2003.

[5] J. F. Hall, T. H. Heaton, M. W. Halling, and D. J. Wald, "Near-source ground motion and its effects on flexible buildings," *Earthquake Spectra*, vol. 11, no. 4, pp. 569–605, 1995.

[6] T. Imai, "P- and S-wave velocities of the ground in Japan," in *Proceedings of the 9th International Conference on Soil Mechanics and Foundation Engineering*, pp. 257–260, Tokyo, Japan, 1977.

[7] Y. Ohta and N. Goto, "Empirical shear wave velocity equation in terms of characteristic soil indexes," *Earthquake Engineering and Structural Dynamics*, vol. 6, no. 2, pp. 167–187, 1978.

[8] T. Imai and K. Tonouchi, "Correlation of N-value with S-wave velocity and shear modulus," in *Proceedings of the 2nd European Symposium of Penetration Testing*, pp. 57–72, A. A. Balkema Publishers, Amsterdam, The Netherlands, 1982.

[9] BSSC, *NEHRP Recommended Seismic Provisions for New Buildings and other Structures*, Building Seismic Safety Council, Federal Emergency Management Agency, Washington, DC, USA, 2009.

[10] PEER, Pacific Earthquake Engineering Research Center, "PEER Strong Motion Database," http://peer.berkeley.edu/smcat/.

[11] P. G. Somerville, "Engineering characteristics of near fault ground motion," in *Proceedings of the Seminar on Utilization of Strong Motion Data (SMIP '97)*, Los Angeles, Calif, USA, 1997.

[12] I. M. Idriss and J. I. Sun, "A Computer program for conducting equivalent linear seismic response analysis of horizontally layered soil deposits," in *Users Manual For SHAKE91*, Center for Geotechnical Modeling, Department of Civil and Environmental Engineering, University of California, Davis, Davis, Calif, USA, 1992.

[13] S. L. Kramer, *Geotechnical Earthquake Engineering*, Prentice Hall, Upper Saddle River, NJ, USA, 1996.

[14] H. H. M. Hwang, H. Lin, and J. R. Huo, "Site coefficients for design of buildings in eastern United States," *Soil Dynamics and Earthquake Engineering*, vol. 16, no. 1, pp. 29–40, 1997.

16

Propagation of Rayleigh Wave in a Two-Temperature Generalized Thermoelastic Solid Half-Space

Baljeet Singh

Department of Mathematics, Post Graduate Government College, Sector 11, Chandigarh 160011, India

Correspondence should be addressed to Baljeet Singh; bsinghgc11@gmail.com

Academic Editors: Y.-J. Chuo, D. Kass, and X. Perez-Campos

The Rayleigh surface wave is studied at a stress-free thermally insulated surface of an isotropic, linear, and homogeneous two-temperature thermoelastic solid half-space in the context of Lord and Shulman theory of generalized thermoelasticity. The governing equations of a two-temperature generalized thermoelastic medium are solved for surface wave solutions. The appropriate particular solutions are applied to the required boundary conditions to obtain the frequency equation of the Rayleigh wave. Some special cases are also derived. The speed of Rayleigh wave is computed numerically and shown graphically to show the dependence on the frequency and two-temperature parameter.

1. Introduction

Lord and Shulman [1] and Green and Lindsay [2] extended the classical dynamical coupled theory of thermoelasticity to generalized thermoelastic theories. These theories treat heat propagation as a wave phenomenon rather than a diffusion phenomenon and predict a finite speed of heat propagation. Ignaczak and Ostoja-Starzewski [3] explained in detail the above theories in their book *Thermoelasticity with Finite Wave Speeds*. The representative theories in the range of generalized thermoelasticity are reviewed by Hetnarski and Ignaczak [4]. Wave propagation in thermoelasticity has many applications in various engineering fields. Some problems on wave propagation in coupled or generalized thermoelasticity are studied by various researchers, for example, Deresiewicz [5], Sinha and Sinha [6], Sinha and Elsibai [7, 8], Sharma et al. [9], Othman and Song [10], Singh [11, 12], and many more.

Gurtin and Williams [13, 14] proposed the second law of thermodynamics for continuous bodies in which the entropy due to heat conduction was governed by one temperature, that of the heat supply by another temperature. Based on this law, Chen and Gurtin [15] and Chen et al. [16, 17] formulated a theory of thermoelasticity which depends on two distinct temperatures, the conductive temperature Φ and the thermodynamic temperature T. The two-temperature

theory involves a material parameter $a^* > 0$. The limit $a^* \to 0$ implies that $\Phi \to T$, and the classical theory can be recovered from two-temperature theory. The two-temperature model has been widely used to predict the electron and phonon temperature distributions in ultrashort laser processing of metals. According to Warren and Chen [18], these two temperatures can be equal in time-dependent problems under certain conditions, whereas Φ and T are generally different in particular problems involving wave propagation. Following Boley and Tolins [19], they studied the wave propagation in the two-temperature theory of coupled thermoelasticity. They showed that the two temperatures, T and Φ, and the strain are represented in the form of a traveling wave plus a response, which occurs instantaneously throughout the body. Puri and Jordan [20] studied the propagation of harmonic plane waves in two-temperature theory. Quintanilla and Jordan [21] derived exact solutions of two initial-boundary value problems in the two-temperature theory with dual-phase-lag delay. Youssef [22] developed a theory of two-temperature generalized thermoelasticity. Kumar and Mukhopadhyay [23] extended the work of Puri and Jordan [20] in the context of the linear theory of two-temperature generalized thermoelasticity developed by Youssef [22]. Magaña and Quintanilla [24] showed the uniqueness and growth of solutions in two-temperature generalized thermoelastic theories. Recently, Youssef [25] formulated a theory of two-temperature thermoelasticity without energy dissipation.

In this paper, the Youssef [22] theory is followed for the theoretical study of the Rayleigh wave at the thermally insulated stress-free surface of an isotropic two-temperature thermoelastic solid half-space. The frequency equation of the Rayleigh wave is obtained. The frequency equation is also approximated by assuming small thermal coupling. The dependence of numerical values of the speed of the Rayleigh wave on material parameters, frequency, and two-temperature parameters is shown graphically for a particular material of the model.

2. Basic Equations

We consider an isotropic, linear, and homogeneous two-temperature thermoelastic solid half-space. Following Youssef [22], the governing equations for a two-temperature generalized thermoelastic half-space with one relaxation time are as follows:

(i) the stress-strain-temperature relations

$$\sigma_{ij} = c_{ijkl}e_{kl} - \gamma_{ij}\left(T - \Phi_0\right),\tag{1}$$

(ii) the displacement-strain relation

$$e_{ij} = \frac{1}{2}\left(u_{i,j} + u_{j,i}\right),\tag{2}$$

(iii) the equation of motion

$$\rho\ddot{u}_i = \sigma_{ji,j} + \rho F_i,\tag{3}$$

(iv) the energy equation

$$-q_{i,i} = \rho T_0 \dot{S},\tag{4}$$

(v) the modified Fourier's law

$$-K_{ij}\Phi_{,j} = q_i + \tau_0 \dot{q}_i,\tag{5}$$

(vi) the entropy-strain-temperature relation

$$\rho S = \frac{\rho c_E}{T_0}\theta + \gamma_{ij}e_{ij}.\tag{6}$$

Here, γ_{ij} are the coupling parameters, T is the mechanical temperature, $\Phi_0 = T_0$ is the reference temperature, $\theta = T - T_0$ with $|\theta/T_0| \ll 1$, σ_{ij} is the stress tensor, e_{kl} is the strain tensor, c_{ijkl} is the tensor of elastic constants, ρ is the mass density, q_i is the heat conduction vector, K_{ij} is the thermal conductivity tensor, c_E is the specific heat at constant strain, u_i are the components of the displacement vector, S is the entropy per unit mass, τ_0 is the thermal relaxation time, which will ensure that the heat conduction equation will predict finite speeds of heat propagation, and Φ is the conductive temperature and satisfies the relation

$$\Phi - T = a^*\Phi_{,ii},\tag{7}$$

where $a^* > 0$ is the two-temperature parameter.

3. Analytical 2D Solution

We consider a homogeneous and isotropic thermoelastic medium of an infinite extent with Cartesian coordinates system (x, y, z), which is previously at uniform temperature. The origin is taken on the plane surface, and z-axis is taken normally into the medium $(z \geq 0)$. The surface $z = 0$ is assumed to be stress-free and thermally insulated. The present study is restricted to the plane strain parallel to x-z plane, with the displacement vector $\mathbf{u} = (u_1, 0, u_3)$. With the help of (1)–(3), we obtain the following two components of the equation of motion:

$$(\lambda + 2\mu)\,u_{1,11} + (\lambda + \mu)\,u_{3,13} + \mu u_{1,33} - \gamma\theta_{,1} = \rho\ddot{u}_1,$$
$$(\lambda + 2\mu)\,u_{3,33} + (\lambda + \mu)\,u_{1,13} + \mu u_{3,11} - \gamma\theta_{,3} = \rho\ddot{u}_3.\tag{8}$$

Equations (4)–(6) lead to the following heat conduction equation:

$$K\left(\Phi_{,11} + \Phi_{,33}\right) = \rho c_E\left(\dot{\theta} + \tau_0\ddot{\theta}\right)$$
$$+ \gamma T_0\left(\dot{u}_{1,1} + \tau_0\ddot{u}_{1,1}\right) + \gamma T_0\left(\dot{u}_{3,3} + \tau_0\ddot{u}_{3,3}\right),\tag{9}$$

and (7) becomes

$$\Phi - T = a^*\left(\Phi_{,11} + \Phi_{,33}\right).\tag{10}$$

The displacement components u_1 and u_3 are written in terms of potentials q and ψ as

$$u_1 = \frac{\partial q}{\partial x} - \frac{\partial \psi}{\partial z}, \qquad u_3 = \frac{\partial q}{\partial z} + \frac{\partial \psi}{\partial x}.\tag{11}$$

Using (11) in (8)–(10), we obtain

$$(\lambda + 2\mu)\left(\frac{\partial^2 q}{\partial x^2} + \frac{\partial^2 q}{\partial z^2}\right) - \gamma\left[\Phi - a^*\left(\frac{\partial^2\Phi}{\partial x^2} + \frac{\partial^2\Phi}{\partial z^2}\right)\right]$$
$$= \rho\frac{\partial^2 q}{\partial t^2},\tag{12}$$

$$K\left(\Phi_{,11} + \Phi_{,33}\right)$$
$$= \rho c_E\left(\frac{\partial\Phi}{\partial t} + \tau_0\frac{\partial^2\Phi}{\partial t^2}\right)$$
$$- a^*\rho c_E\left(1 + \tau_0\frac{\partial}{\partial t}\right)\frac{\partial}{\partial t}\left(\frac{\partial^2\Phi}{\partial x^2} + \frac{\partial^2\Phi}{\partial z^2}\right)\tag{13}$$
$$+ \gamma T_0\left(1 + \tau_0\frac{\partial}{\partial t}\right)\frac{\partial}{\partial t}\left(\frac{\partial^2 q}{\partial x^2} + \frac{\partial^2 q}{\partial z^2}\right),$$

$$\mu\left(\frac{\partial^2\psi}{\partial x^2} + \frac{\partial^2\psi}{\partial z^2}\right) = \rho\frac{\partial^2\psi}{\partial t^2}.\tag{14}$$

For thermoelastic surface waves in the half-space propagating in x-direction, the potential functions Φ, q, and ψ are taken in the following form:

$$(\Phi, q, \psi) = \left(\widehat{\Phi}(z), \widehat{q}(z), \widehat{\psi}(z)\right) \exp i (\eta x - \chi t), \quad (15)$$

where $\chi^2 = \eta^2 c^2$, η is the wave number, and c is the phase velocity.

Substituting (15) with (12) and (13) and eliminating $\widehat{q}$, $\widehat{\Phi}$, we obtain the following auxiliary equation:

$$D^4 - AD^2 + B = 0, \quad (16)$$

where $D = d/dz$,

$$A = 2\eta^2 + \eta^2 c^2 \left[\frac{c_1^2 + \epsilon + \left(\overline{K} - a^* \chi^2\right)}{a^* \epsilon \chi^2 - c_1^2 \left(\overline{K} - a^* \chi^2\right)} \right],$$

$$B = \eta^4 + \eta^4 c^2 \left[\frac{c_1^2 + \epsilon + \left(\overline{K} - a^* \chi^2\right)}{a^* \epsilon \chi^2 - c_1^2 \left(\overline{K} - a^* \chi^2\right)} \right]$$

$$- \eta^4 c^4 \left[\frac{1}{a^* \epsilon \chi^2 - c_1^2 \left(\overline{K} - a^* \chi^2\right)} \right], \quad (17)$$

$$\epsilon = \frac{\overline{\gamma}^2 T_0}{c_E \tau^*}, \qquad \overline{K} = \frac{K}{\rho c_E \tau^*}, \qquad \tau^* = \tau_0 + \frac{i}{\chi},$$

$$\overline{\gamma} = \frac{\gamma}{\rho}, \qquad c_1^2 = \frac{\lambda + 2\mu}{\rho}.$$

With the help of (16) and keeping in mind that $\widehat{q}, \widehat{\Phi} \to 0$ as $z \to \infty$ for surface waves, the solutions q, Φ are written as

$$q = \left[A \exp \left(-\eta \beta_1 z\right) + B \exp \left(-\eta \beta_2 z\right) \right] \exp \iota (\eta x - \chi t),$$

$$\Phi = \left[\zeta_1 A \exp \left(-\eta \beta_1 z\right) + \zeta_2 B \exp \left(-\eta \beta_2 z\right) \right] \exp \iota (\eta x - \chi t), \quad (18)$$

where

$$\beta_1^2 = \frac{1}{2\eta^2} \left[A + \sqrt{A^2 - 4B} \right],$$

$$\beta_2^2 = \frac{1}{2\eta^2} \left[A - \sqrt{A^2 - 4B} \right], \quad (19)$$

$$\zeta_i = \frac{\chi^2 - c_1^2 \left(1 - \beta_i^2\right) \eta^2}{\overline{\gamma} \left[1 + a^* \eta^2 \left(1 - \beta_i^2\right)\right]}, \quad (i = 1, 2). \quad (20)$$

Substituting (15) with (14) and keeping in mind that $\widehat{\psi} \to 0$ as $z \to \infty$ for surface waves, we obtain the following solution:

$$\psi = C \exp \left[-\eta \beta_3 z + \iota \left(\eta x - \chi t\right)\right], \quad (21)$$

where

$$\beta_3^2 = 1 - \frac{c^2}{c_2^2}, \qquad c_2^2 = \frac{\mu}{\rho}. \quad (22)$$

4. Derivation of Frequency Equation

The mechanical and thermal conditions at the thermally insulated surface $z = 0$ are as follows:

(i) vanishing of the normal stress component

$$\sigma_{zz} = 0, \quad (23)$$

(ii) vanishing of the tangential stress component

$$\sigma_{zx} = 0, \quad (24)$$

(iii) vanishing of the normal heat flux component

$$\frac{\partial \Phi}{\partial z} = 0, \quad (25)$$

where

$$\sigma_{zz} = \lambda \left(\frac{\partial^2 q}{\partial x^2} + \frac{\partial^2 q}{\partial z^2} \right) + 2\mu \left(\frac{\partial^2 \psi}{\partial x \partial z} \right)$$

$$+ 2\mu \frac{\partial^2 q}{\partial z^2} - \gamma \left[\Phi - a^* \left(\frac{\partial^2 \Phi}{\partial x^2} + \frac{\partial^2 \Phi}{\partial z^2} \right) \right], \quad (26)$$

$$\sigma_{zx} = \mu \left[2 \frac{\partial^2 q}{\partial x \partial z} - \frac{\partial^2 \psi}{\partial z^2} + \frac{\partial^2 \psi}{\partial x^2} \right].$$

Making use of solutions (18) and (21) for q, Φ, and Ψ in (23) to (25) and eliminating A, B, and C, we obtain the following equation:

$$4\mu \eta^2 \beta_1 \beta_2 \beta_3 \left(\zeta_2 - \zeta_1\right) - \left(1 + \beta_3^2\right)$$

$$\times \left[\eta^2 \left\{-\lambda \left(\zeta_2 \beta_2 - \zeta_1 \beta_1\right) + (\lambda + 2\mu) \beta_1 \beta_2 \left(\zeta_2 \beta_1 - \zeta_1 \beta_2\right)\right\} \right.$$

$$\left. -\rho c_1^2 \zeta_1 \zeta_2 \left(\beta_2 - \beta_1\right) \left\{1 + a^* \eta^2 \left(1 + \beta_1 \beta_2\right)\right\} \right] = 0,$$

$$(27)$$

which is the frequency equation of thermoelastic Rayleigh wave in a two-temperature generalized thermoelastic medium in the context of the Lord and Shulman [1] theory.

5. Special Cases

5.1. Small Thermal Coupling ($\epsilon \ll 1$). For most of materials, ϵ is small at a normal temperature. For $\epsilon \ll 1$ and using (19), we obtain the following approximated relations as

$$\beta_1^2 + \beta_2^2$$
$$\equiv 2 - c^2 \left\{ \frac{c_1^2 + (\overline{K} - a^* \chi^2)}{c_1^2 (K - a^* \chi^2)} \right\}$$
$$\times \left[1 + \epsilon \left\{ \frac{1}{c_1^2 + (\overline{K} - a^* \chi^2)} + \frac{a^* \chi^2}{c_1^2 (\overline{K} - a^* \chi^2)} \right\} \right],$$

$$\beta_1^2 \beta_2^2$$
$$\equiv 1 - c^2 \left\{ \frac{c_1^2 + (\overline{K} - a^* \chi^2)}{c_1^2 (\overline{K} - a^* \chi^2)} \right\}$$
$$\times \left[1 + \epsilon \left\{ \frac{1}{c_1^2 + (\overline{K} - a^* \chi^2)} + \frac{a^* \chi^2}{c_1^2 (\overline{K} - a^* \chi^2)} \right\} \right]$$
$$+ \frac{c^4}{c_1^2 (\overline{K} - a^* \chi^2)} \left[1 + \epsilon \frac{a^* \chi^2}{c_1^2 (\overline{K} - a^* \chi^2)} \right].$$

(28)

With the help of these approximations, we can approximate β_1, β_2, and the coupling coefficients ζ_1 and ζ_2. Finally, the frequency equation (27) can be approximated. Further, if we consider $\epsilon \to 0$, $a^* \to 0$, we obtain $\beta_1 \equiv \sqrt{1 - c^2/c_1^2}$ and $\beta_2 \equiv \sqrt{1 - c^2/\overline{K}}$.

For numerical purpose, we put $c^2 = c^{*2} + \epsilon(\xi_1 + i\xi_2)$, where c^* is the classical Rayleigh wave velocity, and ξ_1 and ξ_2 are two reals, then

$$\eta = \frac{\chi}{c^*} \left(1 - \frac{\epsilon \xi_1}{2c^{*2}} - i \frac{\epsilon \xi_2}{2c^{*2}} \right).$$

(29)

The velocity of propagation is equal to $(c^* + \epsilon \xi_1/2c^*)$, and the amplitude attenuation factor is equal to $\exp[\epsilon \chi \xi_2 x/2c^{*3}]$ with $\xi_2 < 0$.

5.2. Small Frequency ($\chi \ll 1$). For $\chi \ll 1$ and using (19), we obtain the following approximated relations as

$$\beta_1^2 + \beta_2^2 \equiv 2 - c^2 \left\{ \frac{c_1^2 + \epsilon + \overline{K}}{c_1^2 \overline{K}} \right\}$$
$$\times \left[1 + \chi^2 \left\{ \frac{(1+\epsilon)a^*}{c_1^2 \overline{K}} - \frac{a^*}{c_1^2 + \epsilon + \overline{K}} \right\} \right],$$

$$\beta_1^2 \beta_2^2 \equiv 1 - c^2 \left\{ \frac{c_1^2 + \epsilon + \overline{K}}{c_1^2 \overline{K}} \right\}$$
$$\times \left[1 + \chi^2 \left\{ \frac{(1+\epsilon)a^*}{c_1^2 \overline{K}} - \frac{a^*}{c_1^2 + \epsilon + \overline{K}} \right\} \right]$$
$$+ \frac{c^4}{c_1^2 \overline{K}} \left[1 + \chi^2 \frac{(1+\epsilon)a^*}{c_1^2 \overline{K}} \right].$$

(30)

With the help of these approximations, we can approximate β_1, β_2, and the coupling coefficients ζ_1 and ζ_2. Finally, the frequency equation (27) can be approximated. Also, if we consider $\chi \to 0$, $\epsilon \to 0$, we obtain $\beta_1 \equiv \sqrt{1 - c^2/c_1^2}$ and $\beta_2 \equiv \sqrt{1 - c^2/\overline{K}}$.

5.3. Isotropic Elastic Case. If we neglect thermal parameters, then the frequency equation (27) reduces to

$$\left(2 - \frac{c^2}{c_2^2} \right)^2 = 4 \sqrt{1 - \frac{c^2}{c_1^2}} \sqrt{1 - \frac{c^2}{c_2^2}},$$

(31)

which is the frequency equation of Rayleigh wave for an isotropic elastic case.

6. Numerical Example

The speed of propagation is computed for the following material parameters of aluminium metal: $\lambda = 5.775 \times 10^{11}$ Dyn·cm^{-2}, $\mu = 2.646 \times 10^{11}$ Dyn·cm^{-2}, $\rho = 2.7$ g·cm^{-3}, $c_E = 0.236$ Cal·g^{-1} · °C^{-1}, $K = 0.492$ Cal·cm^{-1}·s^{-1} · °C^{-1}, $\gamma = 0.05(3\lambda + 2\mu)$, $T_0 = 27°$C, and $c^* = 0.9554$.

The nondimensional speed of Rayleigh wave is shown graphically against the frequency (2–20 Hz) in Figure 1, when the two-temperature parameter $a^* = 0.5$. With the increase in frequency, it increases very sharply to its maximum value near $\chi = 4.3$ Hz, and then it decreases slowly for the higher frequency range. The nondimensional speed of Rayleigh wave is also shown graphically against the two-temperature parameter a^* in Figure 2 for $\chi = 10$ Hz. With the increase in the value of two-temperature parameter, it increases very sharply to its maximum value at $a^* = 0.47$. Thereafter, it decreases sharply to its minimum value.

7. Conclusion

The appropriate solutions of all the governing equations of a two-temperature generalized thermoelastic medium are applied at the boundary conditions at a thermally insulated free surface of a half-space to obtain the frequency equation of Rayleigh wave in the context of the Lord and Shulman [1] theory. The frequency equation is approximated for the case of small thermal coupling and small frequency and reduced for isotropic elastic case. From the frequency equation of Rayleigh wave, it is observed that the phase speed of Rayleigh wave depends on various material parameters including the two-temperature parameter. The dependence of numerical

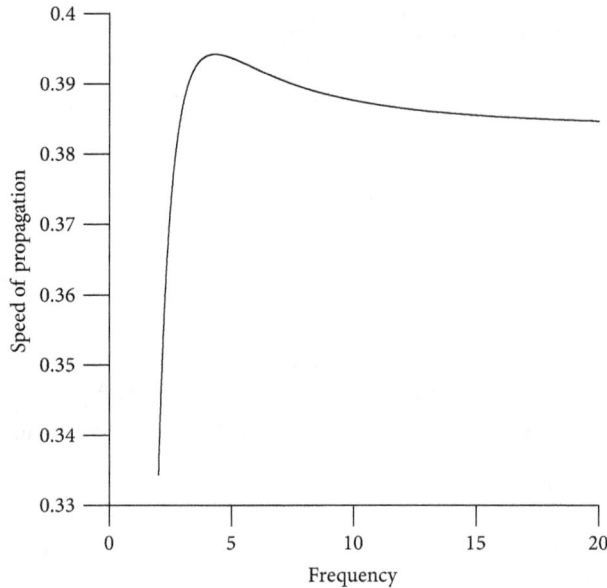

FIGURE 1: Variations of the nondimensional speed of propagation of Rayleigh wave against the frequency (χ) when $a^* = 0.5$.

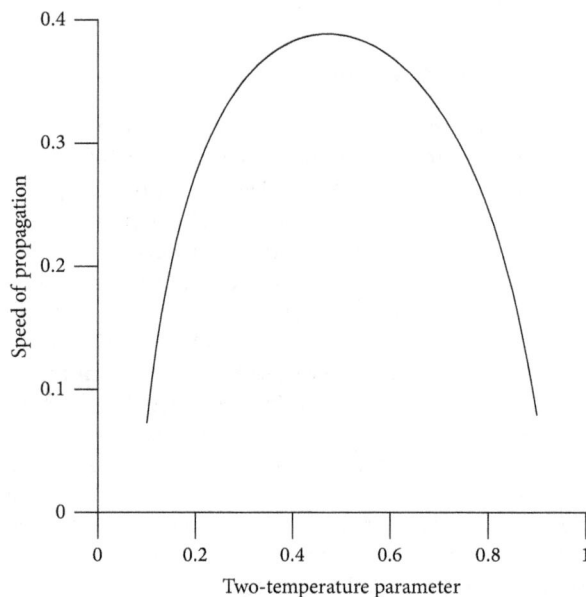

FIGURE 2: Variations of the nondimensional speed of propagation of Rayleigh wave against the two-temperature parameter (a^*) when $\chi = 10$ Hz.

values of speed of propagation on the frequency and two-temperature parameter is shown graphically for a particular material representing the model. The present problem is of geophysical interest, particularly in investigations concerned with earthquakes and other phenomena in seismology. The applications range from geophysical problems to quantitative nondestructive evaluation of mechanical structures and acoustic tomography for medical purposes. The problem

though is theoretical, but it can provide useful information for experimental researchers working in the field of geophysics and earthquake engineering and seismologist working in the field of mining tremors and drilling into the crust of the earth.

References

[1] H. W. Lord and Y. Shulman, "A generalized dynamical theory of thermoelasticity," *Journal of the Mechanics and Physics of Solids*, vol. 15, no. 5, pp. 299–309, 1967.

[2] A. E. Green and K. A. Lindsay, "Thermoelasticity," *Journal of Elasticity*, vol. 2, no. 1, pp. 1–7, 1972.

[3] J. Ignaczak and M. Ostoja-Starzewski, *Thermoelasticity with Finite Wave Speeds*, Oxford University Press, Oxford, UK, 2009.

[4] R. B. Hetnarski and J. Ignaczak, "Generalized thermoelasticity," *Journal of Thermal Stresses*, vol. 22, no. 4, pp. 451–476, 1999.

[5] H. Deresiewicz, "Effect of boundaries on waves in a thermoelastic solid: reflexion of plane waves from a plane boundary," *Journal of the Mechanics and Physics of Solids*, vol. 8, no. 3, pp. 164–172, 1960.

[6] A. N. Sinha and S. B. Sinha, "Reflection of thermoelastic waves at a solid half-space with thermal relaxation," *Journal of Physics of the Earth*, vol. 22, no. 2, pp. 237–244, 1974.

[7] S. B. Sinha and K. A. Elsibai, "Reflection of thermoelastic waves at a solid half-space with two relaxation times," *Journal of Thermal Stresses*, vol. 19, no. 8, pp. 749–762, 1996.

[8] S. B. Sinha and K. A. Elsibai, "Reflection and refraction of thermoelastic waves at an interface of two semi-infinite media with two relaxation times," *Journal of Thermal Stresses*, vol. 20, no. 2, pp. 129–145, 1997.

[9] J. N. Sharma, V. Kumar, and D. Chand, "Reflection of generalized thermoelastic waves from the boundary of a half-space," *Journal of Thermal Stresses*, vol. 26, no. 10, pp. 925–942, 2003.

[10] M. I. A. Othman and Y. Song, "Reflection of plane waves from an elastic solid half-space under hydrostatic initial stress without energy dissipation," *International Journal of Solids and Structures*, vol. 44, no. 17, pp. 5651–5664, 2007.

[11] B. Singh, "Effect of hydrostatic initial stresses on waves in a thermoelastic solid half-space," *Applied Mathematics and Computation*, vol. 198, no. 2, pp. 494–505, 2008.

[12] B. Singh, "Reflection of plane waves at the free surface of a monoclinic thermoelastic solid half-space," *European Journal of Mechanics, A/Solids*, vol. 29, no. 5, pp. 911–916, 2010.

[13] M. E. Gurtin and W. O. Williams, "On the Clausius-Duhem inequality," *Zeitschrift für angewandte Mathematik und Physik*, vol. 17, no. 5, pp. 626–633, 1966.

[14] M. E. Gurtin and W. O. Williams, "An axiomatic foundation for continuum thermodynamics," *Archive for Rational Mechanics and Analysis*, vol. 26, no. 2, pp. 83–117, 1967.

[15] P. J. Chen and M. E. Gurtin, "On a theory of heat conduction involving two temperatures," *Zeitschrift für angewandte Mathematik und Physik ZAMP*, vol. 19, no. 4, pp. 614–627, 1968.

[16] P. J. Chen, M. E. Gurtin, and W. O. Williams, "A note on non-simple heat conduction," *Zeitschrift für angewandte Mathematik und Physik*, vol. 19, no. 6, pp. 969–970, 1968.

[17] P. J. Chen, M. E. Gurtin, and W. O. Williams, "On the thermodynamics of non-simple elastic materials with two temperatures," *Zeitschrift für angewandte Mathematik und Physik*, vol. 20, no. 1, pp. 107–112, 1969.

[18] W. E. Warren and P. J. Chen, "Wave propagation in the two temperature theory of thermoelasticity," *Acta Mechanica*, vol. 16, no. 1-2, pp. 21–33, 1973.

[19] B. A. Boley and I. S. Tolins, "Transient coupled thermoplastic boundary value problems in the half-space," *Journal of Applied Mechanics*, vol. 29, no. 4, pp. 637–646, 1962.

[20] P. Puri and P. M. Jordan, "On the propagation of harmonic plane waves under the two-temperature theory," *International Journal of Engineering Science*, vol. 44, pp. 1113–1126, 2006.

[21] R. Quintanilla and P. M. Jordan, "A note on the two temperature theory with dual-phase-lag delay: some exact solutions," *Mechanics Research Communications*, vol. 36, no. 7, pp. 796–803, 2009.

[22] H. M. Youssef, "Theory of two-temperature-generalized thermoelasticity," *IMA Journal of Applied Mathematics*, vol. 71, no. 3, pp. 383–390, 2006.

[23] R. Kumar and S. Mukhopadhyay, "Effects of thermal relaxation time on plane wave propagation under two-temperature thermoelasticity," *International Journal of Engineering Science*, vol. 48, no. 2, pp. 128–139, 2010.

[24] A. Magaña and R. Quintanilla, "Uniqueness and growth of solutions in two-temperature generalized thermoelastic theories," *Mathematics and Mechanics of Solids*, vol. 14, no. 7, pp. 622–634, 2009.

[25] H. M. Youssef, "Theory of two-temperature thermoelasticity without energy dissipation," *Journal of Thermal Stresses*, vol. 34, no. 2, pp. 138–146, 2011.

17

Day-to-Day Variability of H and Z Components of the Geomagnetic Field at the African Longitudes

T. N. Obiekezie,[1] **S. C. Obiadazie,**[1] **and G. A. Agbo**[2]

[1] Department of Physics and Industrial Physics, Nnamdi Azikiwe University, P.M.B. 5025, Awka, Nigeria
[2] Industrial Physics Department, Ebonyi State University, P.M.B. 053, Abakaliki, Nigeria

Correspondence should be addressed to T. N. Obiekezie; as27ro@yahoo.com

Academic Editors: G. Mele, F. Monteiro Santos, and A. Streltsov

The Day-to-day variability of the geomagnetic field elements at the African longitudes has been studied for the year 1987 using geomagnetic data obtained from four different African observatories. The analysis was carried out on solar quiet days using hourly values of the Horizontal, H, and vertical, Z, geomagnetic field values. The results of this study confirm that Sq is a very changeable phenomenon, with a strong day-to-day variation. This day-to-day variation is seen to be superimposed on magnetic disturbances of a magnetospheric origin.

1. Introduction

Changes in the magnetic environment of the Earth are of interest to those studying space weather and climate change, particularly in the upper atmosphere. The upper atmosphere is ionized by the Sun's ultraviolet and X-radiation to create the ionosphere, and the free ions and electrons are moved by winds arising from the heating effects of the Sun. The currents in the ionosphere have magnetic effect on the ground and are monitored using magnetometers on the Earth surface. The records of any observatory show that on some days there are regular variations on the magnetic record while on other days the variation is irregular. The daily variations of the geomagnetic field when solar-terrestrial disturbances are absent are called solar quiet (Sq) variations [1]. These Sq variations are caused mainly by electrical currents in the upper atmosphere, at altitudes of about 110 km above the Earth surface [2]. Studies on solar quiet daily variation of the Earth's magnetic field show that Sq on one day can be different from Sq of the next day in amplitude, phase, and focal latitude [3–5]. This change in Sq between two adjacent days is the day-to-day variability in Sq between the two days. This day-to-day variability has been highly attributed to changes in the ionospheric dynamo currents, which depend on the ionospheric conductivity and tidal winds, varying with solar radiation and ionospheric conditions [6, 7].

Hasegawa [8] examined the day-to-day changes of the quiet day variation and the ionospheric current systems for the second polar year and suggested that some or all the day-to-day variabilities in solar quiet daily variation (Sq) are due to variability in the positions of the foci of the ionospheric current systems rather than changes in the distribution of ionization and conductivity. Studies conducted by [9–16] clearly showed that the variability of Sq occurred at all hours of the day.

Rabiu et al. [17] from their comprehensive study of Sq day-to-day variability at Addis Ababa, an equatorial electrojet station, found out that the daytime (0700–2000 hours) magnitudes of Sq H (horizontal component of the earth's magnetic field) and Z (vertical component of the earth's magnetic field) were greater than the nighttime (2000–0700 hours) for all the months they studied. They also found that the day-to-day variability peaks during the daytime mostly around the local noon within the range of 1000–1400 hours for all the months in the two Sq elements, H and Z. Their findings are therefore in agreement with the diurnal variation pattern of Sq in the earlier works of [18, 19] which showed that the maximum intensity of Sq occurs around local noon. The diurnal variation of day-to-day variability, which followed the variation pattern of Sq, can be attributed to the variability of the ionospheric process and physical structures such as conductivity and wind structures,

TABLE 1: The stations and their locations.

S/n	Stations	Abbreviations	Location in Africa	Geographic		Geomagnetic	
				Lat. (°)	Long. (°)	Lat. (°)	Long (°)
1	Hermanus	HMN	South	−34.42	9.23	−33.67	83.35
2	Addis Ababa	AAB	East	9.03	38.77	5.16	111.38
3	Mbour	MBR	West	14.40	343.02	20.36	57.16
4	Bangui	BANG	Central	4.43	18.57	4.34	90.75

which are responsible for Sq variation. Studying the Sq variability in Indian equatorial electrojet sector, Okeke et al. [13] noted that changes in the electric field control the phase and randomness of the variabilities, while the magnitude of the ionospheric conductivity controls the magnitude of the variabilities. Campbell [20], Rabiu [21], and Obiekezie and Obiadazie [22] found consistent nighttime variation in the horizontal magnetic field component at midlatitudes and attributed same to distant magnetospheric sources.

The similarities between the diurnal variation patterns of day-to-day variability and Sq in the earlier works of many researchers such as Onwumechili [18], Matsushita [19], and Fambitakoye and Mayaud [23] suggested that the root cause of Sq may also explains the day-to-day variability effect. Several causes of Sq have been identified which can thus explain the day-to-day variability. Onwumechili and Ezema [24] concluded that the diurnal variation of Sq of H arises in daytime which is consistent with atmospheric dynamo theory of the geomagnetic daily variation.

Accurate determination of Sq variability has found applications in the determination of the Earth's electrical conductivity, in determining the baselines for quantifying of magnetospheric disturbances, and in improving the satellite main-field modeling. Although so much work has been done on the Sq variability, Hibberd [25] noted that the phenomenon of day-to-day variability is poorly understood. Thus, monitoring the day-to-day variability could provide very important contributions to the understanding of the atmospheric dynamics affecting the Sq variability. The aim of this paper therefore is to analyze the day-to-day variability in the H and Z geomagnetic field elements *at* four different locations along the African longitudes and also deduce the mechanisms responsible for the observed variations and variabilities. The result of this research is expected to add to the few existing ones in the African region thus bridging the gap between Africa and the other parts of the world where much work had been done.

2. Data and Method of Analysis

A set of observatories located in the East, West, Central, and Southern Africa supplied the dataset used in this work. The geographic and geomagnetic locations of the stations can be seen in Table 1.

Magnetically quiet days from ten internationally quiet days (IQDs) in each month for the year 1987 were selected. The ten International Quiet Days (IQD) are the ten quietest days of the month according to the classification of planetary magnetic index, Kp. The local time (LT) for all the four

stations was employed throughout the analysis. The baseline values (H_o and Z_o) were calculated as the average of the values of the hours flanking the midnight plus the midnight values

$$H_o = \frac{H_{22} + H_{23} + H_{00} + H_{01}}{4}$$
$$Z_o = \frac{Z_{22} + Z_{23} + Z_{00} + Z_{01}}{4},$$

(1)

where H_{00}, Z_{00}, H_{01}, Z_{01}, H_{22}, Z_{22}, and H_{23}, Z_{23}, represent the hourly values of H and Z at 00, 01, 22, and 23 hours LT, respectively.

The Sq amplitude ΔH_t, ΔZ_t for any hour t is the difference between hourly values H_t, Z_t and the baseline value, H_0, Z_0. Thus,

$$\Delta H_t = H_t - H_0,$$
$$\Delta Z_t = Z_t - Z_0,$$

(2)

where t = 0 to 23 hours.

The hourly departures were then corrected for noncyclic variation (Δ_C). This noncyclic variation (Δ_C) defined as a phenomenon in which the value at 00LT is different from the value at 23LT, [26, 27]

$$\Delta_C = \frac{V_0 - V_{23}}{23}.$$

(3)

The linearly adjusted values at the hours are

$$V_0 + 0\Delta_C, V_1 + 1\Delta_C, V_2 + 2\Delta_C, V_3 + 3\Delta_C \cdots V_{23} + 23\Delta_C.$$

(4)

In other words,

$$S_T(V) = V_T + t\Delta_C,$$

(5)

where t = 0 to 23 hrs and V can be either H or Z.

The hourly departures corrected for noncyclic variation on quiet days give the solar quiet daily variation in H and Z denoted as Sq(H) and Sq(Z).

The variabilities of these hourly amplitudes for the hour t from the day i to the next day $i + 1$ for all hours of the day are

$$H_{DD} = H_{t_i} - H_{t_{i+1}},$$
$$Z_{DD} = Z_{t_i} - Z_{t_{i+1}}.$$

(6)

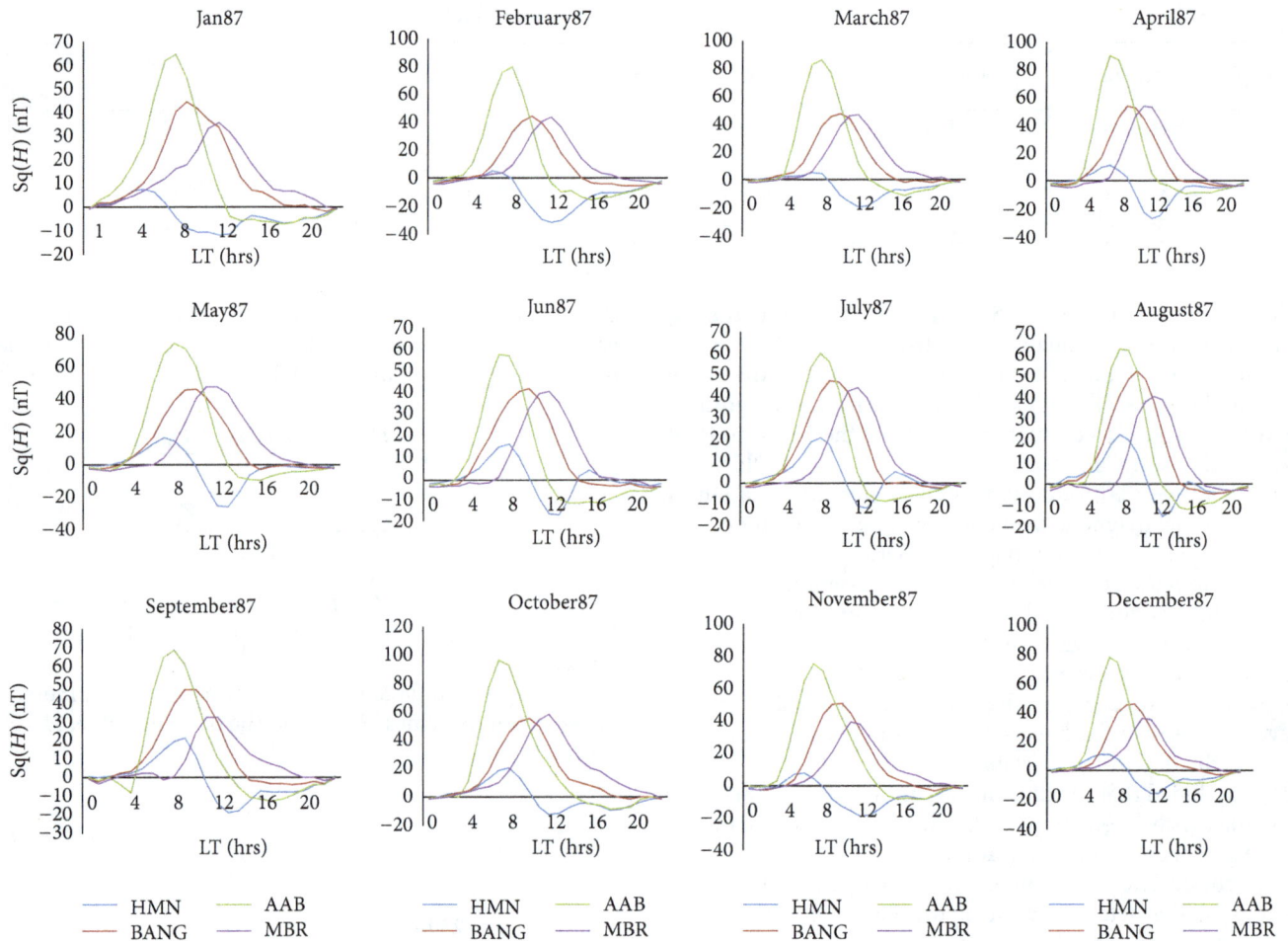

FIGURE 1: Monthly variations in Sq(H).

3. Results and Discussion

Figures 1 and 2 show the monthly variations in the Sq(H) and Sq(Z), respectively, for the four stations. It is evident from Figure 1 that the amplitude of Sq(H) in AAB is higher than that in all other stations. The amplitude curves present the same shape for AAB, BANG, and MBR but different from HMN. The morphology of the curves for AAB, BANG, and MBR shows a regular increase in amplitude in the morning hours which picks around 8.00 hrs for AAB, 10.00 hrs for BANG, and 12.00 hrs for MBR and a gradual decrease from the peak value down to the night value. A kind of phase difference is observed between AAB, BANG, and MBR. This phase shift may be attributed to the differences in their latitudinal locations. AAB, BANG, and MBR stations are located in the equatorial region, AAB (0.18° dip latitude) is located within the equatorial electrojet (EEJ) zone; thus, from Figure 1 the amplitudes of AAB is seen to be higher than Bang, and MBR Sq(H) amplitudes. These high amplitudes could be as a result of influence of the Equatorial electrojet current. The EEJ current is an east-west current which is seen flowing positive in the morning thus causing an enhancement in the Sq(H) values of stations within the EEJ region.

The morphology of the amplitude curve for HMN which is seen different from the other stations is seen to be having a minimum when the others are having a maximum. *H amplitude in HMN* decreases from dawn to about noon, when a minimum is achieved; it later increases up to dusk. This variance could be attributed to the hemispherical difference between HMN and the other three stations. HMN is in the southern hemisphere while the other three stations are on the northern hemisphere. It has been established that the regular daily variations are mainly caused by electric currents flowing at approximately 100 km altitude in the ionosphere (external source current). The ionospheric currents typically form two global horizontal current vortices at the sunlit side of the Earth, one flowing clockwise in the southern hemisphere and the other flowing counterclockwise in the northern hemisphere.

The amplitude curves for Sq(Z) are seen to be conspicuously opposite that of Sq(H) for MBR and HMN stations. The Z variation at Mbour decreases from morning hours at about sunrise, a minimum at about local noon, and a gentle rise towards sunset period. Z variations for HMN show a maximum at about noon, given the clockwise sense of the southern hemisphere current circulation. The variations in Z

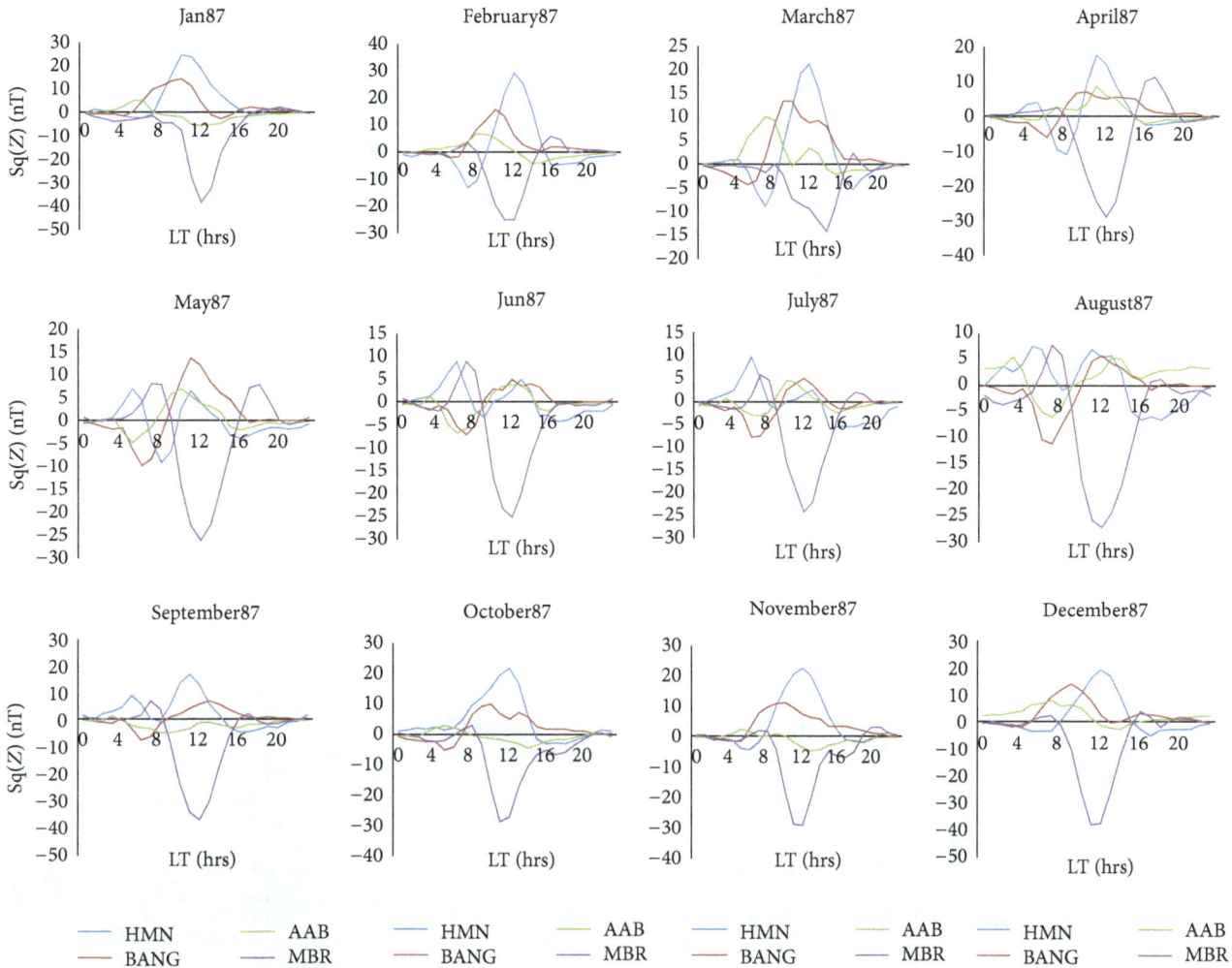

FIGURE 2: Monthly variations in Sq(Z).

at the AAB station show a maximum between the periods from 0800 to 1200 hours LT within December and Feb–August. This result therefore is suggesting that there is a presence of counterelectrojet during these hours. Alex et al. [28] found "Z" variation to be in phase with the H variation and suggested that it could be the cancellation of EEJ. Maximum values were observed almost in all the months in BANG which suggest an anomaly in the current pattern. This is actually expected to be minimal as it is in line with the counterclockwise sense of the northern hemisphere current circulation. This observed anomaly could be said to be the reversal of the atmospheric dynamo electric field as it is found to occur after sunrise with a peak around local noon when the significant increase in E-region ionization must have been formed. Another possible explanation for this anomalous behavior in Bangui can be found in a combination of factors affecting the induced currentsignals. Obiekezie and Okeke [29] found that, at ground level, the induced currents signals have an opposite sign compared with the external signal.

A minimum nocturnal variation is observed in the two components Sq(H) and Sq(Z) for all the stations. This nighttime variation could be attributed to distant current of nonionospheric origin. This night variation is in line with the earlier works of [14–16, 21, 22, 30–32]. Obiekezie and Okeke [16] attributed the observed nighttime variations to currents flowing in the magnetosphere (such as the ring currents) in which most of it filter into the ionosphere at night even during magnetic quiet periods.

The variabilities of Sq hourly amplitudes for the hour, t from the day i to the next day $i + 1$ for all hours of the day called the day-to-day variability, were investigated only for consecutive IQDs. The day-to-day variabilities of Sq(H) and Sq(Z) are as shown in Figures 3 and 4. For the four stations, seven consecutive IQDs in the month of September were chosen. The variability occurrence was a dawn to dusk phenomenon, although more noticeable in the daytime but turns very mild during the night in both Sq(H) and Sq(Z) in all the stations. It could be seen from Figures 3 and 4 that the variabilies between two paired consecutive days are quite different from any other two paired subsequent consecutive days. For example, on the 3/4 September, 5/6 September, and 6/7 September, the variations are seen to be remarkably different from one another. Observing this figure on the 3/4 September at AAB, the day-to-day variability

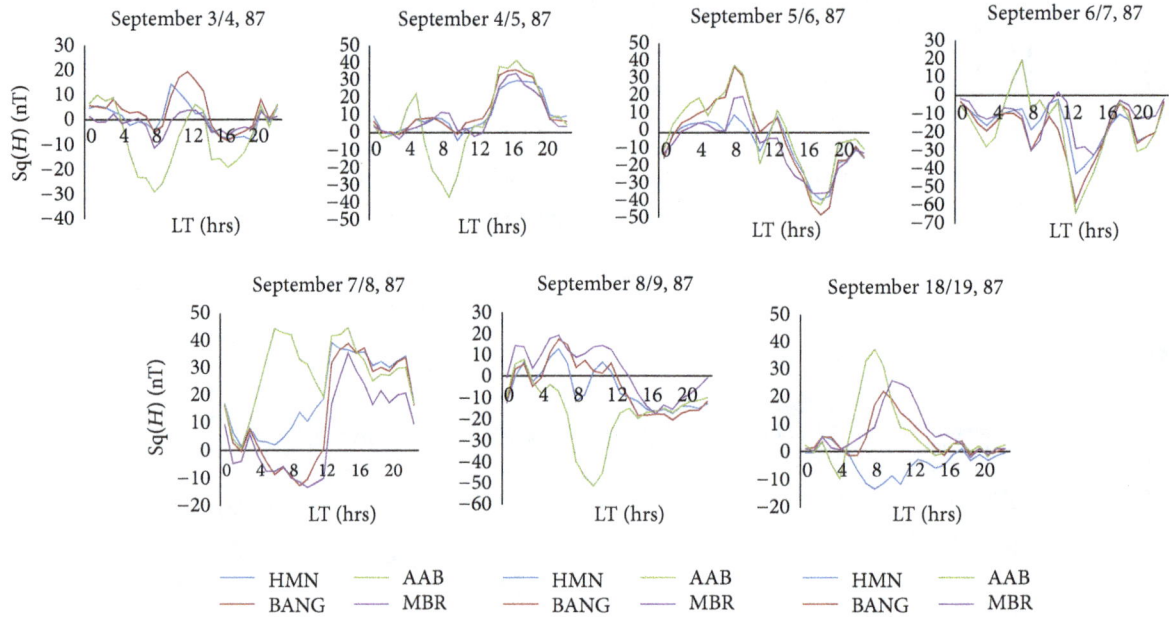

FIGURE 3: Day-to-day variations in Sq(H) for the month of September.

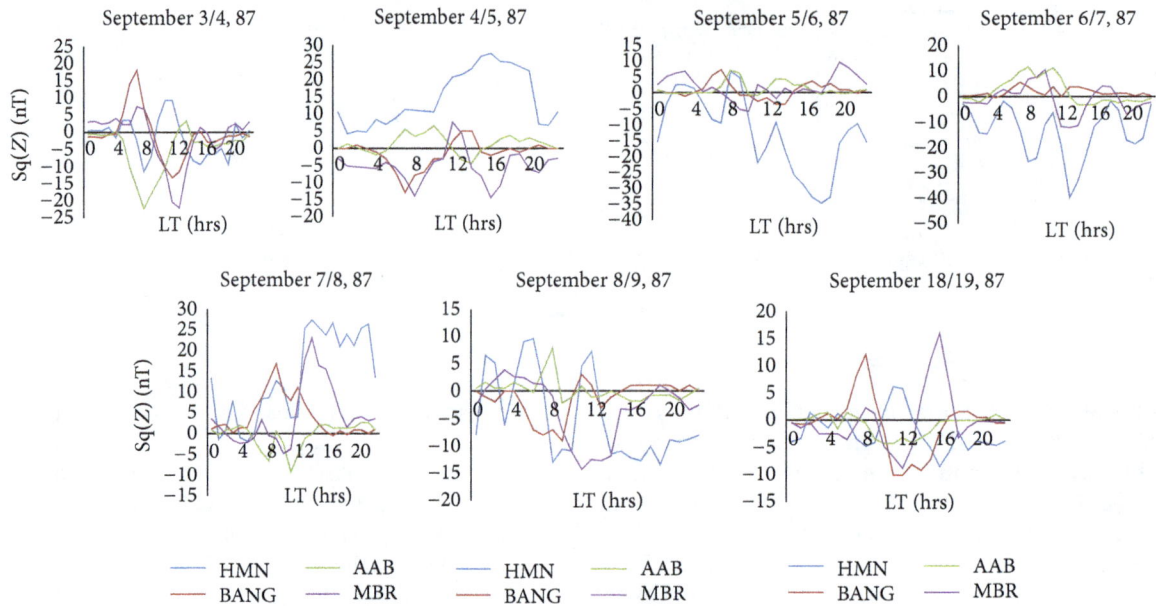

FIGURE 4: Day to day variations in Sq(Z) for the month of September.

in H was seen to have maximum amplitude of about 10 nT and a minimum amplitude up to −30 nT, while on the 4/5, at the same station, the maximum amplitude was about 40 nT and the minimum was about −37 nT while on the 6/7 the maximum was 37 nT and the minimum was about −40 nT. The maxima and minima were seen to be occurring at different times; thus, there exists phase variations. Significant differences in amplitude as well as in phase can be seen in the other three stations. These amplitude and phase variations are seen not to have a definite pattern; they are seen to be random. This result is in line with Okeke et al. [13] who noted that

changes in the electric field control the phase and randomness of the variabilities, while the magnitude of the ionospheric conductivity controls the magnitude of the variabilities.

It could be seen from Figures 3 and 4 also that the variabilities were maximum during daylight hours and mild at night suggesting that the root cause of Sq could be responsible for its day-to-day variability.

The observed day-to-day variability in Sq(H) on 18/19 September followed the exact pattern of variation of Sq(H) observed in Figure 1. These two days may be said to be very quiet suggesting that the observed randomness on the other consecutive days may actually be due to magnetic disturbances of a magnetospheric origin. Following Okeke et al. [13] who noted that changes in electric field control the phase and randomness, we are suggesting that it is the magnetic disturbances that give rise to the change in the electric field. These magnetic disturbances are actually found to affect the determination of a true Sq variation.

The day-day amplitudes in Sq(Z) at Addis Ababa are found to be small compared to all other stations in all the consecutive days except on 6/7 September. this fact can be related to the observed counterelectrojet in Sq(Z) at Addis Ababa because the widths of both the electrojet and the counterelectrojet are nearly equivalent, thus causing the amplitude to be greatly reduced. This finding is in agreement with the works of [33, 34].

4. Conclusion

The results of this study confirm that Sq is a very changeable phenomenon, with a strong day-to-day variation, and that it is superimposed on magnetic disturbances of a magnetospheric origin. It could be seen from the results that the root cause of Sq could be responsible for its day-to-day variability. It is found that, for the Equatorial Electrojet stations, when the widths of both the electrojet and the counter-electrojet are nearly equivalent, the amplitude of variation is greatly reduced. Since the magnetic disturbances are of magnetospheric origin, monitoring the day-to-day variability could provide very important contributions to the knowledge of the ionospheric dynamics as it could be the key to investigate the magnetospheric/ionospheric interaction which actually affects the determination of the true Sq variation.

References

[1] W. H. Campbell, "The regular geomagnetic-field variations during quiet solar conditions," in *Geomagnetism*, J. Jacobs, Ed., vol. 3, pp. 386–460, Academic Press, San Diego, Calif, USA, 1989.

[2] S. Chapman, "The solar and lunar diurnal variation of the earth magnetism," *Philosophical Transactions of the Royal Society A*, vol. 218, pp. 1–118, 1919.

[3] P. N. Mayaud, "Analyse morphologique de la variabilité jourà-jour de la variation journalière "régulière" SR du champ magnétique terrestre, II—Le système de courants CM (Régions non-polaires)," *Annals of Geophysics*, vol. 21, pp. 514–544, 1965.

[4] E. C. Butcher and G. M. Brown, "The variability of Sq(H) on normal quiet days," *The Geophysical Journal of the Royal Astronomical Society*, vol. 64, no. 2, pp. 527–537, 1981.

[5] A. Palumbo, "Lunar and solar daily variations of the geomagnetic field at Italian stations," *Journal of Atmospheric and Terrestrial Physics*, vol. 43, no. 7, pp. 633–642, 1981.

[6] V. W. J. H. Kirchhoff and L. A. Carpenter, "The day-to-day variability in ionospheric fields and currents," *Journal of Geophysical Research*, vol. 81, pp. 2737–2742, 1976.

[7] J. M. Torta, "Behavior of the quiet day ionospheric current system in the European region," *Journal of Geophysical Research A*, vol. 102, no. 2, Article ID 96JA03463, pp. 2483–2494, 1997.

[8] M. Hasegawa, "Geomagnetic Sq current system," *Journal of Geophysical Research*, vol. 65, pp. 1437–1447, 1960.

[9] C. A. Onwumechili, "A study of the equatorial electrojetI: an experimental study," *Journal of Atmospheric and Solar-Terrestrial Physics*, vol. 13, pp. 222–234, 1959.

[10] C. A. Onwumechili, "Geomagnetic variations in the equatorial zone," in *Physics of Geomagnetic Phenomena*, S. Matsushita and W. H. Campbell, Eds., vol. 1, pp. 425–507, Academic press, New York, NY, USA, 1967.

[11] A. B. Rabiu, *Day-to-day variability in the geomagnetic variations at the low latitude observatory of Muntilupa in the Philippines [M.S. thesis]*, University of Nigeria, Nsukka, Nigeria, 1992.

[12] F. N. Okeke, *Day-to-day variability in the geomagnetic variations in the equatorial zone [Ph.D. thesis]*, University of Nigeria, Nsukka, Nigeria, 1995.

[13] F. N. Okeke, C. A. Onwumechili, and B. A. Rabiu, "Day-to-day variability of geomagnetic hourly amplitudes at low latitudes," *Geophysical Journal International*, vol. 134, no. 2, pp. 484–500, 1998.

[14] T. N. Obiekezie and G. A. Agbo, "Day to day variability of Sq(H) variation in the Indian sector," *JANS*, vol. 3, pp. 81–85, 2008.

[15] T. N. Obiekezie, "Geomagnetic field variations at dip equatorial latitudes of West Africa International," *Journal of Physical Sciences*, vol. 7, no. 36, pp. 5372–5377, 2012.

[16] T. N. Obiekezie and F. N. Okeke, "Solar quiet day ionospheric source current in the West African region," *Journal of Advanced Research*, vol. 4, no. 3, pp. 303–305, 2013.

[17] A. B. Rabiu, A. I. Mamukuyomi, and E. O. Joshua, "Variability of equatorial ionosphere inferred from geomagnetic field measurements," *The Bulletin of the Astronomical Society of India*, vol. 35, pp. 607–618, 2007.

[18] A. Onwumechili, "Fluctuations in the geomagnetic horizontal field near the magnetic equator," *Journal of Atmospheric and Terrestrial Physics*, vol. 17, no. 4, pp. 286–294, 1960.

[19] S. Matsushita, "Dynamo currents, winds and electric fields," *Radio Science*, vol. 4, p. 771, 1969.

[20] W. H. Campbell, "Occurrence of AE and Dst geomagnetic index levels and the selection of the quietest days in the year," *Journal of Geophysical Research*, vol. 84, p. 875, 1979.

[21] A. B. Rabiu, "Night time geomagnetic variations at middle latitudes Nigerian," *Journal of Physics*, vol. 85, pp. 35–37, 1996.

[22] T. N. Obiekezie and S. C. Obiadazie, "The variability of H component of geomagnetic field at the African sector," *Physical Review & Research International*, vol. 3, no. 2, pp. 154–160, 2013.

[23] O. Fambitakoye and P. N. Mayaud, "Equatorial electrojet and regular daily variation SR-I. A determination of the equatorial electrojet parameters," *Journal of Atmospheric and Terrestrial Physics*, vol. 38, no. 1, pp. 1–17, 1976.

[24] C. A. Onwumechili and P. O. Ezema, "On the course of the geomagnetic daily variation in low latitudes," *Journal of Atmospheric and Terrestrial Physics*, vol. 39, no. 9-10, pp. 1079–1086, 1977.

[25] F. H. Hibberd, "Day-to-day variability of the S_q geomagnetic field variation," *Australian Journal of Physics*, vol. 34, no. 1, pp. 81–90, 1981.

[26] E. Vestine, *The Geomagnetic Field, Its Description and Analysis*, Carnegie Institute, Washington, DC, USA, 1947.

[27] A. B. Rabiu, *Geomagnetic field variations at the middle latitudes [Ph.D. thesis]*, University of Nigeria, Nsukka, Nigeria, 2000.

[28] S. Alex, B. D. Kadam, and R. G. Rastogi, "A new aspect of daily variations of the geomagnetic field at low latitude," *Journal of Atmospheric and Terrestrial Physics*, vol. 54, no. 7-8, pp. 863–869, 1992.

[29] T. N. Obiekezie and F. N. Okeke, "External S_q currents in the West African region," *Moldavian Journal of the Physical Sciences*, vol. 9, no. 1, pp. 117–122, 2010.

[30] T. N. Obiekezie and F. N. Okeke, "Variations of geomagnetic H, D and Z field intensities on Quiet days at West African latitudes Moldavian," *Journal of Physical Science*, vol. 8, no. 3-4, pp. 366–372, 2009.

[31] W. H. Campbell, "An external current representation of the quiet night side geomagnetic field level changes," *Journal of Geomagnetism and Geoelectricity*, vol. 36, pp. 257–265, 1984.

[32] G. A. Agbo, A. O. Chikwendu, and T. N. Obiekezie, "Variability of daily horizontal component of geomagnetic field component at low and middle latitudes," *Indian Journal of Scientific Research*, vol. 1, no. 2, pp. 1–8, 2010.

[33] P. Gouin and P. N. Mayaud, "A propos de l'existence possible d'un contre-électrojet aux longitudes magnétiques équatoriales," *Annals of Geophysics*, vol. 23, pp. 41–47, 1967.

[34] P. N. Mayaud, "Comment on "the ionospheric disturbances dynamo" by M. Blanc and A.D. Richmond," *Journal of Geophysical Research*, vol. 87, pp. 6353–6355, 1967.

18

Study of Rainfall from TRMM Microwave Imager Observation over India

Anoop Kumar Mishra[1] and Rajesh Kumar[2]

[1] Research Center for Environmental Changes, Academia Sinica, Taipei 115, Taiwan
[2] Remote Sensing & Aerial Surveys, Mission-1B, Geological Survey of India, Govt. of India, Ministry of Mines, Bangalore, India

Correspondence should be addressed to Anoop Kumar Mishra, daksha112@gmail.com

Academic Editors: G. Chisham and A. Streltsov

This paper presents a technique to estimate precipitation over Indian land (6–36°N, 65–99°E) at 0.25° × 0.25° spatial grid using tropical rainfall measuring mission (TRMM) microwave imager (TMI) observations. It adopts the methodology recently developed by Mishra (2012) to monitor the rainfall over the land portion. Regional scattering index (SI) developed for Indian region and polarization corrected temperature (PCT) have been utilized in this study. These proxy rain variables (i.e., PCT and SI) are matched with rainfall from precipitation radar (PR) to relate rain rate with PCT, SI, and their combination. Retrieval techniques have been developed using nonlinear relationship between rain and proxy variables. The results have been compared with the observations (independent of training data set) from PR. Results have also been validated with the observations from automatic weather station (AWS) rain gauges. It is observed from the validation results that nonlinear algorithm using single variable SI underestimates the low rainfall rates (below 20 mm/h) but overestimates the high rain rates (above 20 mm/h). On the other hand, algorithm using PCT overestimates the high rain rates (above 25 mm/h). Validation results with rain gauges show a CC of 0.68 and RMSE of 4.76 mm when both SI and PCT are used.

1. Introduction

High rainfall events in India have increased by 50% during past 50 years [1]. Hence an accurate prediction of high rainfall event is essential for prevention and management of disasters. In India, there is a low spatial density of automatic rain gauges and Doppler Weather Radars (DWRs). So satellite-based rainfall estimates are highly valuable for synoptic situations. Rainfall estimations based on infrared (IR) measurements from satellite have large errors because IR radiances from cloud tops have only indirect and weak relationship with surface rainfall [2, 3]. On the other hand, satellite microwave measurements provide more direct estimates of rainfall. Microwave algorithms are generally classified as statistical or physical [4]. Statistical algorithms (e.g., [5]) use observed data to derive an empirical relationship between brightness temperature and precipitation. Physical algorithm (e.g., [6–9]) uses a data base of radiative transfer calculations based on atmospheric profiles, which are compared with

an observed set of brightness temperature. The higher the number of channels used, the greater the chances of finding an accurate hydrometeor profile in data base. Precipitation estimation algorithms from microwaves using radiative transfer models have been developed by Wu and Weinmann [10] and Kummerow et al. [11]. Most of the microwave rainfall techniques have been based on tropical, subtropical, or continental convective rainfall concentrating on accurate retrieval of higher rainfall rates. Grody [12] used microwave observations to suggest that scattering index (SI) is a good indicator of rainfall. On the basis of Grody's algorithm, Ferraro and Marks [13] developed a global scattering index separately for land and ocean to estimate the rainfall. Chen et al. [14] used SI for rainfall retrieval over Taiwan. Mishra et al. [15] found that global scattering indices were highly variable for different regions and seasons and hence developed a regional scattering index for the Indian land and associated oceanic regions. A nonlinear relationship was developed between SI and rain rate, separately for the Indian land

and associated oceanic regions. Comparison with rain gauge observations showed a correlation coefficient of 0.68 and root mean square error of 9.35 mm/h over Indian land. The validation with rain gauge observations confirmed that regional based scattering indices were able to retrieve the rainfall over Indian region better than the global scattering index developed by Ferraro and Marks [13]. Grody [16] introduced the surface independent polarization-corrected temperature (PCT) to estimate rainfall. Spencer et al. [17] and Kidd and Barret [18] used PCT imagery at 85 GHz microwave brightness temperature to analyze the rainfall signatures. Kumar et al. [19] utilized PCT to explore the precipitation feature over Indian region from special sensor microwave/imager (SSM/I) observations. Nativi et al. [20] utilized PCT, SI, polarization difference, and combination of brightness temperatures to estimate the rainfall from SSM/I observations. Tsai et al. [21] used PCT depression to estimate rainfall over Taiwan during Typhoon by the application of nonlinear regression. These studies showed the possibilities for rainfall estimation at the highest possible resolution over nonscattering surface by identification of depressed PCT values. Biscaro and Morales [22] applied the probability matching method (PMM) developed by Calheiros and Zawadzki [23] to derive the relation between PCT and rain rate. The main concept behind the PMM was to relate two independent variables through their probability frequency. Liu and Curry [24] proposed the use of both scattering and emission signatures from microwave observations to estimate the rainfall. Todd and Bailey [25] used depression in PCT to estimate the rainfall rate by the application of nonlinear regression. Yao et al. [26] utilized the PCT and SI from TRMM microwave observations to estimate the rainfall over Tibetan Plateau assuming a linear relationship between the radar rain rate, PCT, and SI. They observed that high rainfall rates were underestimated most of the time by the application of linear algorithm. Zhao et al. [27] used both SI and PCT to monitor the high rainfall events over China on the basis of linear regression between rain rate and combination of SI and PCT using TMI observation. Mahesh et al. [28] used both SI and PCT to estimate the rainfall using artificial neural network. Recently, heavy rainfall events during cyclonic cases over Indian ocean using SI and PCT were studied by Mishra [29]. He found that nonlinear algorithm performs better than linear algorithm in retrieving the rainfall over oceanic region (especially heavy rainfall). In the present study, TRMM data are analyzed to estimate the rainfall over Indian land based on nonlinear algorithm using PCT, SI, and combination of them.

2. Data Used

TRMM data is used in this study. The tropical rainfall measuring mission (TRMM) is a joint space mission between National Aeronautics and Space Administration (NASA) and the Japan Aerospace Exploration Agency (JAXA) designed to monitor and study tropical rainfall. It was launched in 1997 into a near circular orbit. Along with other sensors it carries TRMM microwave imager (TMI) and first space borne precipitation radar (PR). In the present study surface rainfall

data of Precipitation Radar Version 6 (2A25) [30] is used during the years 2008–2011. Brightness temperature data of TMI at 19, 22, and 85 GHz channels are also utilized for the same period in this study. In the present study, the Indian Space Research Organization-Designed Automatic Weather Station (AWS) rain gauge data is used. Among many other sources, AWS has a tipping bucket rain gauge with unlimited rain measuring capacity. At present, more than 300 AWSs are deployed in clusters working all over specific regions in India. Temporal resolutions of these rain gauge observations are 15 minutes. For validation of present algorithm, AWS rain gauge data during 2009 and 2010 are used. The point measurement by AWS rain gauges is averaged within $0.25° \times 0.25°$ spatial grid. Area of study and AWS distribution over India is shown in Figure 1.

3. Methodology

Rain signatures based on scattering and polarization properties of the microwave radiation from the hydrometeors are identified. Region-specific scattering index developed by Mishra et al. [15] was utilized for scattering-based rain signature. This scattering index for the Indian region is given by following equation:

$$SI = \left(448.6809 + (-1.5456 \times T_{19V}) + (-0.6020 \times T_{22V}) + \left(0.0055 \times (T_{22V})^2\right)\right) - T_{85V},$$
(1)

where T_f is vertically polarized brightness temperature in °K at frequency "f."

Another rain signature based on polarization and scattering property of the microwaves at 85 GHz brightness temperature proposed by Spencer et al. [17] is defined as follows:

$$PCT = (1 + 0.818)T_{85V} - 0.818 T_{85H},$$
(2)

where T_{85V} and T_{85H} are vertically and horizontally polarized brightness temperatures at 85 GHz, respectively.

Now, these variables SI and PCT are matched with rainfall rates from PR in $0.25° \times 0.25°$ grid box over Indian land. Coastal areas are excluded from the area of study.

Figures 2 and 3 show the relationship between rainfall rate versus SI and PCT, respectively. Lines drawn over scatters are lines of best fit. For this purpose, total number of 5764 collocated data points from PR rain rate and SI, PCT, and combination of SI and PCT during the years 2008–2011 are used.

Nonlinear equations using SI, PCT, and their combination with rainfall rate (mm/h) from PR are as follows:

$$Rain = 0.0001 \times (SI)^{2.7496},$$
(3)

$$Rain = -5.02 + (585.39 \times \exp(-0.0168 \times PCT)),$$
(4)

$$Rain = -6.75 + (89.15 \times \exp(-0.0088 \times PCT)) + \left(0.0001 \times (SI)^{2.5977}\right).$$
(5)

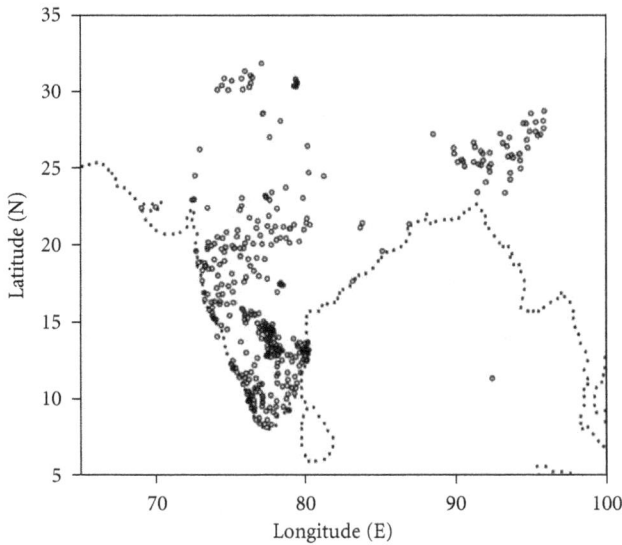

FIGURE 1: Area of study and density of AWS rain gauges.

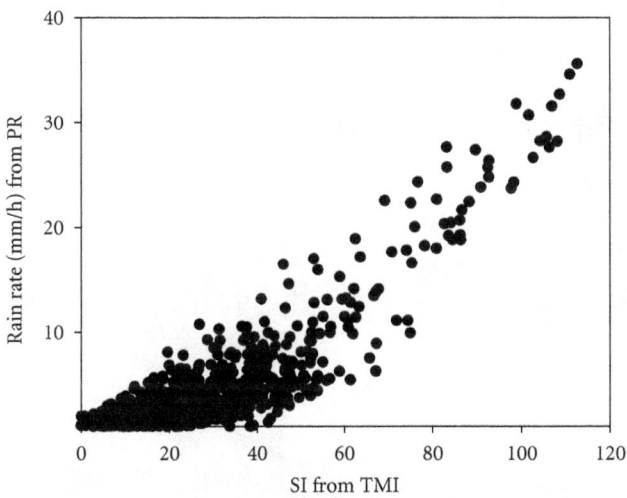

FIGURE 2: Scatter plot between rain rate (mm/h) from PR and SI from TMI.

TABLE 1: Error statistics of the training data sets.

Regression equation	Data points	Correlation coefficients	Root mean square error (mm/h)
Nonlinear equation using SI (3)	5764	0.762	2.061
Nonlinear equation using PCT (4)	5764	0.786	1.952
Nonlinear equation using SI and PCT (5)	5764	0.842	1.764

Correlation coefficients of 0.76, 0.78, and 0.84 and root mean square error of 2.06, 1.95, and 1.76 mm/h are observed for the above equations (3), (4), and (5), respectively, (see Table 1).

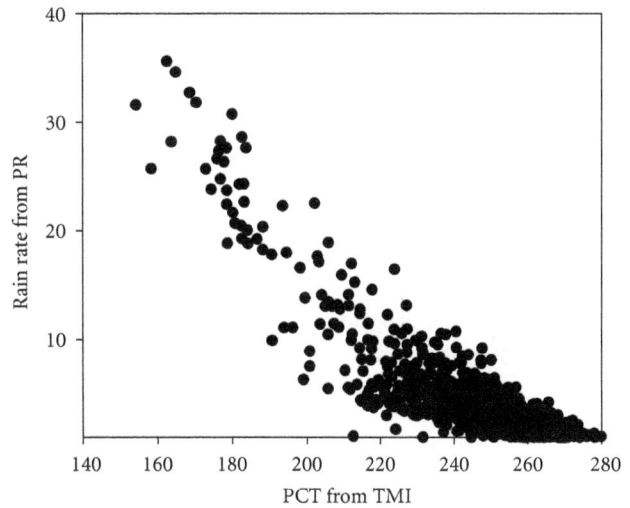

FIGURE 3: Scatter plot between rain rate (mm/h) from PR and PCT from TMI.

4. Results and Discussions

Regression equations (3)–(5) are used to study the rainfall over land portion of the Indian region. We have tested the algorithm by comparing the rainfall obtained from above equation with that from independent (of training data sets) PR observations. For this purpose, PR observations were averaged at $0.25° \times 0.25°$ grid box. Total 6842 independent data sets (different from the points used for the training purpose) during the years 2008–2011 are selected. Figures 4(a)–4(c) show scatter plots between PR rain rate and rain rates retrieved using nonlinear regression ((3), (4), and (5)). Lines drawn over scatters are lines of best fit. From Figure 4(a), it may be observed that nonlinear regression using SI only underestimates the lighter rain rates (1–10 mm/h). It may also be noted that higher rain rates (above 25 mm/h) are overestimated using SI only. Careful observation of Figure 4(b) reveals that rainfall using PCT compares well with PR at low-to-moderate rainfall rates (below 15 mm/h) but overestimates the high rainfall rates (above 25 mm/h). It may further be observed from Figure 4(c) that algorithm using both SI and PCT compares well with PR observations at each rainfall range. Algorithm using both SI and PCT performs better than those using either SI or PCT for precipitation observations over Indian region. Error statistics of this comparison are listed in Table 2. It may be observed from Table 2 that statistical comparison with PR shows a CC of 0.72, 0.74, 0.78 RMSE of 3.18, 2.76, 1.89 mm/h and bias of 2.17, 1.96, 1.18 mm/h using SI, PCT, and combination of SI and PCT, respectively. It may be concluded that algorithm using both SI and PCT shows superior comparison with PR than that using either SI or PCT.

We analyzed the accuracy of this technique to monitor the rainfall at different ranges using bin analysis. For this purpose, independent PR observations are averaged in seven ranges, that is, 0–5, 5–10, 10–15, 15–20, 20–25, 25–30, and 30 to above 30 (mm/h). The total number of observations in each bin is almost constant, so that statistical weight is same.

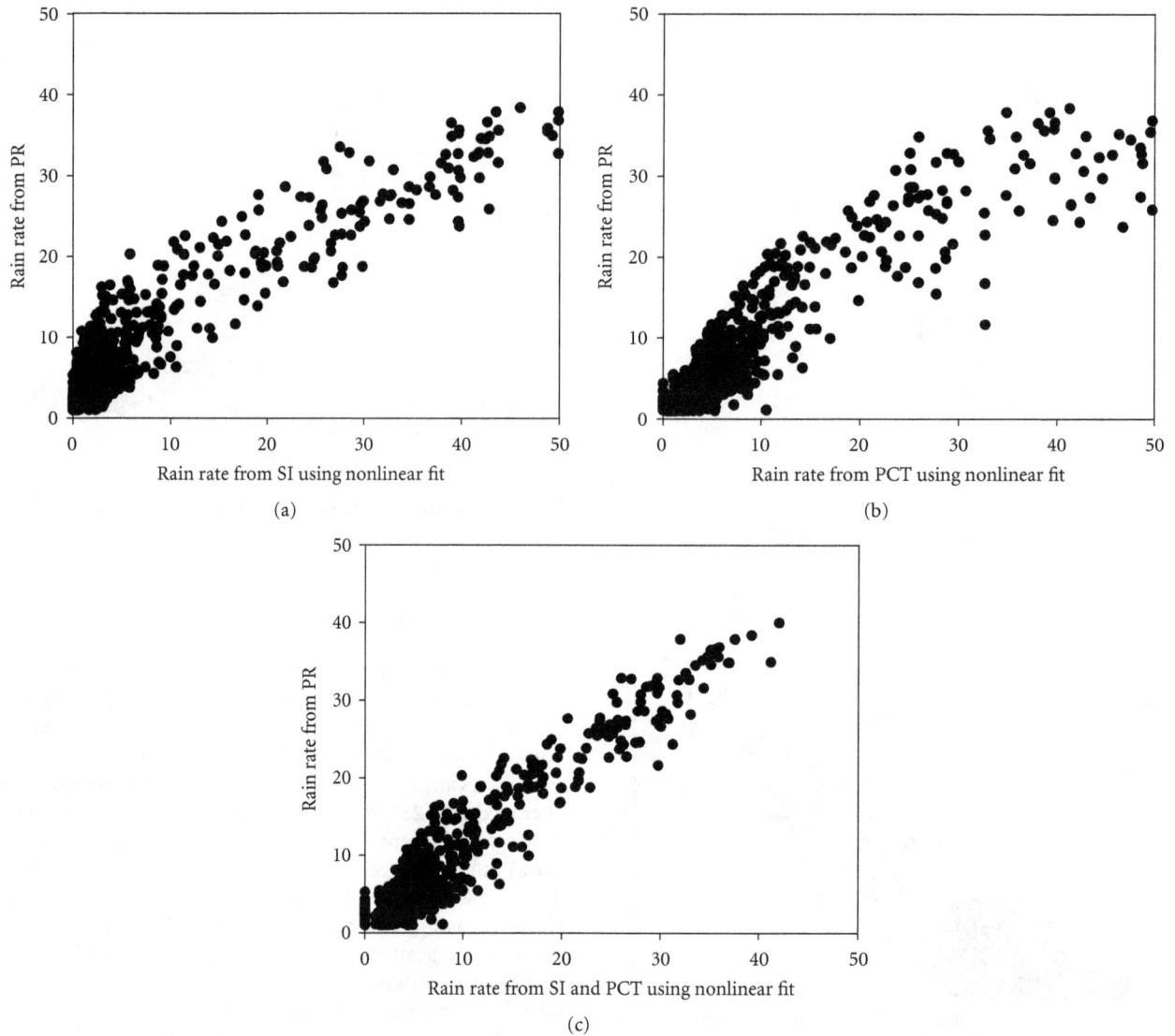

FIGURE 4: (a) Scatter plot between rain rate (mm/h) from PR and that from nonlinear regression using SI, (b) same as Figure 4(a), but using PCT only, and (c) same as Figure 4(a), but using both SI and PCT.

TABLE 2: Statistical comparison of algorithms using nonlinear regression with PR observations (validation data sets).

Algorithm applied	No. of data points	Correlation coefficients	Root mean square error (RMSE) (mm/h)	Bias (mm/h)
Nonlinear regression using SI only	6842	0.722	3.186	2.171
Nonlinear regression using PCT only	6842	0.741	2.763	1.962
Nonlinear regression using both SI and PCT	6842	0.784	1.896	1.182

Figure 5 shows scatter plot of RMSE and averaged rainfall rates. It is observed that algorithm using both SI and PCT shows the least error.

Finally, rainfall rates obtained from linear and nonlinear algorithms are validated using AWS rain gauges. The total number of 2516 data points is selected for this purpose during high rainfall events of 2009 and 2010. Figures 6(a)–6(c) represent scatter plots between rain rate from rain gauge and that from nonlinear regression using SI, PCT, and both SI and PCT, respectively. Lines drawn over scatters are lines of best fit. It may be concluded from these figures that nonlinear regression using multiproxy rain variable shows better

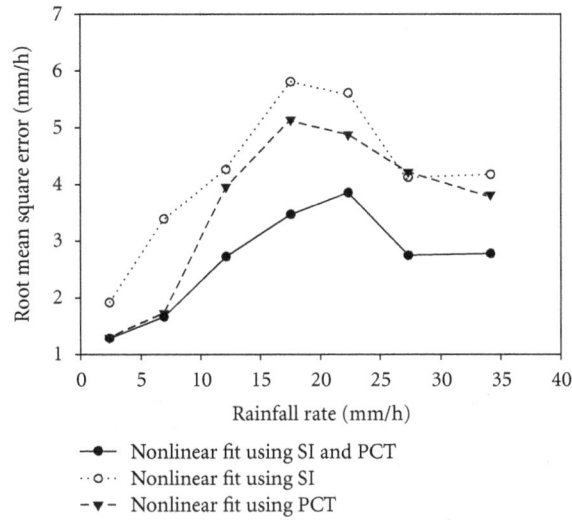

FIGURE 5: Scatter plot between binned rain rate (mm/h) from PR and RMSE obtained from binned rain rate (mm/h) from PR and those from SI, PCT, and combination of SI and PCT using linear and nonlinear regression.

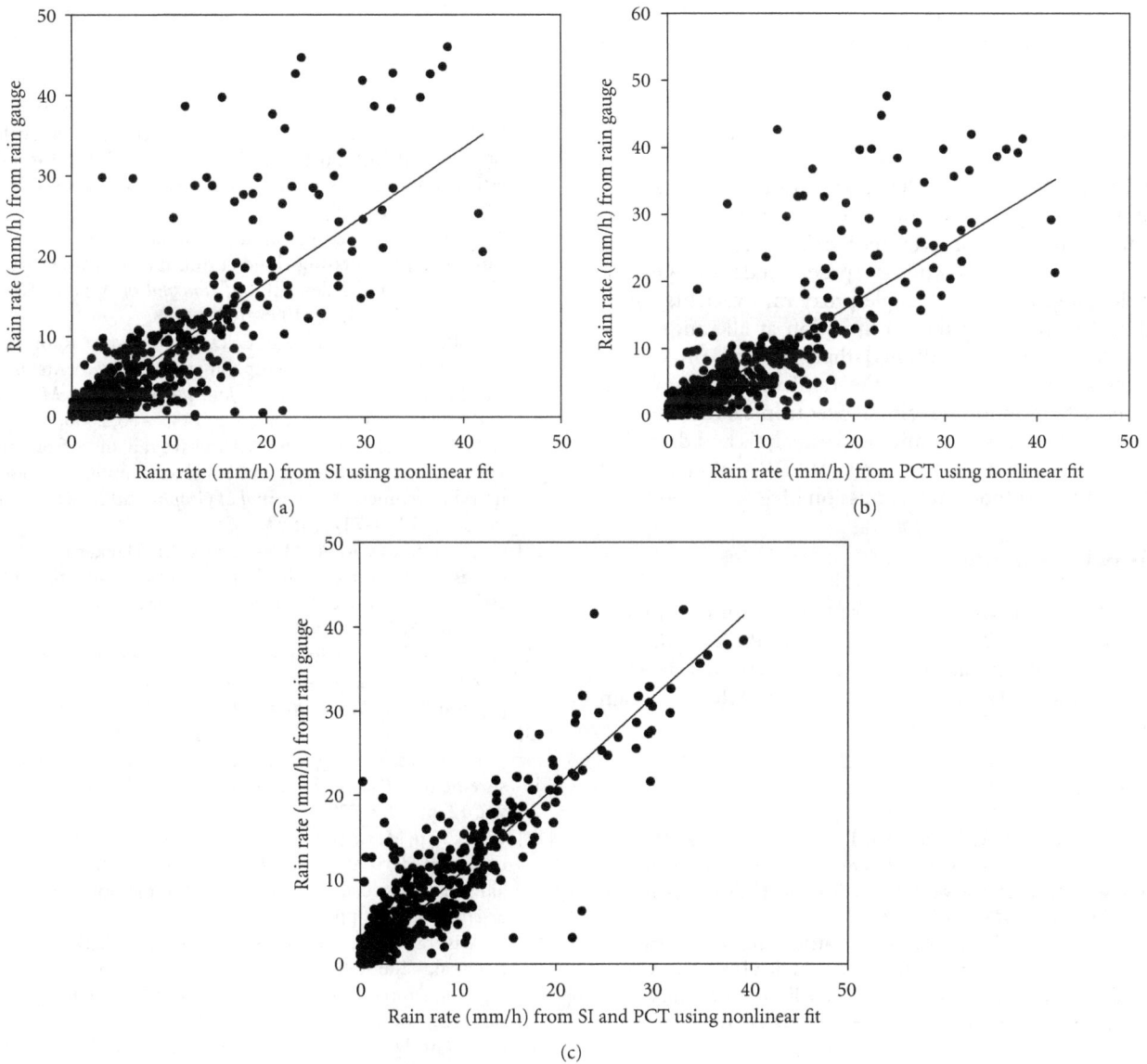

FIGURE 6: (a) Scatter plot between rain rate (mm/h) from rain gauge and that from nonlinear regression using SI, (b) same as Figure 6(a), but using PCT, and (c) same as Figure 6(a), but using SI and PCT.

TABLE 3: Statistical comparison of algorithms using nonlinear regression with AWS rain gauge observations.

Algorithm applied	No. of data points	Correlation coefficients	Root mean square error (RMSE) (mm/h)	Bias (mm/h)
Nonlinear regression using SI only	2516	0.654	7.542	6.653
Nonlinear regression using PCT only	2516	0.673	6.763	4.563
Nonlinear regression using both SI and PCT	2516	0.682	4.763	3.987

comparison with gauges as compared to that using single variable. Error statistics are listed in Table 3. Correlation coefficients (CC) of 0.65, 0.67, and 0.68, RMSE of 7.542, 6.763, and 4.763, and bias of 6.65, 4.56, and 3.98 mm/h are observed with rain gauge from the measurement of SI, PCT, and combination of SI and PCT, respectively, using nonlinear regression.

It may be inferred from validation results that multiple rain signatures are required to monitor the rainfall over Indian land region.

5. Conclusion

The technique described in the present study emphasizes the use of multiple signatures rain from microwave observations to monitor the rainfall over Indian land. Region-specific SI and PCT are used for this purpose. Validation with PR and rain gauge shows that single proxy rain variable (SI or PCT) cannot represent the precipitation at all ranges. The present technique has indicated that multiple proxy rain variables successfully monitor the rainfall. This algorithm can be used to monitor the rainfall over tropical region using more frequent passes from microwave analysis and detection of rain and atmospheric structures (MADRASs) channel of recently launched Indo-French mission Megha-Tropiques.

Acknowledgments

TRMM data from NASA/GSFC used in this study is thankfully acknowledged. Useful discussions with Professor J. Srinivasan are thankfully acknowledged. The authors also acknowledge the MOSDAC for providing the ISRO AWS rain gauge data.

References

[1] B. N. Goswami, V. Venugopal, D. Sangupta, M. S. Madhusoodanan, and P. K. Xavier, "Increasing trend of extreme rain events over India in a warming environment," *Science*, vol. 314, no. 5804, pp. 1442–1445, 2006.

[2] A. Mishra, R. M. Gairola, A. K. Varma, and V. K. Agarwal, "Remote sensing of precipitation over Indian land and oceanic regions by synergistic use of multi-satellite sensors," *Journal of Geophysical Research*, vol. 115, Article ID D08106, 12 pages, 2010.

[3] A. K. Mishra, R. M. Gairola, A. K. Varma, and V. K. Agarwal, "Improved rainfall estimation over the Indian region using

satellite infrared technique," *Advances in Space Research*, vol. 48, no. 1, pp. 49–55, 2011.

[4] A. Mishra, *Remote Sensing of Precipitation from Satellite Infrared and Microwave Observations Over Indian Tropics*, Lambert Academic Publication, Berlin, Germany, 2012.

[5] R. F. Adler, G. J. Huffman, and P. R. Keehn, "Global tropical rain estimates from microwave-adjusted geosynchronous IR data," *Remote Sensing Reviews*, vol. 11, no. 1–4, pp. 125–152, 1994.

[6] K. F. Evans, J. Turk, T. Wong, and G. L. Stephens, "A Bayesian approach to microwave precipitation profile retrieval," *Journal of Applied Meteorology*, vol. 34, no. 1, pp. 260–278, 1995.

[7] C. Kummerow, Y. Hong, W. S. Olson et al., "The evolution of the Goddard profiling algorithm (GPROF) for rainfall estimation from passive microwave sensors," *Journal of Applied Meteorology*, vol. 40, no. 11, pp. 1801–1820, 2001.

[8] A. Mugnai and E. A. Smith, "Radiative transfer to space through a precipitating cloud at multiple microwave frequencies. Part1: model description," *Journal of Applied Meteorology,*, vol. 27, no. 9, pp. 1055–1073, 1988.

[9] N. Viltard, C. Burlaud, and C. D. Kummerow, "Rain retrieval from TMI brightness temperature measurements using a TRMM PR-based database," *Journal of Applied Meteorology and Climatology*, vol. 45, no. 3, pp. 455–466, 2006.

[10] R. Wu and J. A. Weinman, "Microwave radiances from precipitating clouds containing aspherical ice, combined phase, and liquid hydrometeors," *Journal of Geophysical Research*, vol. 89, no. 5, pp. 7170–7178, 1984.

[11] C. Kummerow, R. A. Mack, and I. M. Hakkarinen, "A self-consistency approach to improve microwave rainfall rate estimation from space," *Journal of Applied Meteorology*, vol. 28, no. 9, pp. 869–884, 1989.

[12] N. C. Grody, "Classification of snow cover and precipitation using the special sensor Microwave Imager," *Journal of Geophysical Research*, vol. 96, no. 4, pp. 7423–7435, 1991.

[13] R. R. Ferraro and G. F. Marks, "The development of SSM/I rain rate retrieval algorithms using ground based radar measurements," *Journal of Atmospheric and Oceanic Technology*, vol. 12, pp. 755–770, 1995.

[14] W. J. Chen, M. D. Tsai, G. R. Liu, and M. H. Chang, "The study of rainfall derived from TRMM Microwave Imager data over Taiwan land—using Scattering Index method," *Atmospheric Sciences*, vol. 33, pp. 277–300, 2005.

[15] A. Mishra, R. M. Gairola, A. K. Varma, A. Sarkar, and V. K. Agarwal, "Rainfall retrieval over Indian land and oceanic regions from SSM/I microwave data," *Advances in Space Research*, vol. 44, no. 7, pp. 815–823, 2009.

[16] N. C. Grody, "Precipitation monitoring over land from satellites by microwave radiometry," in *Proceedings of the International Geoscience and Remote Sensing Symposium (IGARSS*

'84), ESA SP-215, pp. 417–423, Strasbourg, France, August 1984.

[17] R. W. Spencer, H. M. Goodman, and R. E. Hood, "Precipitation retrieval over land and ocean with the SSM/I: identification and characteristics of the scattering signal," *Journal of Atmospheric and Oceanic Technology*, vol. 6, no. 2, pp. 254–273, 1989.

[18] C. Kidd and E. C. Barret, "The use of passive microwave imagery in rainfall monitoring," *Remote Sensing Reviews*, vol. 4, no. 2, pp. 415–450, 1990.

[19] R. Kumar, R. M. Gairola, A. Mishra, A. K. Varma, and I. M. L. Das, "Evaluation of precipitation features in high-frequency SSM/I measurements over Indian land and oceanic regions," *IEEE Geoscience and Remote Sensing Letters*, vol. 6, no. 3, pp. 373–377, 2009.

[20] S. Nativi, E. C. Barrett, and M. J. Beaumont, "Microwave monitoring of rainfall: intercomparisons of data from the Chilbolton radar and the DMSP-SSM/I," *Meteorological Applications*, vol. 4, no. 2, pp. 101–114, 1997.

[21] M. D. Tsai, W. J. Chen, and J. L. Wang, "Implementing polarization-corrected temperature into typhoon rainfall estimates over Taiwan from TRMM/TMI data," *Terrestrial, Atmospheric and Oceanic Sciences*, vol. 21, no. 4, pp. 697–712, 2010.

[22] T. S. Biscaro and C. A. Morales, "Continental passive microwave-based rainfall estimation algorithm: application to the Amazon Basin," *Journal of Applied Meteorology and Climatology*, vol. 47, no. 7, pp. 1962–1981, 2008.

[23] R. V. Calheiros and I. Zawadzki, "Reflectivity-rain rate relationships for radar hydrology in Brazil," *Journal of Climate & Applied Meteorology*, vol. 26, no. 1, pp. 118–132, 1987.

[24] G. Liu and J. A. Curry, "An investigation of the relationship between emission and scattering signals in SSM/I data," *Journal of the Atmospheric Sciences*, vol. 55, no. 9, pp. 1628–1643, 1998.

[25] M. C. Todd and J. O. Bailey, "Estimates of rainfall over the United Kingdom and surrounding seas from the SSM/I using the polarization corrected temperature algorithm," *Journal of Applied Meteorology*, vol. 34, no. 6, pp. 1254–1265, 1995.

[26] Z. Yao, W. Li, Y. Zhu, B. Zhao, and Y. Chen, "Remote sensing of precipitation on the Tibetan Plateau using the TRMM Microwave Imager," *Journal of Applied Meteorology*, vol. 40, no. 8, pp. 1381–1392, 2001.

[27] B. Zhao, Z. Yao, W. Li et al., "Rainfall retrieval and flooding monitoring in China using TRMM Microwave Imager (TMI)," *Journal of the Meteorological Society of Japan*, vol. 79, no. 1, pp. 301–315, 2001.

[28] C. Mahesh, S. Prakash, V. Sathiyamoorthy, and R. M. Gairola, "Artificial neural network based microwave precipitation estimation using scattering index and polarization corrected temperatures," *Atmospheric Research*, vol. 102, no. 3, pp. 358–364, 2011.

[29] A. Mishra, "Estimation of heavy rainfall during cyclonic storms from microwave observations using nonlinear approach over Indian Ocean," *Natural Hazards*, vol. 63, no. 2, pp. 673–683, 2012.

[30] T. Iguchi, T. Kozu, R. Meneghini, J. Awaka, and K. I. Okamoto, "Rain-profiling algorithm for the TRMM precipitation radar," *Journal of Applied Meteorology*, vol. 39, no. 12, pp. 2038–2052, 2000.

19

Theoretical Study on the Flow Generated by the Strike-Slip Faulting

Chi-Min Liu,[1,2] **Ray-Yeng Yang,**[2,3] **and Hwung-Hweng Hwung**[2,4]

[1] *Division of Mathematics, General Education Center, Chienkuo Technology University, Changhua City 500, Taiwan*
[2] *International Wave Dynamics Research Center, National Cheng Kung University, Tainan 701, Taiwan*
[3] *Tainan Hydraulic Laboratory and Research Center of Ocean Environment and Technology, National Cheng Kung University, Tainan 701, Taiwan*
[4] *Department of Hydraulic and Ocean Engineering, National Cheng Kung University, Tainan 701, Taiwan*

Correspondence should be addressed to Chi-Min Liu; cmliu@ctu.edu.tw

Academic Editors: G. Casula, E. Liu, and K. Maamaatuaiahutapu

The flow driven by the strike-slip faulting is theoretically analyzed in this paper. The surface of the strike-slip fault is generally near vertical, and the corresponding plates move in horizontal directions during the faulting. The focus of present paper is on the flow at the early stage when the faulting is activated. Standard procedures for deriving the exact solution of the induced flow are first demonstrated. Based on the derived solution, flows generated by three kinds of faulting are examined to observe and compare the evolution of velocity profiles and the corresponding kinetic energy. The results show that the flow energy rapidly decays as the speed of the moving plates begins to slow down. Moreover, mathematical methods proposed in this study provide a useful basis for related studies on not only geophysics, but also fluid mechanics, industry manufacturing, heat-conduction problems, and other possible applications.

1. Introduction

It is well known that many earthquakes are attributed to the fault motion. In general, there are two kinds of faults, the dip-strike and the slip-strike faults, which indicate that the relative movements on the fault plane are approximately vertical and horizontal, respectively. For oceanographers and ocean engineers, a great deal of attention is focused on not only the faulting properties but also the induced flow above the plates. From the viewpoint of energy, the dip-strike and slip-strike faults, respectively, result in the transfer of potential energy and kinetic energy from the moving plates to the fluid. As the dip-strike faulting usually transfers a considerable amount of potential energy into the above fluid, a tsunami wave is simultaneously generated on the ocean surface. Due to the catastrophic damage by the tsunami wave, related studies either on the generation or the propagation of tsunamis are quite abundant (see Ben-Menahem and Rosenman [1], Hammack [2], Okada [3], Dutykh and Dias [4, 5], and Saito and Furumura [6] for details). However, for the flow induced by the slip-strike faulting, it seems to receive much less attention from scientists.

For the sake of modeling the flow driven by the strike-slip faulting, Zeng and Weinbaum [7] may be the first who theoretically studied the flow induced by relatively moving half plates. The two half plates are assumed to move either in a constant speed or a harmonic oscillation. The steady-state solution to flow velocity is then obtained by employing some mathematical techniques. However, the steady-state solution cannot exactly capture the flow at the early stage because the transient component of the flow is totally ignored in their analysis. As the duration of the strike-slip faulting is usually short, Liu [8] theoretically solved the same problems by applying different mathematical methods and then acquired the exact solutions which can successfully describe the flow not only at larger times but also at smaller times. Later, Liu [9] further studied similar problems, while either the mass influx or outflux is allowed on the moving plates.

Though aforementioned papers have preliminary studied the topic considered in the present paper, some important

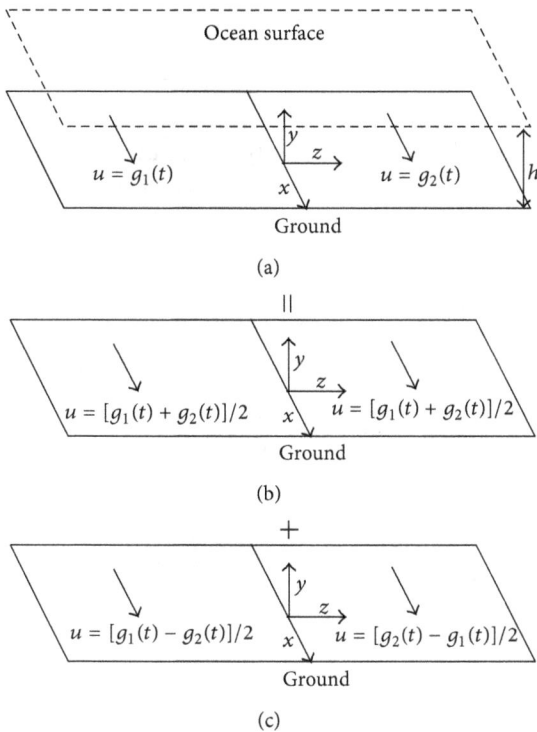

FIGURE 1: Definition sketches for (a) the original problem, (b) the first subproblem of one-dimensional flow, and (c) the second subproblem of two-dimensional flow.

aspects, however, have not yet been well investigated. First, standard procedures for solving the flow driven by arbitrary motion of plates have not been established even though the cases of the so-called extended Stokes' problems were investigated by Liu [8]. Besides, the kinetic energy which plays a more important role in assessing the influence of the flow on neighboring environment has not been calculated in previous studies. To this end, the primary goal of this study is to study these aspects. This paper is organized as follows. In Section 2, mathematical techniques and standard procedures are developed to derive the exact solution of the flow generated arbitrarily moving half plates. Based on the proposed methods, flows induced by three faulting types are, respectively, solved and the corresponding exact solutions are given in Section 3. The velocity evolution and kinetic energy are also analyzed. Finally, important results and future works are briefly concluded in Section 4.

2. Theoretical Analysis

For a viscous flow driven by the moving plates, the difficulty of solving the exact solution greatly depends on how complex the plates shift is. The most well-known case is the Stokes' problems [10, 11] describing a flow generated by a single plate moving in either a constant speed or a harmonic oscillation. The Stokes' problem depicts a one-dimensional flow. One only needs to apply the Laplace transform to acquire the exact solution which provides an accurate prediction of flow at any time. The problem considered is depicted in Figure 1(a).

The fault surface is assumed to be located at the x-y plane and extends from $x = -\infty$ to $x = \infty$ and from $y = 0$ to $y = -\infty$. A viscous fluid of depth h above the x-z plane remains motionless for $t < 0$. When $t > 0$, the flow is suddenly driven by two plates moving in the opposite directions along the x coordinate. The speeds of the negative-z and positive-z plates are represented by arbitrarily time-dependent functions, $g_1(t)$ and $g_2(t)$. Since the induced flow is of two spatial dependences y and z, more mathematical techniques besides the integral transform used in solving Stokes' problems are required to derive the solution. The most important technique is to suitably decompose the original problem into two subproblems, as shown in Figures 1(b) and 1(c). For the first subproblem, the induced flow is one-dimensional since the motion of two plates is uniform and equal to $[g_1(t) + g_2(t)]/2$. Based on the solutions of the classical Stokes' first and second problems, the exact solution is readily acquired by using the Fourier analysis. Therefore, the following analysis will focus on analyzing the second subproblem. In this problem that two plates move in the same speed, $[g_1(t)-g_2(t)]/2$, but in the opposite direction implying the induced flow will be antisymmetrical with respect to the x-y plane. Therefore, only the analysis of the flow in the positive-z domain is required to understand the whole flow. The momentum equation along the x direction, boundary conditions and initial condition in the positive-z domain are shown as follows:

$$u_t = \nu\left(u_{yy} + u_{zz}\right),\qquad(1)$$

$$u\left(y = 0\right) = g\left(t\right),\qquad(2)$$

$$u_y\left(y = h\right) = 0,\qquad(3)$$

$$u\left(z = 0\right) = 0,\qquad(4)$$

$$u\left(t = 0\right) = 0,\qquad(5)$$

where ν is the kinematic viscosity, the subscript denotes the differentiation, and

$$g\left(t\right) \equiv \frac{1}{2}\left[g_2\left(t\right) - g_1\left(t\right)\right]\qquad(6)$$

denotes the plate speed. It is noted that (4) results from that the induced flow is antisymmetrical with respect to the x-z plane. In addition, for the sake of employing the integral transform with suitable boundary conditions, (3) is replaced by

$$u\left(y = 2h\right) = g\left(t\right).\qquad(7)$$

To obtain the exact solution, several mathematical techniques are required and will be briefly demonstrated after. Firstly, the Laplace transform,

$$\hat{u}\left(y,z,s\right) = \int_0^\infty u\left(y,z,t\right)\cdot e^{-st}ds,\qquad(8)$$

is applied to the previous PDE system with the help of the initial condition, (5). Next, one has to shift the transformed variable $\hat{u}$ to a new variable u^* where $u^* = \hat{u} - \hat{g}(s)$. The purpose of this shifting is to make the resulting boundary conditions $u^*(y = 0)$ and $u^*(y = 2h)$ be zero. With the help of the shifted boundary conditions, the Fourier sine transform defined as

$$\tilde{u}(s, n, z) = \int_0^{2h} u^*(s, y, z) \sin\left(\frac{n\pi}{2h} y\right) dy \qquad (9)$$

is subsequently performed to the PDE system. The final result is

$$\tilde{u}_{zz} - \left(\frac{s}{\nu} + \frac{n^2\pi^2}{4h^2}\right)\tilde{u} = \frac{2h}{n\pi\nu}\left(1 - (-1)^n\right) s\hat{g}(s), \qquad (10)$$

with boundary conditions

$$\tilde{u}(z = 0) = -\frac{2h}{n\pi}\left(1 - (-1)^n\right) \cdot \hat{g}(s), \qquad (11)$$

$$\tilde{u}(z = \infty) = \text{finite value.} \qquad (12)$$

Now the solution to $\tilde{u}$ can be determined by solving (10) to (12). It reads

$$\tilde{u} = \frac{2h\left(1 - (-1)^n\right)}{n\pi\nu\left(4h^2 s\nu^{-1} + n^2\pi^2\right)}$$

$$\times \left[\left(4h^2 - 4h^2 s\nu^{-1} - n^2\pi^2\right) \right. \qquad (13)$$

$$\left. \times \exp\left(-\sqrt{\frac{s}{\nu} + \frac{n^2\pi^2}{4h^2}} \cdot z\right) - 4h^2 \right] \cdot s\hat{g}(s).$$

After taking the inverse spatial and time transforms, the exact solution of the flow velocity u in the positive-z domain can be obtained. As for the negative-z domain, the solution is given by the relation $u(-z) = -u(z)$ due to the antisymmetrical concept.

3. Flow Development and Kinetic Energy

From the previous section, the flow velocity can be exactly calculated at any spatial position and at any time. In this section, the evolution of flow velocity and the corresponding kinetic energy will be explored to understand the fundamental characteristics of the induced flows. For the sake of simplification, only the second subsystem displayed in Figure 1(c) is investigated herein. For simulating the possible strike-slip faulting and the induced flow, three faulting types are considered:

(A) uniform motion:

$$g(t) = u_0, \quad \text{at } 0 \le t \le t_0, \qquad (14)$$

(B) linear decay:

$$g(t) = u_0\left(1 - \frac{t}{t_0}\right), \quad \text{at } 0 \le t \le t_0, \qquad (15)$$

(C) exponential decay:

$$g(t) = u_0 \exp(-\alpha t), \quad \text{at } 0 \le t \le t_0, \qquad (16)$$

where u_0 is the constant velocity and t_0 the duration of the faulting. The exact solutions of faulting types (A), (B) and (C) are respectively shown

$$U = 1 + \sum_{n=1}^{\infty} 2 \cdot \frac{1 - (-1)^n}{n\pi} \sin\left(\frac{n\pi Y}{2}\right)$$

$$\cdot \left\{ -\exp\left(-\frac{n^2\pi^2 T}{4}\right) \text{erf}\left(\frac{Z}{2\sqrt{T}}\right) \right.$$

$$\left. - \int_0^T \frac{Z}{2\sqrt{\pi(T')^3}} \exp\left(-\frac{n^2\pi^2 T'}{4} - \frac{Z^2}{4T'}\right) dT' \right\},$$

$$U = 1 - \frac{T}{T_0} + \sum_{n=1}^{\infty} 2 \cdot \frac{1 - (-1)^n}{n\pi} \sin\left(\frac{n\pi Y}{2}\right)$$

$$\cdot \left\{ \frac{4}{n^2\pi^2 T_0} - \left(1 + \frac{4}{n^2\pi^2 T_0}\right) \exp\left(-\frac{n^2\pi^2 T}{4}\right) \text{erf}\left(\frac{Z}{2\sqrt{T}}\right) \right.$$

$$+ \int_0^T \left(\frac{T - T'}{T_0} - 1 - \frac{4}{n^2\pi^2 T_0}\right) \frac{Z}{2\sqrt{\pi(T')^3}}$$

$$\left. \times \exp\left(-\frac{n^2\pi^2 T'}{4} - \frac{Z^2}{4T'}\right) dT' \right\},$$

$$U = \exp(-AT) + \sum_{n=1}^{\infty} 2 \cdot \frac{1 - (-1)^n}{n\pi} \sin\left(\frac{n\pi Y}{2}\right)$$

$$\cdot \left\{ \frac{-A}{A - (n^2\pi^2/4)} \exp(-AT) + \frac{n^2\pi^2/4}{A - (n^2\pi^2/4)} \right.$$

$$\times \exp\left(-\frac{n^2\pi^2}{4} T\right) \text{erf}\left(\frac{Z}{2\sqrt{T}}\right)$$

$$+ \frac{n^2\pi^2/4}{A - (n^2\pi^2/4)} \int_0^T \frac{Z}{2\sqrt{\pi(T')^3}}$$

$$\left. \times \exp\left[-A\left(T - T'\right) - \frac{n^2\pi^2 T'}{4} - \frac{Z^2}{4T'}\right] dT' \right\}, \qquad (17)$$

where the dimensionless variables written in capitals are scaled as

$$U = \frac{u}{u_0}, \quad Y = \frac{y}{h}, \quad Z = \frac{z}{h},$$

$$T = \frac{\nu t}{h^2}, \quad T_0 = \frac{\nu t_0}{h^2}, \quad A = \frac{h^2\alpha}{\nu}. \qquad (18)$$

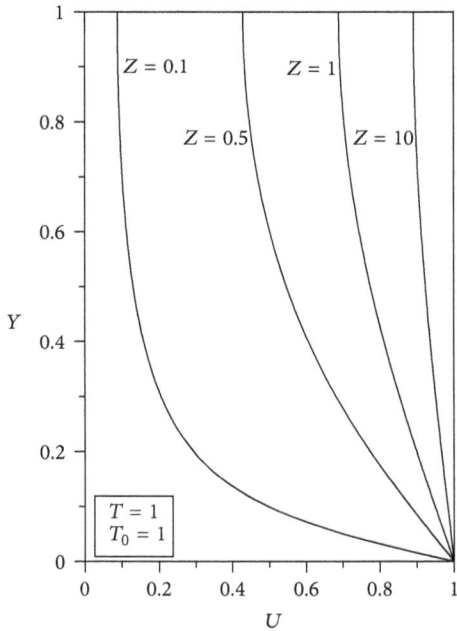

FIGURE 2: Velocity profiles at various Z sections at $T = 1$ for the faulting type A.

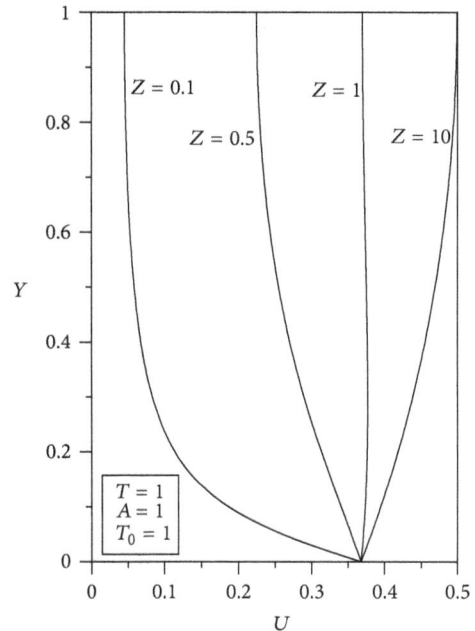

FIGURE 4: Velocity profiles at various Z sections at $T = 1$ for the faulting type C.

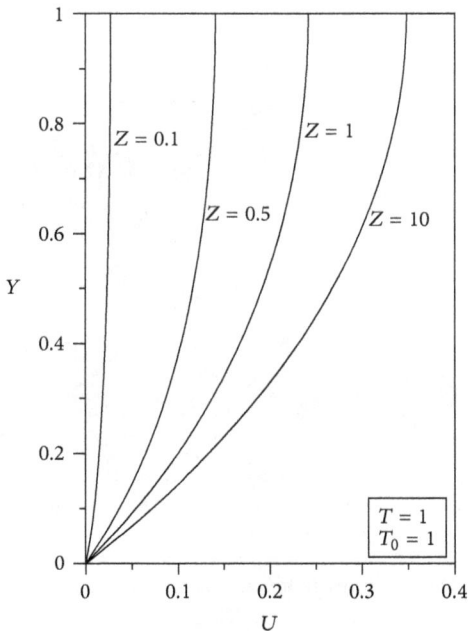

FIGURE 3: Velocity profiles at various Z sections at $T = 1$ for the faulting type B.

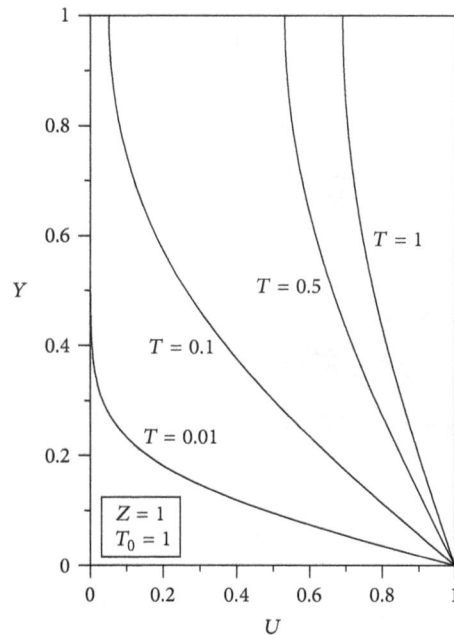

FIGURE 5: Velocity evolution at $Z = 1$ for the faulting type A.

Without loss of generality, the duration of faulting T_0 is assumed to be unity in the following analyses. The velocity profiles at various Z sections at a fixed time $T = 1$ (the end time of the faulting) are first calculated. Results of faulting types (A), (B), and (C) are, respectively, drawn in Figures 2, 3, and 4. It is clearly seen that the velocity profile at smaller Z section develops more slowly than that at larger Z section due to the influence of the reverse flow in the

negative-Z domain. This demonstrates a truth that the fluid near the fault moves more slowly than that far from the fault. Next, Figure 5 depicts the flow development during the faulting ($0 \leq T \leq 1$) for the faulting type (A). At a fixed section $Z = 1$, the velocity profile gradually grows and approaches the plate velocity since the energy is continuously transferred from the plate into the fluid. However, the flow developments of the faulting types (B) and (C) are quite different from that of faulting type (A), as shown in

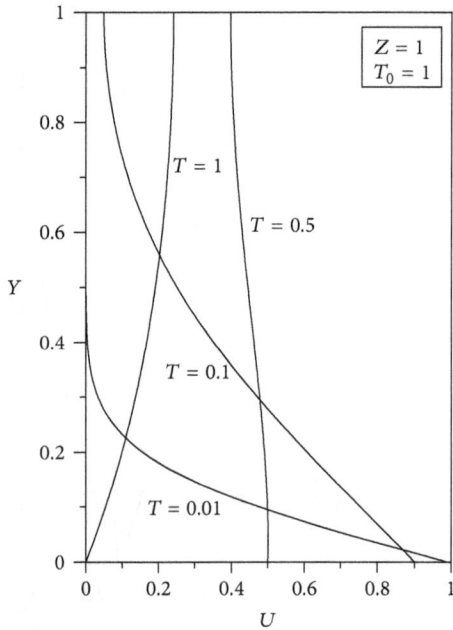

FIGURE 6: Velocity evolution at $Z = 1$ for the faulting type B.

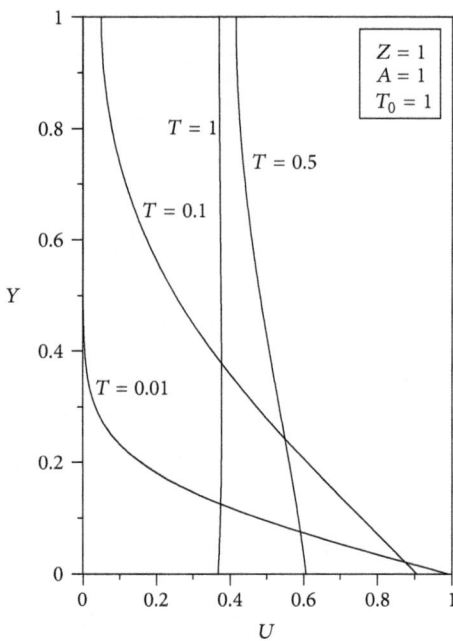

FIGURE 8: Relation between kinetic energy and time for the faulting type A.

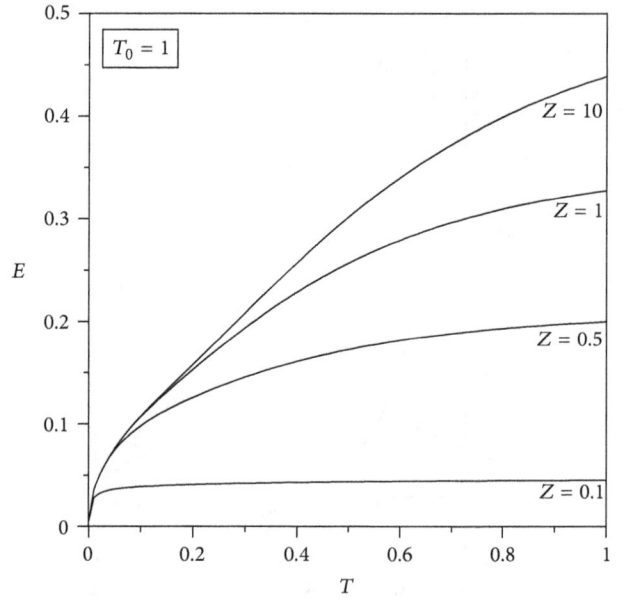

FIGURE 7: Velocity evolution at $Z = 1$ for the faulting type C.

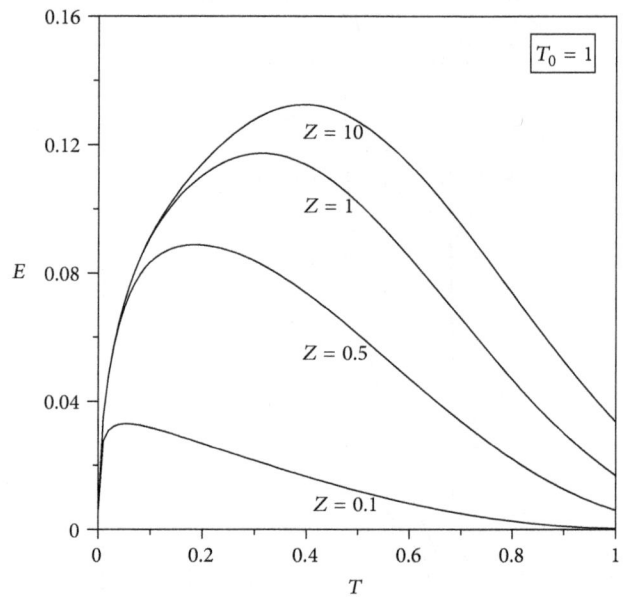

FIGURE 9: Relation between kinetic energy and time for the faulting type B.

Figures 6 and 7. Since the plate speeds of faulting types (B) and (C) decay linearly and exponentially, the velocity will rapidly grow at the early stage of the faulting and then quickly decay due to the effects of viscosity.

Next, the following integral

$$E(Z,T) \equiv \frac{1}{2} \int_0^1 U^2 dy \qquad (19)$$

is defined to measure the kinetic energy per area. In comparison with velocity profiles, exploring the kinetic energy

can provide a much clearer understanding of the potential impact of the earthquake-induced flow on the neighboring region. Figure 8 shows the development of the kinetic energy of faulting type (A) at various Z sections during the faulting. Similar to the velocity development, the kinetic energy at larger Z section is greater than that at smaller Z section. As the energy is continuously transferred into the fluid, the kinetic energy will gradually grow and approach a finite value before the faulting ends. As for results of faulting types (B) and (C), the kinetic energy will reach a maximum value and then decay during the faulting, as shown in Figures 9 and 10.

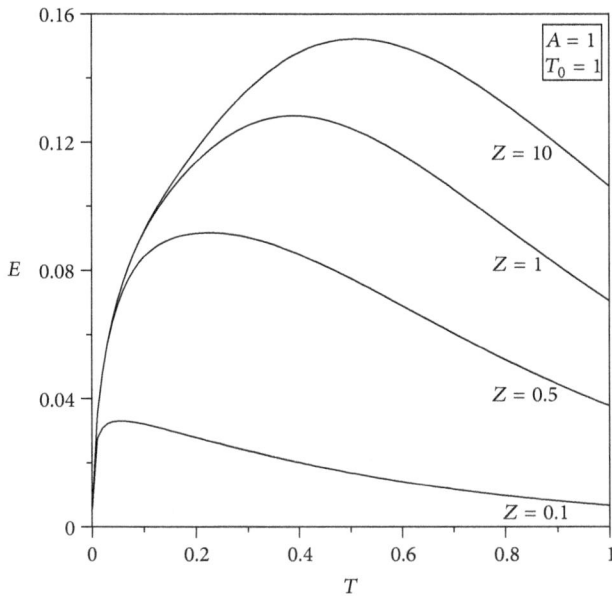

FIGURE 10: Relation between kinetic energy and time for the faulting type C.

Previous results indicate that the faster the faulting decays, the faster the flow energy attenuates.

4. Conclusions

The flow driven by the strike-slip faulting is theoretically analyzed in this paper. The focus is on the flow at the early stage when the faulting is activated. Standard mathematical procedures for deriving the exact solutions are provided. Three types of faulting are also investigated to calculate the velocity evolution and the corresponding kinetic energy. Results show that the kinetic energy rapidly decays as the speed of the moving plates begins to slow down. Based on present results, some suggestions and future works are briefly demonstrated later.

Though the flow after the faulting is not solved in this paper, it is not difficult to calculate it by using mathematical methods described later. The velocity profiles at $T = T_0$ (the end time of the faulting) have to be regarded as a new initial condition. The subsequent flow can be solved by applying integral transforms to the new PDE system. It seems quite straightforward and easy to perform the calculation. However, as the initial condition is no longer zero (e.g, substituting $T = T_0$ into (17) as new initial conditions), it generally results in the difficulty in performing the integral transform and its inversion. Due to the rapid decay of kinetic energy during the faulting, as shown in Figures 9 and 10, it is reasonable to predict that the flow after the faulting will also decay rapidly since no external energy comes in. It indicates that the influence of the induced flow on neighboring structures and environment is much greater at the early stage of the faulting than that at longer times.

In addition to the faulting types considered in this paper, the flow induced by other types of faulting can be readily analyzed. Based on the present paper and that of Liu [8], the flow driven by arbitrary motion of two half plates can be theoretically calculated by using the Fourier analysis. Moreover, the procedures and mathematical techniques introduced herein can be applied to the case of non-Newtonian fluids. One needs to modify the constitutive equation (momentum equation) by including the rheological properties of fluids. As there exist a wide variety of non-Newtonian fluids, other mathematical techniques for dealing with additional rheological terms appearing in the constitutive equation may be necessary to derive the exact solution. The case of an Oldroyd-B fluid for similar problems was studied by Liu [12] who additionally employed the technique of the series expansion due to the unavailability of the inversed transforms.

Finally, mathematical methods used in this study can be applied to other fields. For example, the PDE system of the heat-conduction problem will be completely identical to (1) by modifying the velocity to temperature and the viscosity to conduction coefficient. In conclusion, the present study contributes to not only the theoretical solution of the earthquake-induced flow but also the related analyses in fluid mechanics, industry manufacturing, chemical engineering, and other fields.

Acknowledgment

The research funding provided by National Science Council of Taiwan with Contracts NSC 101-2221-E-270-001-MY2 and NSC 102-2911-I-006-302 is acknowledged.

References

[1] A. Ben-Menahem and M. Rosenman, "Amplitude patterns of tsunami waves from submarine earthquakes," *Journal of Geophysical Research*, vol. 77, no. 17, pp. 3097–3128, 1972.

[2] J. Hammack, "A note on tsunamis: their generation and propagation in an ocean of uniform depth," *Journal of Fluid Mechanics*, vol. 60, pp. 769–799, 1973.

[3] Y. Okada, "Surface deformation due to shear and tensile faults in a half-space," *Bulletin of the Seismological Society of America*, vol. 75, no. 4, pp. 1135–1154, 1985.

[4] D. Dutykh and F. Dias, "Tsunami generation by dynamic displacement of sea bed due to dip-slip faulting," *Mathematics and Computers in Simulation*, vol. 80, no. 4, pp. 837–848, 2009.

[5] D. Dutykh and F. Dias, "Energy of tsunami waves generated by bottom motion," *Proceedings of the Royal Society A*, vol. 465, no. 2103, pp. 725–744, 2009.

[6] T. Saito and T. Furumura, "Three-dimensional tsunami generation simulation due to sea-bottom deformation and its interpretation based on the linear theory," *Geophysical Journal International*, vol. 178, no. 2, pp. 877–888, 2009.

[7] Y. Zeng and S. Weinbaum, "Stokes problems for moving half-planes," *Journal of Fluid Mechanics*, vol. 287, pp. 59–74, 1995.

[8] C.-M. Liu, "Complete solutions to extended Stokes' problems," *Mathematical Problems in Engineering*, vol. 2008, Article ID 754262, 18 pages, 2008.

[9] C.-M. Liu, "Extended stokes' problems for relatively moving porous half-planes," *Mathematical Problems in Engineering*, vol. 2009, Article ID 185965, 10 pages, 2009.

[10] G. G. Stokes, "On the effect of the internal friction of fluids on the motion of pendulums," *Transactions of the Cambridge Philosophical Society*, vol. 9, pp. 8–106, 1851.

[11] C.-M. Liu and I.-C. Liu, "A note on the transient solution of Stokes' second problem with arbitrary initial phase," *Journal of Mechanics*, vol. 22, no. 4, pp. 349–354, 2006.

[12] C.-M. Liu, "Another approach to the extended Stokes' problems for the Oldroyd-B fluid," *ISRN Applied Mathematics*, vol. 2012, Article ID 274914, 14 pages, 2012.

20

The Effects of Marine Cloud Brightening on Seasonal Polar Temperatures and the Meridional Heat Flux

Ben Parkes,[1] Alan Gadian,[1] and John Latham[2, 3]

[1] NCAS, SEE, University of Leeds, Leeds LS2 9JT, UK
[2] MMM, National Center for Atmospheric Research, Boulder, CO 80307-3000, USA
[3] SEAS, University of Manchester, Manchester M13 9PL, UK

Correspondence should be addressed to Alan Gadian, alan@env.leeds.ac.uk

Academic Editors: S. Verma and G. Zhang

Marine cloud brightening (MCB) is one of several proposed solar radiation management (SRM) geoengineering schemes designed to ameliorate some of the undesirable effects of climate change, for example polar ice loss and associated increased sea levels. Satellite measurements over the last 40 years show a general reduction in polar sea ice area and thickness which is attributed to climate change. In our studies, HadGEM1, a fully coupled climate model, is used to predict changes in surface temperatures and ice cover as a result of implementing MCB in a double carbon dioxide concentration atmosphere. The meridional heat flux (MHF) is the mechanism within the earth system for the transport of energy from tropical to polar regions. This poleward transport of heat in a double carbon dioxide atmosphere amplifies the effects in polar regions, where it has a significant impact on both temperatures and ice cover. The results from this work show that MCB is capable of roughly restoring control temperatures and ice cover (where control is defined as 440 ppm carbon dioxide, a predicted 2020 level) in a double carbon dioxide atmosphere scenario. This work presents the first results on the impact of MCB on the MHF and the ability of the MCB scheme to restore the MHF to a control level.

1. Introduction

Global warming is a major feature of climate change, and many publications have shown that it is most pronounced at high latitudes with the Arctic and Antarctic showing considerable heating compared to the rest of the world [1]. This additional heating of the polar regions is known as polar amplification and results in temperature changes far above the global average. Several mechanisms have been proposed to explain polar amplification. One is a sea ice-albedo feedback proposed by Curry et al. in 1996 [2]. A reduction of sea ice cover exposes the sea surface, which is of a lower albedo, so that there is more absorption of solar radiation and concomitant additional warming. This heating of polar waters further increases the sea ice loss. The Arctic has been shown to be more susceptible than the Antarctic to changes in temperature; resulting in larger effects in the northern hemisphere [3]. This result is repeated in several papers on MCB (e.g., [4–7]). The work in [4, 7] modifies clouds in tropical regions and finds an associated local cooling to 1.5 m air temperature, however Figure 3 of [4] and Figure 4(d) of [7] show that MCB also preferentially cools the Arctic. The aim of this study is to investigate the preferential cooling of the Arctic and if possible suggest a mechanism to explain how seeding in the tropics leads to a cooling in the Arctic. MCB is one of several proposed Solar Radiation Management (SRM) geoengineering ideas designed to reduce some of the impacts of climate change. MCB utilises the first and second aerosol indirect effects on clouds ([8, 9], resp.). Exploitation of the first indirect effect is based upon increasing the cloud droplet number concentration (CDNC). A higher number of smaller droplets increase the optical thickness of clouds which therefore reflect more shortwave radiation. The second indirect effect prolongs cloud lifetime as smaller droplets take longer to coalesce into droplets large enough to precipitate out of the clouds.

Several approaches are taken to investigate MCB and its impacts, from atmosphere only modelling work [11, 12],

TABLE 1: Simulations produced using HadGEM1 to investigate the climate effects of MCB.

Experiment name	Carbon dioxide fraction (ppm)	Seeding scheme
CON	440	None
2CO2	440 + 1%/year, held at 560	None
MCB3	440 + 1%/year, held at 560	Three regions
MCBA	440 + 1%/year, held at 560	All ocean

to coupled atmosphere slab ocean models [4, 6], to fully coupled global climate models (GCMs) [4, 5, 7, 13]. A GCM is used to compare the effects of seeding three regions of persistent marine stratocumulus clouds individually or as a group [4, 7, 13]. These simulations impose a cloud droplet number concentration (CDNC) of $N = 375\,cm^3$ and found that MCB is capable of substantially reducing the polar impact of increasing carbon dioxide. The impacts of MCB on both annual and seasonal polar ice coverage are investigated in [5–7, 13]. Previous studies predict an accompanying rainfall reduction in the Amazon region [4, 5, 7, 13], the amount of which varies from model to model. These differences appear to be related to differences in seeding strategy but are not relevant for this paper.

Most GCM studies assume a fixed value for the CDNC and do not assess the technological requirements for attaining this value [4–7, 11, 13]. A seeding mechanism was proposed by Salter et al. in 2008 [14] whereby autonomous GPS guided ships would use solar power to create large numbers of 2 nm seawater droplets which would act as cloud condensation nuclei. The ships were designed to move perpendicular to trade winds allowing the seawater particles to spread over a large area.

2. Experiment Description

HadGEM1—used in our computations—is the Hadley centre Global Environment Model developed by the UK Met Office as version 6.1 of the Unified Model [15]. HadGEM1 is the combination and coupling of the HadGAM and HadGOM atmosphere and ocean models. The model atmosphere has a horizontal resolution of 1.875° longitude by 1.25° latitude and 38 vertical levels of increasing size to a maximum height of 39 km with 10 levels between 0 km and 2 km. The dynamics and radiation schemes within the atmospheric model are described in [16, 17]. The atmosphere is coupled to an ocean which has a resolution of 1° square between the poles and 30°. Between 30° and the equator the meridional resolution of the ocean increases smoothly to (1/3)°. The ocean has 40 smoothly increasing depths, 10 m near the surface, to 345 m at 5.3 km [18].

The model has been modified to have a fixed CDNC in the three regions of low-level marine stratocumulus; these regions are of the coasts of California, Peru and Namibia as shown in Figure 1 of Jones et al. 2009 [4]. The second geoengineering simulation seeds the entire marine atmosphere. In the unseeded regions, and the CON and 2CO2 experiments, the CDNC is unmodified and the original model derivation of CDNC is used. In a modified

region the CDNC is set to be $375\,cm^3$ at all model levels between 0 km and 3 km and held static for the duration of the simulation. This bypasses the normal method used by the model to calculate the CDNC. The model average CDNC for marine regions at 1 km is roughly $60\,cm^3$ which is much lower than the $375\,cm^3$ that is used to simulate MCB.

Four cases were run to investigate the effects of MCB on polar temperatures. The setup of the four simulations is shown in Table 1. A comparison of CON and 2CO2 shows the effects of increasing carbon dioxide levels, while comparing MCB3 or MCBA with 2CO2 shows the differences resulting from seeding. Comparing MCB3 or MCBA with CON shows the combined impacts of increasing carbon dioxide while using MCB. Each simulation was run for 70 years with the final 20 years used for analysis. The initial state for the model was copied from an existing model run that had been simulated between 1860 and 2020.

3. Calculation of the Meridional Heat Flux

Incoming shortwave solar radiation warms the tropics more than the poles, while emitted longwave radiation cools the whole planetary surface. In the absence of a fluid atmosphere or ocean, the only method to balance the earth's surface temperature would be by conduction. With an atmosphere or ocean, transport of energy polewards is possible, and this is defined as the Meridional Heat Flux (MHF). Large-scale dynamics and eddies transport energy polewards as shown analytically and numerically in [19], while the oceanic energy transport is largely driven by the thermohaline circulation and ocean currents. The total MHF can be calculated from the top-of-atmosphere radiative balance [10, 20]. MHF can be estimated in total or in oceanic and atmospheric components [10, 20]. In previous work [20] the ERBE dataset was used to find the total MHF with the NCEP and ECMWF models used to find the contribution from the atmosphere and ocean.

The method used to calculate the MHF in this work replicates that used in both Trenberth and Caron 2001 [20], and Wunsch 2005 [10]. The MHF is calculated from monthly average radiative flux difference at the top of the atmosphere. For each latitude band around the globe, the radiative flux difference is summed to give a total flux difference at each band. These total flux differences are then multiplied by the area in each band, to calculate an energy flux out of the atmosphere for each band. The fluxes are then accumulated from the South pole to the North pole with the final sum defining the MHF [10].

Figures 1(a) and 1(b) show the annual average radiative balance at the top of the atmosphere from the ERBE dataset and our computed HadGEM1 results. These values are multiplied by the area in each latitude band to give the contribution of each band to the total MHF. To produce Figure 1(d) the data from Figure 1(b) was regridded, from the HadGEM1 model grid to the 2.5° square grid used by the ERBE dataset. The regridding enables a direct comparison between the results in Figures 1(c) and 1(d). If the original grid spacing were retained the HadGEM results would show a larger number of smaller fluxes. Figure 1(e) shows the summation of the results in Figure 1(c) from the South pole to the North pole. Figure 1(f) is generated using the same method as Figure 1(e) with data that has not been regridded to the lower ERBE resolution. It can be seen that the MHF values derived from HadGEM1 were compared well with the dataset and show less of an imbalance when the calculation direction is reversed from the North to the South.

4. Results

The results in Figure 2 show the change in average summer and winter surface temperature and in many cases are reflected in the sea ice fraction plots shown in Figure 3. Figures 3(a) and 3(b) show the effects of doubling atmospheric carbon dioxide concentration on northern polar sea ice fraction. The warming values found in Figures 2(a) and 2(b) agree with those found in Figure 1(a) of [5] and Figure 4(b) of [7] where, again, the doubling of atmospheric carbon dioxide leads to a disproportionate warming in the polar regions. This warming of the climate results in a loss of 3.6×10^6 km^2 of Arctic sea ice and a further 1.0×10^6 km^2 of Antarctic sea ice. The global average temperature change between the control and double carbon dioxide concentration atmosphere for the work in [4] is $+0.58$ K; in this work the difference is $+0.82$ K.

It can be seen in Figures 3(a) and 3(b) that doubling the carbon dioxide concentration causes a greater increase in temperature difference during the winter, than in summer. We find that seeding the three regions results in a global average polar cooling of 0.8 K as can be seen in Figures 3(c) and 3(d). This cooling acts against the polar amplification and reduces South polar ice loss to 0.79×10^6 km^2 while increasing North polar ice cover by 0.20×10^6 km^2. The results from a comparison between MCBA and CON shown in panels (e) and (f) of Figures 2 and 3, indicate the extensive cooling brought on by all-sea seeding. The majority of the northern hemisphere is subject to a significant reduction in surface temperatures which in turn influences the sea ice coverage. There is however a warming found in the region South of Greenland. Despite this warming, the Arctic sea ice is increased by 2.3×10^6 km^2 and in the Southern hemisphere the increase in sea ice cover is 7.9×10^6 km^2.

We present the first analysis of changes to an MHF as a result of simulating the deployment of MCB in a double carbon dioxide concentration atmosphere. The maximum MHF, in the northern hemisphere, is generally found close to 40 degrees N and the maximum value from the control,

CON is found to be 5.8 PW. The heating from doubling carbon dioxide raises the maximum to 6.1 PW while three region seeding in MCB3 reduces this to 5.7 PW. In MCBA the maximum MHF is reduced to 4.0 PW. These results demonstrate how MCB, even when seeding is applied in three-relatively small maritime regions, can cause an appreciable change in the global MHF.

5. Discussion

The results from our four climate simulations including two MCB scenarios show a strong connection between sea ice fraction and sea surface temperatures. These two quantities are influenced by a sea ice-albedo feedback loop which in turn is influenced by the MHF, which transports heat energy polewards [21]. Furthermore our control simulation (CON) is in good agreement with previous work on the MHF using the ERBE dataset [10, 20]. Polar amplification leads to a polar heating and thus a reduction in sea ice which then possibly starts the positive feedback resulting in further heating and sea ice loss. Thus MCB may be able to target polar regions more effectively than other geoengineering methods [5, 11]. In particular, in the double carbon dioxide scenario, MCB produces a significant reduction in sea ice loss.

We can see from the results in Figures 2(c), 2(d), 3(c), and 3(d) that seeding the selected three regions of stratocumulus clouds returns climate to close to but not exactly the control situation. This agrees with results [5] where different areas of cloud were seeded to a much higher value of N. The ability of MCB to return to close to a control simulation is further reflected in Figure 4 where the MCB3 MHF flux curve is almost overlaid on CON.

The results from a double carbon dioxide atmosphere (no seeding) run are consistent with those of [4–7]. These show that polar regions are warmed significantly more than tropical regions and that this warming has a significant impact upon sea ice cover. When an MCBA seeding scenario is used we see further evidence to support the link between MHF and polar temperatures and also the impact these temperatures have on sea ice fraction. With reduced polar temperatures, ice growth is significant.

This work builds upon previous studies [4–7, 11] which show that MCB cannot reproduce the control climate but can return the polar ice state to that similar to the control. There is a need for further studies to understand better the complexities of marine stratocumulus clouds, to assess the consequences of modifying the patterns of polar temperature and sea ice cover, and to thoroughly examine the (possibly adverse) consequences of MCB deployment. The impact of MCB on precipitation patters has been investigated in several publications [4–7, 13]. Work in Latham et al. (In press) [13] uses simulations similar to those in Table 1 to assess the impact of three region MCB on precipitation patterns.

Further work is needed to estimate the relative sizes of the atmospheric and oceanic contributions to the MHF in a geoengineered atmosphere.

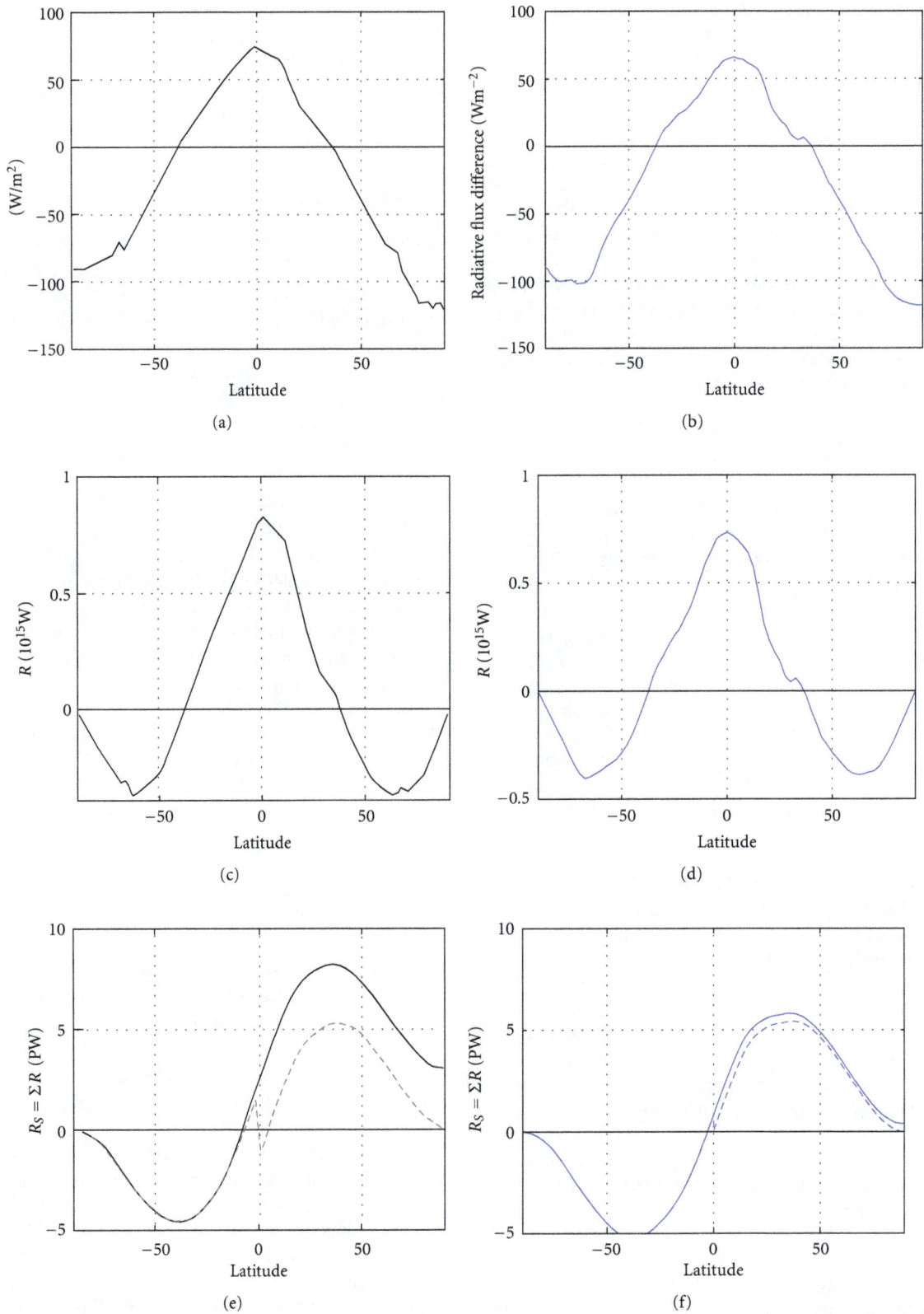

FIGURE 1: Calculation of the MHF from radiative balance values in the ERBE dataset [10] (left) and HadGEM1 (right). (a) and (b) show the annual average radiative balance. (c) and (d) multiply (a) and (b) by the area in each latitude band. (e) and (f) sum these values from 90°S to 90°N to give the MHF. Dotted lines in (e) and (f) show the result from 90°N to 90°S. (a), (c), and (e) are copied from Figure 2(a), 2(b), and 2(c) of Wunsch 2005 [10].

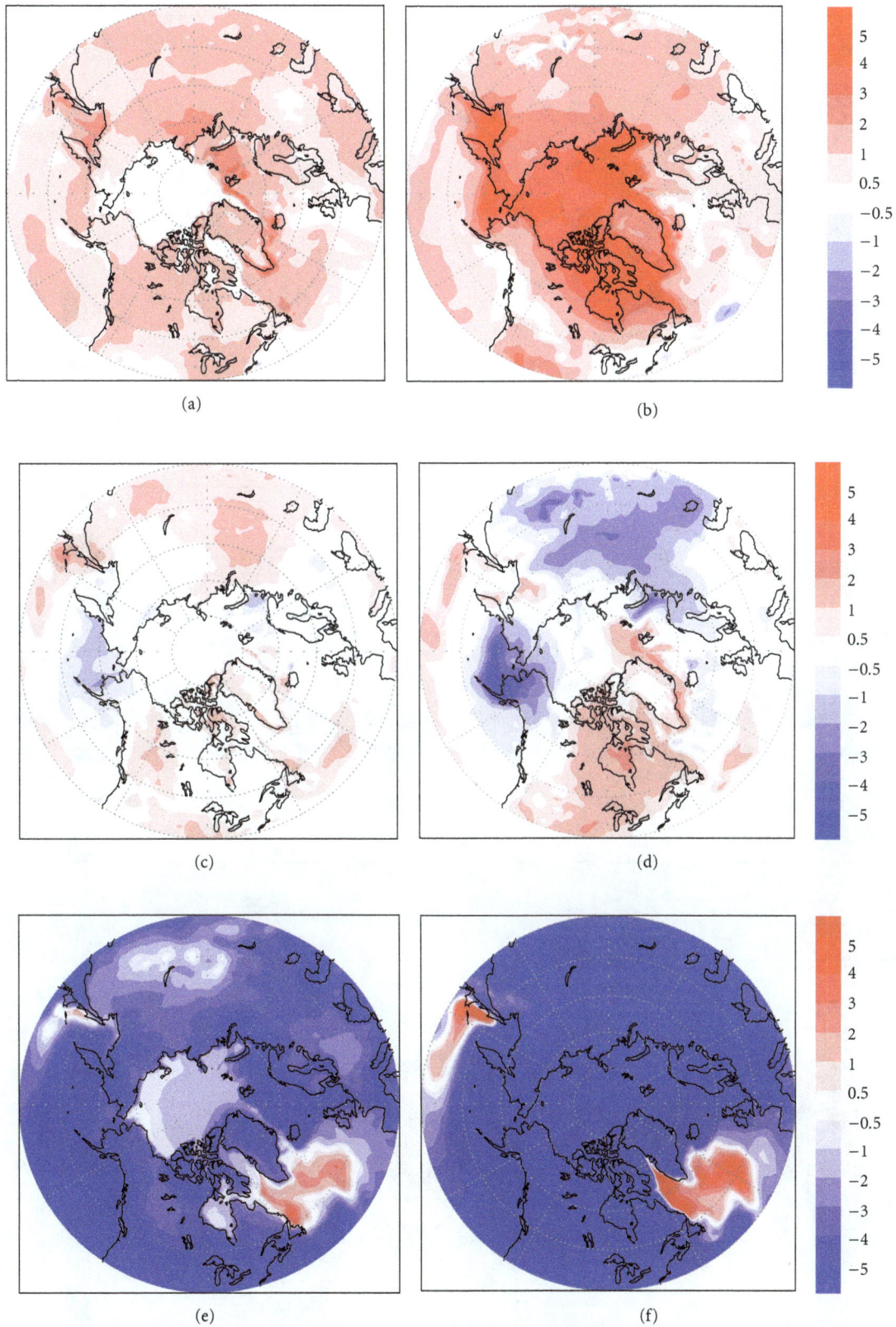

FIGURE 2: Comparison of summer (left) and winter (right) polar surface temperatures (K) for four geoengineering simulations. (a) and (b) show the differences between 2CO2 and CON. (c) and (d) show the differences between MCB3 and CON. (e) and (f) show the differences between MCBA and CON.

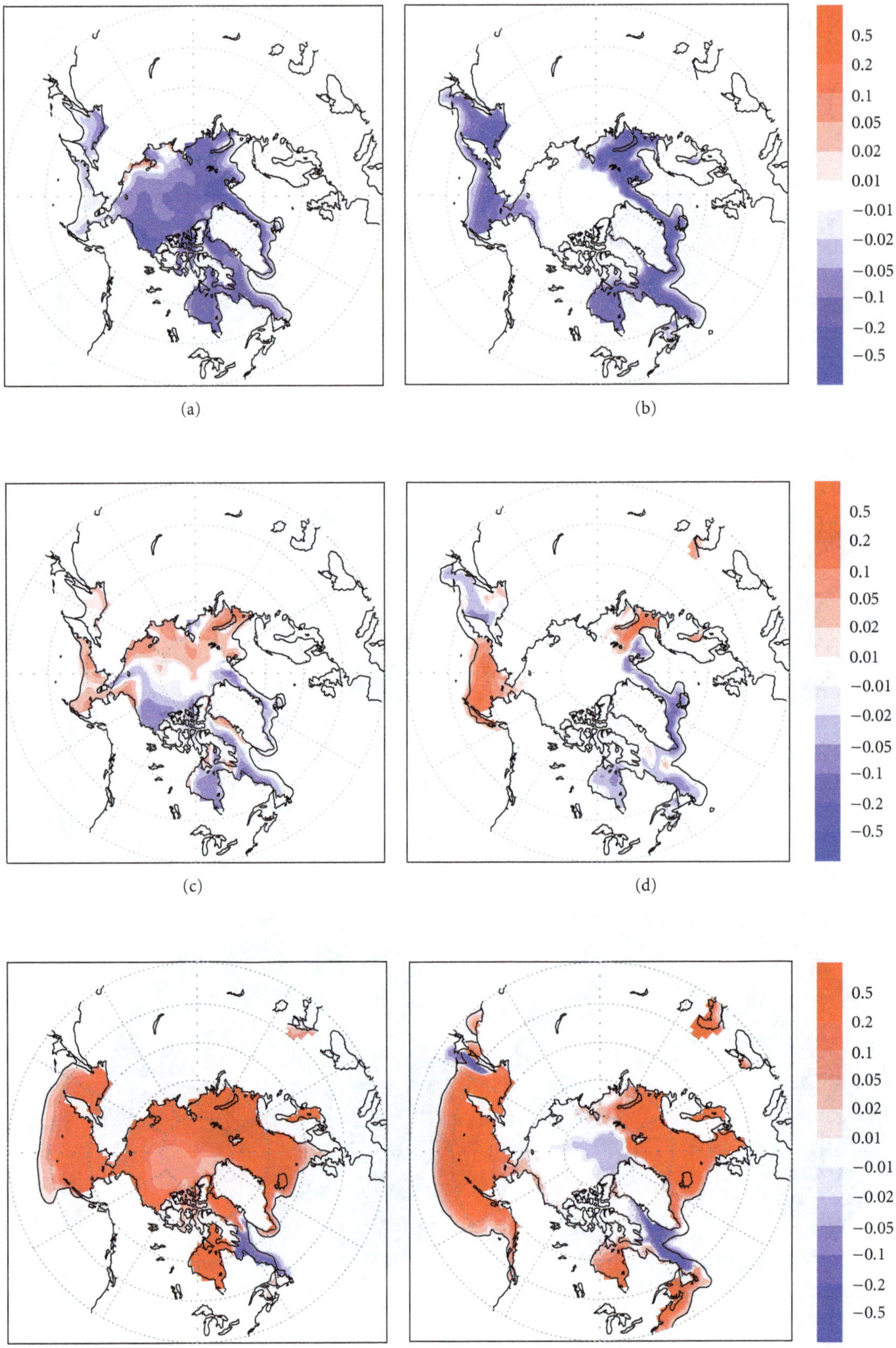

FIGURE 3: As Figure 2 except for sea ice fraction. The black contour shows the limit of sea ice in the CON simulation.

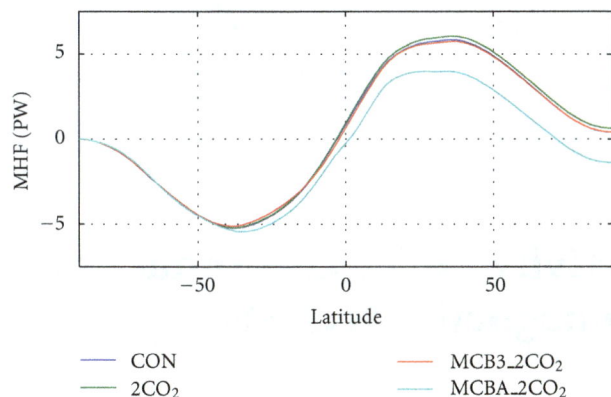

Legend:
— CON
— 2CO₂ (2CO$_2$)
— MCB3_2CO₂ (MCB3_2CO$_2$)
— MCBA_2CO₂ (MCBA_2CO$_2$)

FIGURE 4: Meridional heat flux for four climate scenarios as described in Table 1.

Acknowledgment

We are grateful for the use of NERC, NCAS, HECToR supercomputer resources. Support for elements of this research was provided by the Fund for Innovative Climate and Energy Research, FICER, at the University of Calgary. This does not constitute endorsement of deployment in any form of Cloud Albedo Modification by the funding agency.

References

[1] P. Forster, V. Ramaswamy, P. Artaxo et al., "IPCC synthesis report chapter 2 changes in atmospheric constituents and in radiative forcing climate change 2007: The physical science basis. contribution of working group i to the fourth assessment report of the intergovernmental panel on climate change," in Climate Change 2007: The Physical Science Basis. Contribution of Working Group I to the Fourth Assessment Report of the Intergovernmental Panel on Climate Change, S. Solomon, M. Manning, D. Qin et al., Eds., Cambridge University Press, Cambridge, UK, 2007.

[2] J. A. Curry, W. B. Rossow, D. Randall, and J. L. Schramm, "Overview of arctic cloud and radiation characteristics," Journal of Climate, vol. 9, no. 8, pp. 1731–1764, 1996.

[3] M. M. Holland and C. M. Bitz, "Polar amplification of climate change in coupled models," Climate Dynamics, vol. 21, no. 3-4, pp. 221–232, 2003.

[4] A. Jones, J. Haywood, and O. Boucher, "Climate impacts of geoengineering marinestratocumulus clouds," Journal of Geophysical Research, vol. 114, Article ID D10106, 9 pages, 2009.

[5] P. J. Rasch, J. Latham, and C.-C. J. Chen, "Geoengineering by cloud seeding: influenceon sea ice and climate system," Environmental Research Letters, vol. 4, Article ID 045112, 2010.

[6] G. Bala, K. Caldeira, R. Nemani, L. Cao, G. Ban-Weiss, and H. J. Shin, "Albedo enhancement of marine clouds to counteract global warming: impacts on the hydrological cycle," Climate Dynamics, vol. 37, no. 5-6, pp. 915–931, 2010.

[7] A. Jones, J. Haywood, and O. Boucher, "A comparison of the climate impacts ofgeoengineering by stratospheric so2 injection and by brightening of marine stratocumulus cloud," Atmospheric Science Letters, vol. 291, no. 2, pp. 176–183, 2011.

[8] S. Twomey, "Pollution and the planetary albedo," Atmospheric Environment, vol. 8, no. 12, pp. 1251–1256, 1974.

[9] B. A. Albrecht, "Aerosols, cloud microphysics, and fractional cloudiness," Science, vol. 245, no. 4923, pp. 1227–1230, 1989.

[10] C. Wunsch, "The total meridional heat flux and its oceanic and atmospheric partition," Journal of Climate, vol. 18, no. 21, pp. 4374–4380, 2005.

[11] J. Latham, P. Rasch, C. C. Chen et al., "Global temperature stabilization via controlled albedo enhancement of low-level maritime clouds," Philosophical Transactions of the Royal Society A: Mathematical, Physical and Engineering Sciences, vol. 366, no. 1882, pp. 3969–3987, 2008.

[12] H. Korhonen, K. S. Carslaw, and S. Romakkaniemi, "Enhancement of marine cloud albedo via controlled sea spray injections: a global model study of the influence of emission rates, microphysics and transport," Atmospheric Chemistry and Physics, vol. 10, no. 9, pp. 4133–4143, 2010.

[13] J. Latham, K. Bower, T. Choularton et al., "Marine cloud brightening," Philosophical Transactionsof the Royal Society A. In press.

[14] S. Salter, G. Sortino, and J. Latham, "Sea-going hardware for the cloud albedo method of reversing global warming," Philosophical Transactions of the Royal Society A: Mathematical, Physical and Engineering Sciences, vol. 366, no. 1882, pp. 3989–4006, 2008.

[15] G. M. Martin, M. A. Ringer, V. D. Pope, A. Jones, C. Dearden, and T. J. Hinton, "The physical properties of the atmosphere in the new Hadley Centre Global Environmental Model (HadGEM1). Part 1: Model description and global climatology," Journal of Climate, vol. 19, no. 7, pp. 1274–1301, 2006.

[16] T. Davies, M. J. P. Cullen, A. J. Malcolm et al., "A new dynamical core of the Met Office's global and regional modelling of the atmosphere," Quarterly Journal of the Royal Meteorological Society, vol. 131, no. 608, pp. 1759–1782, 2005.

[17] J. M. Edwards and A. Slingo, "Studies with a flexible new radiation code. I: Choosing a configuration for a large-scale model," Quarterly Journal of the Royal Meteorological Society, vol. 122, no. 531, pp. 689–719, 1996.

[18] T. C. Johns, C. F. Durman, H. T. Banks et al., "The new Hadley Centre Climate Model (HadGEM1): Evaluation of coupled simulations," Journal of Climate, vol. 19, no. 7, pp. 1327–1353, 2006.

[19] J. S. A. Green, "Transfer properties of the large-scale eddies and the general circulation ofthe atmosphere," Quarterly journal of the Royal Meteorological Society, vol. 96, pp. 157–185, 1970.

[20] K. E. Trenberth and J. M. Caron, "Estimates of meridional atmosphere and ocean heat transports," Journal of Climate, vol. 14, no. 16, pp. 3433–3443, 2001.

[21] V. A. Alexeev, P. L. Langen, and J. R. Bates, "Polar amplification of surface warming on an aquaplanet in "ghost forcing" experiments without sea ice feedbacks," Climate Dynamics, vol. 24, no. 7-8, pp. 655–666, 2005.

21

Error Analysis in Measured Conductivity under Low Induction Number Approximation for Electromagnetic Methods

George Caminha-Maciel[1,2] and Irineu Figueiredo[1,3]

[1] *Observatório Nacional (ON), MCT, Rua General José Cristino 77, 20921-400 Rio de Janeiro, RJ, Brazil*
[2] *Universidade Federal do Pampa (UNIPAMPA), Avenida Pedro Anunciação s/n°, 96570-000 Caçapava do Sul, RS, Brazil*
[3] *Universidade Estadual do Rio de Janeiro (UERJ), Rua São Francisco Xavier 524, 20550-900 Rio de Janeiro, RJ, Brazil*

Correspondence should be addressed to George Caminha-Maciel; caminha.maciel@unipampa.edu.br

Academic Editors: G. Chisham, A. De Santis, G. Mele, and A. Tzanis

We present an analysis of the error involved in the so-called low induction number approximation in the electromagnetic methods. In particular, we focus on the EM34 equipment settings and field configurations, widely used for geophysical prospecting of laterally electrical conductivity anomalies and shallow targets. We show the theoretical error for the conductivity in both vertical and horizontal dipole coil configurations within the low induction number regime and up to the maximum measuring limit of the equipment. A linear relationship may be adjusted until slightly beyond the point where the conductivity limit for low induction number ($B = 1$) is reached. The equations for the linear fit of the relative error in the low induction number regime are also given.

1. Introduction

The induction method consists basically in determining subsurface rock conductivities with the help of electromagnetic fields generated by a coil at the Earth's surface and by catching the response to this field from the conducting media under surface by using a reception coil [1–3].

From the Maxwell equations, in particular the Faraday induction law applied to an infinite homogenous half-plane, the subsurface rock conductivity can be estimated through the ratio between the magnetic field measured in the receiving coil and the magnetic field produced at the transmission coil with both at surface. Then we can take laterally distributed measurements along a transect for identifying conductivity-related anomalies. We can also get information on the vertical conductivity structure by varying the coil's dipole configurations (vertical dipole or horizontal dipole) as well as by increasing the instrument height. This information is very useful in several geophysical problems as, for example, water prospecting or mapping pollution plumes.

The basic model for both configurations is described in Figure 1 where a transmission coil Tx with a given alternate electric current at a given frequency is located on the terrain (assumed to be an uniform semiplane) and a receiving coil Rx is located at a short distance s from Tx. The time variation of the magnetic field H_p, called primary magnetic field, produced by the electric current in the transmission coil generates a small alternate current in the soil. This electric current, on its turn, produces a magnetic field H_s, called secondary field, which can be measured at the receiving coil together with the primary field.

In general, the secondary magnetic field H_s is a complicated function of the distance between coils s, of the magnetic permeability of the medium μ—which will be considered exactly the same as for vacuum μ_0 ($4\pi \cdot 10^{-7}$A·m)—of the angular frequency of the oscillatory electric current ω, and of the soil conductivity σ. Under well-established conditions, technically known as "operation with low values of induction number," the secondary magnetic field H_s becomes a simple

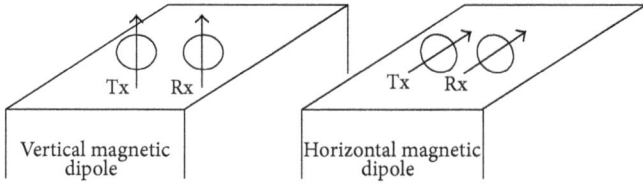

FIGURE 1: Representation of the transmission (Tx) and reception (Rx) coils for both the vertical and the horizontal dipole configurations.

function of these variables when we consider only the quadrature component:

$$\frac{H_s}{H_p} \cong \frac{\omega\mu\sigma}{4}s^2. \tag{1}$$

This result can be used when coils are in the vertical or horizontal dipole orientations. Actually, these two situations are described by different sets of equations for the secondary field; however, these equations give the same result for the component in quadrature of the ratio between the primary and secondary fields under low induction number conditions.

In order to understand the induction number we need to define the "electromagnetic skin depth" δ:

$$\delta = \left(\frac{2}{\omega\mu\sigma}\right)^{1/2}, \tag{2}$$

which is equal to the distance in a conducting medium an electromagnetic front wave has to travel in order to reduce the amplitude to $1/e(\sim 37\%)$ of its value outside the conducting medium.

The induction number B describes the distance between the transmission coil and the reception coil s in units of skin depth δ:

$$B = \frac{s}{\delta}. \tag{3}$$

The recent literature contains some debate on which values of B (and consequently σ) are valid for low induction number approximations [4, 5]. It also contains suggested corrections for nonzero height dependence and nonlinear departure for higher values of conductivity even above the low induction number regime [4].

In this work we introduce an error analysis for the induction method taking as a starting point the set of equations given by McNeill [1] for the field ratios on both coil dipole configurations. We also consider zero elevation of the coils and, as in Beamish [4], we discard medium magnetic and dielectric components as we are working in frequencies lower than 15 kHz. Those equations represent the ratio between primary and secondary fields as the response of a homogeneous earth. We calculate the relative difference between this conductivity value and that obtained through the low induction number approximation.

2. Methods

The equations for the primary and secondary field ratios for vertical and horizontal dipole configurations as given by McNeill [1] are

$$\left(\frac{H_s}{H_p}\right)_V$$

$$= \frac{2}{(\gamma s)^2} \cdot \left\{9 - \left\{9 + 9 \cdot (\gamma s) + 4 \cdot (\gamma s)^2 + (\gamma s)^3\right\} \cdot e^{-\gamma s}\right\}, \tag{4}$$

$$\left(\frac{H_s}{H_p}\right)_H = 2 \cdot \left\{1 - \frac{3}{(\gamma s)^2} + \left[3 + 3 \cdot (\gamma s) + (\gamma s)^2\right] \cdot \frac{e^{-\gamma s}}{(\gamma s)^2}\right\}, \tag{5}$$

where

$$\gamma = \sqrt{(i\omega\mu_0\sigma)},$$
$$\omega = 2\pi f,$$
$$f = \text{frequency (Hz)},$$
$$\mu_0 = \text{permeability of free space},$$
$$i = \sqrt{-1}.$$

Expressions (4) and (5) are multivalued functions of the conductivity, frequency, and intercoil spacing, and as functions of the complex exponential $\sqrt{i}$ they have several branches in the complex plane. Therefore, as a matter of simplification, the induction number, a nondimensional parameter defined as the ratio between intercoil spacing and skin depth, is introduced, which will act as a conformal map simplifying equations.

To do this we begin by writing $x = i\gamma s$ as a function of the induction number $B = s/\delta$:

$$x = \sqrt{\frac{2}{i}}B. \tag{6}$$

Remembering that

$$(1-i)^2 = -2i = \frac{2}{i}, \tag{7}$$

we have

$$x = (1-i)B. \tag{8}$$

By substituting in (4) the following expression for the vertical dipole configuration comes out:

$$\left(\frac{H_s}{H_p}\right)_V = h_V$$

$$= -\frac{18}{(1-i)^2 B^2}\left[1 - \left(1 - (1-i)B + \frac{4}{9}(1-i)^2 B^2\right.\right.$$

$$\left.\left. - \frac{1}{9}(1-i)^3 B^3\right)e^{(1-i)B}\right]. \tag{9}$$

As

$$(1 - i)^2 = -2i, \qquad (1 - i)^3 = -2(1 + i), \qquad (10)$$

then

$$h_V = -\frac{9i}{B^2}\left[1 - \left(1 - B + iB - \frac{8}{9}iB^2 + \frac{2}{9}B^3 + \frac{2}{9}iB^3\right)e^B e^{-iB}\right], \qquad (11)$$

and doing $e^{-iB} = \cos B - i \sin B$, it follows that

$$h_V = -\frac{1}{B^2}\left[9i + \left[(9B - 8B^2 + 2B^3) - (9 - 9B + 2B^3)i\right]\right.$$
$$\left. \times e^B [\cos B - i \sin B]\right]. \qquad (12)$$

By taking only the quadrature of h we have

$$Qh_V = -\frac{1}{B^2}\left[9 - (9B - 8B^2 + 2B^3)e^B \sin B\right.$$
$$\left. - (9 - 9B + 2B^3)e^B \cos B\right]. \qquad (13)$$

Similarly, for the horizontal dipole configuration we obtain

$$Qh_H = -\frac{1}{B^2}\left[-3 + (3B - 2B^2)e^B \sin B + (3 - 3B)e^B \cos B\right]. \qquad (14)$$

The imaginary part (quadrature term) is related to the conductivity measurements under low induction number conditions. The magnitude of the secondary magnetic field is now directly proportional to the conductivity and its phase leads the primary field to 90°. Under low induction number ($B \ll 1$) the expressions (4) and (5) can be expanded in a power series, and considering only the first terms the quadrature part (neglecting the in-phase term) can be written as [1]

$$\left(\frac{H_s}{H_p}\right)_V = \left(\frac{H_s}{H_p}\right)_H = \frac{i\sigma\mu\omega}{4}s^2. \qquad (15)$$

This is the low induction number approximation key expression which leads to the conductivity value from the equipment readings.

The induction number B is a function of the material conductivity σ for a given frequency ω and the separation between coils s. Furthermore, B is a function of the conductivity σ and of the product ($f^{1/2} * s$). For the EM34-3 and EM34-X (Geonics Ltd.) equipment settings all different possibilities of frequency and intercoil spacing give the same product $f^{1/2} * s$ (800 mS$^{1/2}$ * m$^{1/2}$). Since the low induction number condition ($B \ll 1$) implies that we should have $\sigma \ll 2/\mu\omega s^2$, then, considering the EM34-3 instrument that works in the frequencies 6400, 1600, and 400 Hz with distances between coils of 10, 20, and 40 m, respectively, it comes out that the value for the soil conductivity measured by this instrument should be quite lower than 400 mS/m.

We define the error associated with the low induction number approximation as the theoretical deviation from the exact model which is the electromagnetic response for a homogenous half-plane as follows:

$$\varepsilon = \frac{|\sigma' - \sigma|}{\sigma}. \qquad (16)$$

From the induction number definition we have

$$\sigma = \frac{2B^2}{\mu\omega s^2}, \qquad (17)$$

and from (15)

$$\sigma' = \frac{4(Qh)}{\mu\omega s^2}. \qquad (18)$$

In this way, the error associated with the approximation is

$$\varepsilon = \frac{|4(Qh) - 2B^2|}{2B^2} = \left|\frac{2(Qh)}{B^2} - 1\right|. \qquad (19)$$

Substituting those Qh values in expression (19) we obtain the relative theoretical error for the approximation. Figure 2 represents the plot of the error for the whole conductivity range (and a little further) mostly used in practical applications (10^{-2} mS/m–10^3 mS/m) and shows the deviation from the linear response. We add two vertical black lines marking the limits of interest: the low induction number limit ($B = 1$) and the usual limit of the equipment (1000 mS/m).

From the studied lower limit through the whole low induction number regime (considering its limit as 400/10 = 40 mS/m) the relationship is clearly linear, but it also can easily be extended to 100 mS/m or even to 400 (mS)/m ($B = 1$), as it is shown in Figure 2.

Figure 3 shows the absolute apparent conductivity as a function of true conductivity until the linear response limit is surpassed. McNeill [1] shows this plot for a narrower range of conductivity values, and therefore the departure from linearity was not clear. The same linear behaviour is seen in the apparent conductivity (Figure 3) for vertical and horizontal dipole configurations, as seen by the fitted straight line to the logarithmic scaled error curves on the low induction number regime (10^{-2} mS/m–40 mS/m) given by

$$\log(\varepsilon)$$
$$= \left\{ \begin{array}{l} 0.4915 \cdot \log(\sigma_V) + 0.7321 \longrightarrow \text{Vertical dipole} \\ 0.4959 \cdot \log(\sigma_H) + 0.4296 \longrightarrow \text{Horizontal dipole} \end{array} \right\}, \qquad (20)$$

where ε is given in percent (%) and σ in mS/m.

3. Conclusions

It is well known that the vertical dipole configuration is more sensitive to conductivity anomalies in deep, while the horizontal one is more sensitive to near surface conductivity

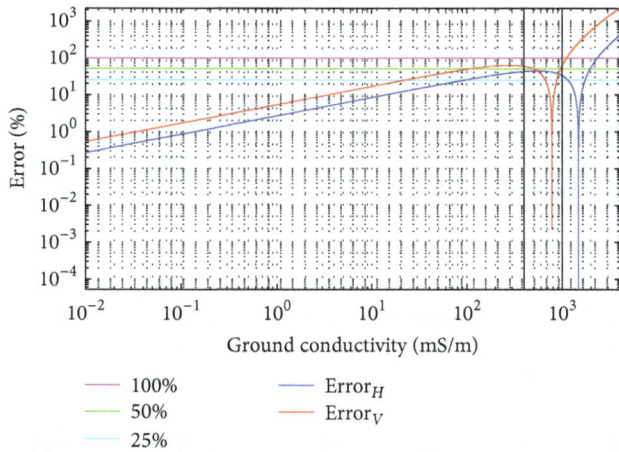

FIGURE 2: Relative percent error (%) as a function of conductivity (mS/m). It also shows error isolines for the error levels of 100%, 50%, and 25%. The vertical black lines mark two main limits: induction number equals to one ($B = 1$, for which $\sigma \sim 400$ mS/m) and the EM34 instrument limit of 1000 mS/m.

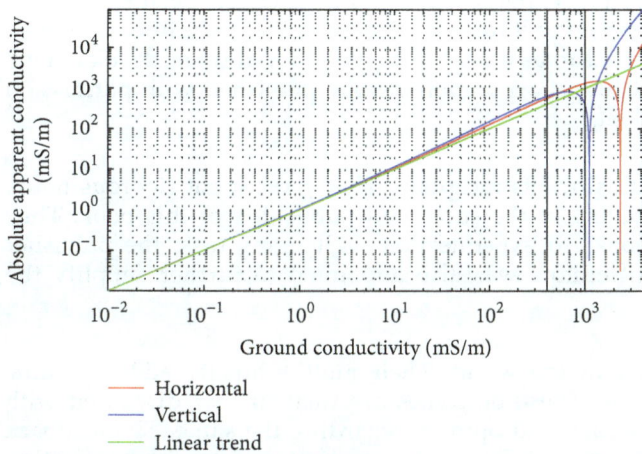

FIGURE 3: Apparent conductivity versus true conductivity in mS/m until slightly above the EM34 instrument limit. The vertical black lines mark two main limits: induction number equals to one ($P = 1$, for which $\sigma \sim 400$ mS/m) and the EM34 instrument limit of 1000 mS/m. This curve shows a slight departure from the linear behaviour after the low induction number limit (~ 40 mS/m) until the limit for $B = 1$ (~ 400 mS/m) and a strong departure from the linear behaviour after that.

variations [1]. However, if we consider only the measured conductivity values, the error is lower for the horizontal than for the vertical dipole configuration as noted in Figure 2. For both configurations conductivity converges to the same values when conductivity tends to zero.

An approximately linear tendency can be observed over the whole low induction number range and up to 400 mS/m. Above this value there are strong deviations from linearity indicating that we are close to singularities of the complex function (not only the quadrature term). Despite a correction procedure suggested by Beamish [4], we believe that under statistical fluctuations of the measured signal the response

could depart from the expected model leading to very high errors in the estimated conductivity due to the proximity of the function singularity.

It is worth noting that the determination of these bounds depends on the configurations (operation frequency and intercoil spacing) of the considered instrument.

Acknowledgments

This work was supported by the Brazilian agency CNPq through a postdoctoral scholarship to George Caminha-Maciel. Thanks are due to E. La Terra for the valuable discussions and M. Ernesto for helping in improving the paper. Any use of product names is for descriptive purposes only and does not imply any endorsement.

References

[1] J. D. McNeill, "Electromagnetic terrain conductivity measurements at low induction numbers," Technical Note TN-6, Geonics Ltd., Mississauga, Canada, 1980.

[2] G. V. Keller and F. C. Firschknecht, *Electrical Methods in Geophysical Prospecting*, Pergamon, New York, NY, USA, 1966.

[3] J. R. Waitt, "A note on the electromagnetic response of a stratified Earth," *Geophysics*, vol. 27, pp. 382–385, 1962.

[4] D. Beamish, "Low induction number, ground conductivity meters: A correction procedure in the absence of magnetic effects," *Journal of Applied Geophysics*, vol. 75, no. 2, pp. 244–253, 2011.

[5] J. B. Callegary, T. P. A. Ferré, and R. W. Groom, "Vertical spatial sensitivity and exploration depth of low-induction-number electromagnetic-induction instruments," *Vadose Zone Journal*, vol. 6, no. 1, pp. 158–167, 2007.

Permissions

The contributors of this book come from diverse backgrounds, making this book a truly international effort. This book will bring forth new frontiers with its revolutionizing research information and detailed analysis of the nascent developments around the world.

We would like to thank all the contributing authors for lending their expertise to make the book truly unique. They have played a crucial role in the development of this book. Without their invaluable contributions this book wouldn't have been possible. They have made vital efforts to compile up to date information on the varied aspects of this subject to make this book a valuable addition to the collection of many professionals and students.

This book was conceptualized with the vision of imparting up-to-date information and advanced data in this field. To ensure the same, a matchless editorial board was set up. Every individual on the board went through rigorous rounds of assessment to prove their worth. After which they invested a large part of their time researching and compiling the most relevant data for our readers. Conferences and sessions were held from time to time between the editorial board and the contributing authors to present the data in the most comprehensible form. The editorial team has worked tirelessly to provide valuable and valid information to help people across the globe.

Every chapter published in this book has been scrutinized by our experts. Their significance has been extensively debated. The topics covered herein carry significant findings which will fuel the growth of the discipline. They may even be implemented as practical applications or may be referred to as a beginning point for another development. Chapters in this book were first published by Hindawi Publishing Corporation; hereby published with permission under the Creative Commons Attribution License or equivalent.

The editorial board has been involved in producing this book since its inception. They have spent rigorous hours researching and exploring the diverse topics which have resulted in the successful publishing of this book. They have passed on their knowledge of decades through this book. To expedite this challenging task, the publisher supported the team at every step. A small team of assistant editors was also appointed to further simplify the editing procedure and attain best results for the readers.

Our editorial team has been hand-picked from every corner of the world. Their multi-ethnicity adds dynamic inputs to the discussions which result in innovative outcomes. These outcomes are then further discussed with the researchers and contributors who give their valuable feedback and opinion regarding the same. The feedback is then collaborated with the researches and they are edited in a comprehensive manner to aid the understanding of the subject.

Apart from the editorial board, the designing team has also invested a significant amount of their time in understanding the subject and creating the most relevant covers. They scrutinized every image to scout for the most suitable representation of the subject and create an appropriate cover for the book.

The publishing team has been involved in this book since its early stages. They were actively engaged in every process, be it collecting the data, connecting with the contributors or procuring relevant information. The team has been an ardent support to the editorial, designing and production team. Their endless efforts to recruit the best for this project, has resulted in the accomplishment of this book. They are a veteran in the field of academics and their pool of knowledge is as vast as their experience in printing. Their expertise and guidance has proved useful at every step. Their uncompromising quality standards have made this book an exceptional effort. Their encouragement from time to time has been an inspiration for everyone.

The publisher and the editorial board hope that this book will prove to be a valuable piece of knowledge for researchers, students, practitioners and scholars across the globe.

List of Contributors

P. A. Alao
Department of Geology, Institute of Earth and Environmental Sciences, University of Freiburg, Albertstraße 23b, 79104 Freiburg, Germany

A. I. Ata
Department of Geology, University of Malaya, Kuala Lumpur, Malaysia

C. E. Nwoke
School of Earth, Atmospheric and Environmental Sciences, The University of Manchester, Oxford Road, Manchester, Greater Manchester M13 9PL, UK

Abu Mallam
Department of Physics, Federal University of Technology P.M.B. 65 Minna, Niger State, Nigeria
Department of Physics, University of Abuja, P.M.B. 117Abuja, Nigeria

Adetona A. Abbass
Department of Physics, Federal University of Technology P.M.B. 65 Minna, Niger State, Nigeria

M. W. Dongmo, L. Y. Kagho, F. B. Pelap, G. B. Tanekou, Y. L. Makenne and A. Fomethe
Laboratory of Mechanics and Modelling of Physical Systems, Department of Physics, University of Dschang, P.O. Box 69, Dschang, Cameroon

D. A. Angus
CiPEG, University of Leeds, Leeds LS2 9JT, UK

J. P. Verdon
School of Earth Sciences, University of Bristol, Bristol BS8 1RJ, UK

Igor V. Mingalev, Konstantin G. Orlov, Victor S. Mingalev and Oleg V. Mingalev
Polar Geophysical Institute, Kola Scientific Center of the Russian Academy of Sciences, Murmansk Region, Apatity 184209, Russia

Natalia M. Astafieva
Space Research Institute of the Russian Academy of Sciences, Moscow 117997, Russia

Valery M. Chechetkin
Keldysh Institute of Applied Mathematics of the Russian Academy of Sciences, Moscow 125047, Russia

Aniekan Martin Ekanem
Department of Physics, Akwa Ibom State University, PMB 1167, Mkpat Enin, Nigeria
British Geological Survey, Murchison House, Edinburgh EH9 3LA, UK
School of Geosciences, University of Edinburgh, Edinburgh EH9 3JW, UK

Main Ian
School of Geosciences, University of Edinburgh, Edinburgh EH9 3JW, UK

Mark Chapman
British Geological Survey, Murchison House, Edinburgh EH9 3LA, UK
School of Geosciences, University of Edinburgh, Edinburgh EH9 3JW, UK

Xiang Yang Li
British Geological Survey, Murchison House, Edinburgh EH9 3LA, UK
CNPC Geophysical Key Laboratory, China University of Petroleum, Beijing 102249, China

Jianxin Wei
CNPC Geophysical Key Laboratory, China University of Petroleum, Beijing 102249, China

J. S. Nandal
Department of Mathematics, M.D. University, Rohtak 124001, India

T. N. Saini
Department of Mathematics, Government College Kalka, Kalka 133302, India

Zoulin Liu and Stephen M. J. Moysey
Department of Environmental Engineering and Earth Science, Clemson University, Clemson, SC 29670, USA

Andy A. Bery
Geophysics Section, School of Physics, Universiti Sains Malaysia, 11800 Penang, Malaysia

Suze Nei P. Guimaraes and Valiya M. Hamza
Observatorio Nacional, Rua General Jose Cristino 77, Rio de Janeiro, CEP 20921-400, RJ, Brazil

Roberta A. Cardoso
Observatorio Nacional-ON/MCT, Rio de Janeiro, Brazil
Superintendence for Petroleum and Gas/DPG, Empresa de Pesquisa Energetica (EPE), Rio de Janeiro, Brazil

Valiya M. Hamza
Observatorio Nacional-ON/MCT, Rio de Janeiro, Brazil

Jean Jacques Nguimbous-Kouoh, Robert Nouayou, Charles Tabod Tabod and Eliezer Manguelle-Dicoum
Department of Physics, Faculty of Science, University of Yaounde I, P.O. Box 812, Yaounde, Cameroon

Eric M. Takam Takougang
Department of Earth Sciences, Simon Fraser University, 8888 University Drive, Burnaby, BC, Canada V5A1S6

D. T. Luong and R. Sprik
Van der Waals-Zeeman Institute, University of Amsterdam, 1098 XH Amsterdam, The Netherlands

N. J. George and E. U. Nathaniel
Department of Physics, Akwa Ibom State University, Ikot Akpaden 53001, Nigeria

S. E. Etuk
Department of Physics, University of Uyo, Uyo 53001, Nigeria

Jagabandhu Dixit, D. M. Dewaikar and R. S. Jangid
Department of Civil Engineering, Indian Institute of Technology Bombay, Mumbai 400076, PIN, India

T. N. Obiekezie and S. C. Obiadazie
Department of Physics and Industrial Physics, Nnamdi Azikiwe University, P.M.B. 5025, Awka, Nigeria

G. A. Agbo
Industrial Physics Department, Ebonyi State University, P.M.B. 053, Abakaliki, Nigeria

Anoop Kumar Mishra
Research Center for Environmental Changes, Academia Sinica, Taipei 115, Taiwan

Rajesh Kumar
Remote Sensing & Aerial Surveys, Mission-1B, Geological Survey of India, Govt. of India, Ministry of Mines, Bangalore, India

Chi-Min Liu
Division of Mathematics, General Education Center, Chienkuo Technology University, Changhua City 500, Taiwan
International Wave Dynamics Research Center, National Cheng Kung University, Tainan 701, Taiwan

Ray-Yeng Yang
International Wave Dynamics Research Center, National Cheng Kung University, Tainan 701, Taiwan
Tainan Hydraulic Laboratory and Research Center of Ocean Environment and Technology, National Cheng Kung University, Tainan 701, Taiwan

Hwung-Hweng Hwung
International Wave Dynamics Research Center, National Cheng Kung University, Tainan 701, Taiwan
Department of Hydraulic and Ocean Engineering, National Cheng Kung University, Tainan 701, Taiwan

Ben Parkes and Alan Gadian
NCAS, SEE, University of Leeds, Leeds LS2 9JT, UK

John Latham
MMM, National Center for Atmospheric Research, Boulder, CO 80307-3000, USA
SEAS, University of Manchester, Manchester M13 9PL, UK

George Caminha-Maciel
Observatorio Nacional (ON), MCT, Rua General Jose Cristino 77, 20921-400 Rio de Janeiro, RJ, Brazil
Universidade Federal do Pampa (UNIPAMPA), Avenida Pedro Anunciacao s/nº, 96570-000 Cacapava do Sul, RS, Brazil

Irineu Figueiredo
Observatorio Nacional (ON), MCT, Rua General Jose Cristino 77, 20921-400 Rio de Janeiro, RJ, Brazil
Universidade Estadual do Rio de Janeiro (UERJ), Rua Sao Francisco Xavier 524, 20550-900 Rio de Janeiro, RJ, Brazil